Materials for Nuclear Waste Immobilization

Materials for Nuclear Waste Immobilization

Special Issue Editors

Michael I. Ojovan
Neil C. Hyatt

MDPI • Basel • Beijing • Wuhan • Barcelona • Belgrade

MDPI

Special Issue Editors
Michael I. Ojovan
University of Sheffield
UK

Neil C. Hyatt
University of Sheffield
UK

Editorial Office
MDPI
St. Alban-Anlage 66
4052 Basel, Switzerland

This is a reprint of articles from the Special Issue published online in the open access journal *Materials* (ISSN 1996-1944) from 2018 to 2019 (available at: https://www.mdpi.com/journal/materials/special_issues/MNWI)

For citation purposes, cite each article independently as indicated on the article page online and as indicated below:

LastName, A.A.; LastName, B.B.; LastName, C.C. Article Title. *Journal Name* **Year**, *Article Number*, Page Range.

ISBN 978-3-03921-845-5 (Pbk)
ISBN 978-3-03921-846-2 (PDF)

Contents

About the Special Issue Editors

Michael I. Ojovan has been Nuclear Engineer of International Atomic Energy Agency (IAEA), visiting Professor of Imperial College London, Associate Reader in Materials Science and Waste Immobilisation of the University of Sheffield, UK, and Leading Scientist of Radiochemistry Department of Lomonosov Moscow State University. M. Ojovan is Chief Editor of journal "Science and Technology of Nuclear Installations", Associate Editor of journal "Innovations in Corrosion and Materials Science", and Editorial Board Member of journals: "npj Materials Degradation" (Nature Partner Journal), "International Journal of Corrosion", "Journal of Nuclear Materials". He has published 12 monographs including the "Handbook of Advanced Radioactive Waste Conditioning Technologies" by Woodhead. He has founded and led the IAEA International Predisposal Network (IPN) and the IAEA International Project on Irradiated Graphite Processing (GRAPA). M. Ojovan is known for the connectivity-percolation theory of glass transition, Sheffield model (two-exponential equation) of viscosity of glasses and melts, condensed Rydberg matter, metallic and glass-composite materials for nuclear waste immobilisation, and self-sinking capsules to investigate Earth' deep interior.

Neil C. Hyatt is a Fellow of the Royal Society of Chemistry, holds the NDA/ Royal Academy of Engineering Chair in Radioactive Waste Management and is Professor of Nuclear Materials Chemistry at The University of Sheffield, UK. He is Adjunct Professor in Nuclear Materials in the School of Mechanical & Materials Engineering at Washington State University. Neil has almost 20 years of professional and academic experience in radioactive waste management and disposal. His current research interests include:

- Design, processing, and performance assessment of glass, ceramic, and cement materials for radioactive waste immobilisation.

- Characterisation, degradation, and remediation of radioactive particulates in the environment.

- Identification and exploitation of forensic signatures in nuclear materials and detonation debris. His research is supported by a current funding portfolio in of excess of £11M, sponsored by EPSRC, EC, NDA, Sellafield Ltd., and others.

Neil was appointed to HM Government's Nuclear Innovation Research Board (2014–20), providing independent advice on nuclear research strategy priorities for the UK; he is Chair of the NIRAB International Strategy Sub-Group. He has served as a technical expert for the International Atomic Energy Agency, and technical advisor and consultant to the Nuclear Decommissioning Authority, Radioactive Waste Management Ltd., and US Nuclear Waste Technical Review Board, amongst others. Neil has guided academic nuclear energy research, education and training strategy at The University of Sheffield, since 2011, serving as Head of the Department of Materials Science & Engineering from 2015–18. Under his leadership, and in partnership with the Nuclear Advanced Manufacturing Research Centre, this activity has grown in size and stature to be one of the top three centres of UK expertise. He held a Royal Academy of Engineering Research Chair, sponsored by the Nuclear Decommissioning Authority, from 2011–16, establishing the Immobilisation Science Laboratory as the leading UK academic research group in radioactive waste management and disposal, with a research order book in excess of £15M, supporting 7 academic staff, 16 PDRAs, and 43 PhDs. He created the

MIDAS laboratory, with investment of £1M from the Department of Energy & Climate Change, as a national centre of excellence in waste management research.

Preface to "Materials for Nuclear Waste Immobilization"

Nuclear energy is clean, reliable and competitive with many useful applications, among which power generation is the most important because it can gradually replace fossil fuels and avoid massive pollution of the environment. Nuclear waste is a useless by-product resulting from the utilization of nuclear energy in both power generation and other applications such as in medicine, industry, agriculture, and research. Safe and effective management of nuclear waste is crucial to ensure the sustainable utilization of nuclear energy. Nuclear waste must be processed to make it safe for storage, transportation, and final disposal, which includes its conditioning; accordingly, it is immobilized and packaged before storage and disposal. The immobilization of waste radionuclides in durable wasteform materials provides the most important barrier to contribute to the overall performance of any storage and/or disposal system. Materials for nuclear waste immobilization are thus at the core of multibarrier systems of isolation of radioactive waste from the environment aimed to ensure the long-term safety of nuclear waste storage and disposal. This Special Issue analyzes the materials currently used, as well as novel materials for nuclear waste immobilization, including technological approaches utilized in nuclear waste conditioning that attempt to ensure the efficiency and long-term safety of storage and disposal systems. It focuses on the advanced cementitious materials, geopolymers, glasses, glass composite materials, and ceramics developed and used in nuclear waste immobilization, with the performance of such materials being considered of the utmost importance.

Michael I. Ojovan, Neil C. Hyatt
Special Issue Editors

materials

MDPI

Editorial
Special Issue: Materials for Nuclear Waste Immobilization

Neil C. Hyatt [1] and Michael I. Ojovan [1,2,3,*]

1   Immobilisation Science Laboratory, Department of Materials Science and Engineering,
     University of Sheffield, Mappin Street, Sheffield S1 3JD, UK; n.c.hyatt@sheffield.ac.uk
2   Department of Radiochemistry, Lomonosov Moscow State University, Moscow 119991, Russia
3   Imperial College London, South Kensington Campus, Exhibition Road, London SW7 2AZ, UK
*   Correspondence: m.ojovan@sheffield.ac.uk

Received: 16 October 2019; Accepted: 1 November 2019; Published: 3 November 2019

Abstract: Nuclear energy is clean, reliable, and competitive with many useful applications, among which power generation is the most important as it can gradually replace fossil fuels and avoid massive pollution of environment. A by-product resulting from utilization of nuclear energy in both power generation and other applications, such as in medicine, industry, agriculture, and research, is nuclear waste. Safe and effective management of nuclear waste is crucial to ensure sustainable utilization of nuclear energy. Nuclear waste must be processed to make it safe for storage, transportation, and final disposal, which includes its conditioning, so it is immobilized and packaged before storage and disposal. Immobilization of waste radionuclides in durable wasteform materials provides the most important barrier to contribute to the overall performance of any storage and/or disposal system. Materials for nuclear waste immobilization are thus at the core of multibarrier systems of isolation of radioactive waste from environment aimed to ensure long term safety of storage and disposal. This Special Issue analyzes the materials currently used as well as novel materials for nuclear waste immobilization, including technological approaches utilized in nuclear waste conditioning pursuing to ensure efficiency and long-term safety of storage and disposal systems. It focuses on advanced cementitious materials, geopolymers, glasses, glass composite materials, and ceramics developed and used in nuclear waste immobilization, with the performance of such materials of utmost importance. The book outlines recent advances in nuclear wasteform materials including glasses, ceramics, cements, and spent nuclear fuel. It focuses on durability aspects and contains data on performance of nuclear wasteforms as well as expected behavior in a disposal environment.

Keywords: nuclear waste; spent nuclear fuel; immobilisation; conditioning; wasteforms; vitrification; glass; ceramics; glass composite materials; durability

---

Materials are at the core of multibarrier systems of isolation of radioactive (nuclear) waste from the environment. Relevant materials are used to ensure long-term safety of handling, storage, transportation, and disposal of nuclear waste. Nuclear waste immobilization is the conversion of waste into a wasteform by solidification, embedding, or encapsulation that reduces the potential for migration or dispersion of radionuclides during operational and disposal stages of waste lifecycle. Immobilization of waste is achieved by its chemical incorporation into the structure of a suitable matrix (typically cement, glass, or ceramic) so it is captured and unable to escape. Chemical immobilization is typically applied to high level waste (HLW). Encapsulation of waste is achieved by physically surrounding it in materials (typically bitumen or cement) so it is isolated, and radionuclides are retained. Physical encapsulation is often applied to intermediate level waste (ILW) but can also be used for HLW, especially where chemical incorporation of radionuclides in the surrounding matrix is also possible.

Within the repository, the wasteform is one part of a multiple engineered barrier system. During storage and transportation, the wasteform is the primary barrier preventing the release of radionuclides into the environment, while during post closure disposal, the wasteform will reduce the release of radionuclides from breached and compromised containers that could result due to corrosion, earthquake, human intrusion, igneous intrusion (volcano), or other disruptive phenomena.

Choosing a suitable wasteform (matrix) to use for nuclear waste immobilization is not easy and its durability is not the sole acceptance criterion. Priority is given to reliable, simple, rugged technologies and equipment, which may have advantages over complex or sensitive equipment and processes. A variety of matrix materials and techniques are available for immobilisation [1–16]. The choice of the immobilization technology depends on the physical and chemical nature of the waste and the acceptance criteria for the storage and disposal facility to which the waste will be consigned.

Factors that are considered primarily when selecting a wasteform material are as follows [3,4]:

- Waste loading—able to accommodate a significant amount of waste (typically 25–45 weight %) to minimize volume;
- Ease of production—accomplished under reasonable conditions;
- Durability—low rate of dissolution to minimize the release of radioactive and chemical constituents;
- Radiation stability—high tolerance to radiation effects from the decay of radioactive constituents;
- Chemical flexibility—able to accommodate a mixture of radioactive and chemical constituents with minimum formation of secondary phases;
- Availability of natural analogues—availability of natural mineral or glass analogues may provide important clues about the long-term performance;
- Compatibility with the intended disposal environment—compatible with the near-field environment of the disposal facility.

A host of regulatory, process, and product requirements has led to the investigation and adoption of a variety of matrices and technologies for waste immobilization. The resistance of the wasteform to aqueous corrosion and release of radionuclides in the disposal environment—chemical durability—is a critical parameter. Figure 1 shows schematically the water durability of main nuclear wasteforms used.

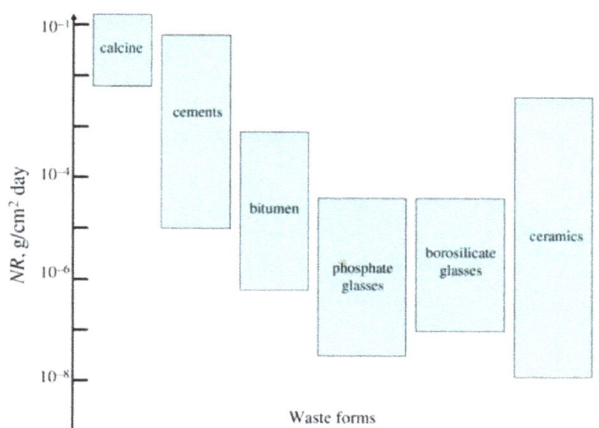

Figure 1. Normalized leaching rates of various wasteforms (after Reference [1]).

The main immobilization technologies that are available commercially and have been demonstrated to be viable are cementation [1–5] and vitrification [1,3,4,8–16], whereas bitumen and polymeric materaisl are used to a smaller extent (see data in [1]) and ceramification is a perspective technology [1,3,6–8,12,14–16]. Table 1 shows generic features and limitations of main wasteforms currently used on industrial scale.

Table 1. Features and limitations of main wasteforms currently used.

Wasteform	Features	Limitations	Secondary Waste
Glasses	Proven method to condition liquid high-level waste (HLW) as well as intermediate-level waste (ILW) and low-level waste (LLW). High flexibility in terms of the glass formulation range. High reliability of the immobilization process. High glass throughput. High durability of the final wasteform. Small volume of the resulting wasteform.	High initial investment and operational costs. Complex technology requiring high qualified personnel. Need to control off-gases. High specific energy consumption.	Off-gases. Filters. Scrub solutions. Used melters.
Ceramics	Possible to incorporate higher levels of actinides than borosilicate glass. Wasteform can be more durable than glass. Expected to be suitable for long term isolation since it simulates natural rocks.	Limited experience. Most efforts have been research-based. The ceramic shall be tailored to nuclear waste composition.	Filters. Off-gases. Scrub solutions.
Glass-composite materials	Combine features of both crystalline and glassy materials. Higher waste loading. Higher compatibility. Higher stability compared glasses.	Limited experience.	Off-gases. Filters. Scrub solutions. Used melters
Cements	Widely used method for variety of LLW and ILW. High flexibility. Low cost. Simplicity of process. Low temperature precludes volatile emissions. High radiation stability, impact, and fire resistance of wasteforms.	Increase of volume (low waste loading). Low retention of some fission and activation products. Poor compatibility with organic materials and high-salt content.	None.
Bitumen	Mostly used for LILW, chemical precipitates, low heat, and low alpha wastes. High flexibility. High compatibility with organic materials. High waste loading. Lower leaching rate compared with cements.	Sensitivity to some components. Low fire resistance.	Filters.
Metals	Extensively proven technology for conditioning of metallic waste. The product is typically homogeneous and stable.	Pre-sorting is usually required due to dedicated melt furnaces and differences in melt temperatures of different metals.	Off-gases. Slag.

The general requirements against one another need optimization for any technological approach considered. For example, ceramics are credited with having higher chemical durability than glasses, however, radionuclides will be released at similar or even higher rates compared with glassy wasteforms (see, e.g., Figure 1) if they are incorporated in the lower durability crystalline phases and intergranular glassy phases.

This book contains 10 dedicated papers prepared by lead researchers covering different aspects of nuclear wasteforms and their expected behavior. They purposely analyze the materials currently used as well as novel materials for nuclear waste immobilization including technological approaches utilized in nuclear waste conditioning pursuing to ensure efficiency and long-term safety of storage and disposal systems, including cementitious materials, glasses, and ceramics. The book outlines recent advances in nuclear wasteform materials including cements, glasses, ceramics, cements, and spent nuclear fuel with focus on durability aspects and presenting data on performance of nuclear wasteforms, as well as expected behavior in a disposal environment.

Funding: This research received no external funding.

Conflicts of Interest: The authors declare no conflict of interest.

References

1. Ojovan, M.I.; Lee, W.E.; Kalmykov, S.N. *An Introduction to Nuclear Waste Immobilisation*, 3rd ed.; Elsevier: Amsterdam, The Netherlands, 2019; p. 497.
2. Abdel Rahman, R.O.; Rahimov, R.Z.; Rahimova, N.R.; Ojovan, M.I. *Cementitious Materials for Nuclear Waste Immobilization*; Wiley: Chichester, UK, 2015; p. 232.
3. Lee, W.E.; Ojovan, M.I.; Jantzen, C.M. *Radioactive Waste Management and Contaminated Site Clean-up: Processes, Technologies and International Experience*; Woodhead: Cambridge, UK, 2013; p. 924.
4. National Research Council. *Waste Forms Technology and Performance: Final Report*; National Academies Press: Washington, DC, USA, 2011; p. 308.
5. Glasser, F. Application of inorganic cements to the conditioning and immobilisation of radioactive wastes. In *Handbook of Advanced Radioactive Waste Conditioning Technologies*; Ojovan, M.I., Ed.; Woodhead: Cambridge, UK, 2011; pp. 67–135, 512.
6. Kinoshita, H. Development of ceramic matrices for high level radioactive waste. In *Handbook of Advanced Radioactive Waste Conditioning Technologies*; Ojovan, M.I., Ed.; Woodhead: Cambridge, UK, 2011; pp. 67–135, 512.
7. Burakov, B.E.; Ojovan, M.I.; Lee, W.E. *Crystalline Materials for Actinide Immobilisation*; Imperial College Press: London, UK, 2010; p. 198.
8. Donald, I.W. *Waste Immobilisation in Glass and Ceramic Based Hosts*; Wiley: Chichester, UK, 2010; p. 507.
9. Jantzen, C.M.; Brown, K.G.; Pickett, J.B. Durable glass for thousands of years. *Int. J. Appl. Glass Sci.* **2010**, *1*, 38–62. [CrossRef]
10. Caurant, D.; Loiseau, P.; Majerus, O.; Aubin-Chevalsdonnet, V.; Bardez, I. Quintas, A. *Glasses, Glass-Ceramics and Ceramics for Immobilization of Highly Radioactive Nuclear Wastes*; Nova Science Publishers: New York, NY, USA, 2009; p. 359.
11. Ojovan, M.I.; Lee, W.E. *New Developments in Glassy Nuclear Wasteforms*; Nova Science Publishers: New York, NY, USA, 2007; p. 131.
12. Lee, W.E.; Ojovan, M.I.; Stennett, M.C.; Hyatt, N.C. Immobilisation of radioactive waste in glasses, glass composite materials and ceramics. *Adv. Appl. Ceram.* **2006**, *105*, 3–12. [CrossRef]
13. Vienna, J.D. Nuclear Waste Glasses. In *Properties of Glass Forming Melts*; Pye, L.D., Joseph, I., Montenaro, A., Eds.; CRC Press: Boca Raton, FL, USA, 2015; pp. 391–404, 512.
14. Stefanovsky, S.V.; Yudintsev, S.V.; Giere, R.; Lumpkin, G.R. Nuclear waste forms. In *Energy, Waste and the Environment: A Geochemical Perspective*; Gieré, R., Stille, P., Eds.; Geological Society of London: London, UK, 2004; pp. 37–63, 688.
15. Ewing, R.C. The Design and Evaluation of Nuclear-waste Forms Clues from Mineralogy. *Canadian Mineralogist* **2001**, *39*, 697–715. [CrossRef]
16. Lutze, W.; Ewing, R.C. *Radioactive Waste Forms for the Future*; Elsevier: Amsterdam, The Netherlands, 1988; p. 778.

materials

MDPI

Article

Magnesium Potassium Phosphate Compound for Immobilization of Radioactive Waste Containing Actinide and Rare Earth Elements

Sergey E. Vinokurov *, Svetlana A. Kulikova and Boris F. Myasoedov

Vernadsky Institute of Geochemistry and Analytical Chemistry, Russian Academy of Sciences, 19 Kosygin st., Moscow 119991, Russia; kulikova.sveta92@mail.ru (S.A.K.); bf@geokhi.ru (B.F.M.)
* Correspondence: vinokurov.geokhi@gmail.com; Tel.: +7-495-939-7007

Received: 17 May 2018; Accepted: 5 June 2018; Published: 8 June 2018

Abstract: The problem of effective immobilization of liquid radioactive waste (LRW) is key to the successful development of nuclear energy. The possibility of using the magnesium potassium phosphate (MKP) compound for LRW immobilization on the example of nitric acid solutions containing actinides and rare earth elements (REE), including high level waste (HLW) surrogate solution, is considered in the research work. Under the study of phase composition and structure of the MKP compounds that is obtained by the XRD and SEM methods, it was established that the compounds are composed of crystalline phases—analogues of natural phosphate minerals (struvite, metaankoleite). The hydrolytic stability of the compounds was determined according to the semi-dynamic test GOST R 52126-2003. Low leaching rates of radionuclides from the compound are established, including a differential leaching rate of ^{239}Pu and ^{241}Am—3.5×10^{-7} and 5.3×10^{-7} g/(cm^2·day). As a result of the research work, it was concluded that the MKP compound is promising for LRW immobilization and can become an alternative material combining the advantages of easy implementation of the technology, like cementation and the high physical and chemical stability corresponding to a glass-like compound.

Keywords: magnesium potassium phosphate compound; actinides; rare earth elements; uranium; plutonium; americium; lanthanum; neodymium; immobilization; leaching

---

1. Introduction

Long-term controlled storage or disposal is one of the key stages of the liquid radioactive waste (LRW) management in terms of radiation safety. The preparation of the LRW for this stage involves the transfer of waste into a stable solidified form using preserving matrices [1,2]. Cementation has found wide use in the nuclear industry for radioactive waste (RW) management of low and intermediate activity levels, in spite of significant disadvantages of the method, especially the relatively low degree of incorporation of waste salts, as well as low hydrolytic stability and frost resistance of cement compound. Vitrification is currently the only high level waste (HLW) management technology that is applied in industry [3]. The disadvantages of the method are low chemical and crystallization resistance of the glass at elevated temperatures, as well as the need to use expensive high temperature melters, the liquidation of which, after the end of a relatively short technical lifetime, represents an unresolved radioecological problem.

Ceramic materials [4], and especially synthetic analogues of natural phosphate minerals [5,6], are considered as an alternative to cement and glass for the immobilization of RW, primarily obtained after the reprocessing of spent nuclear fuel (SNF) and containing long-lived isotopes of highly toxic actinides and rare earth elements (REE).

The mineral-like phosphate materials that were obtained at room temperature in aqueous solution by chemical interaction, as a rule, between metal (II) oxides (MgO, ZnO, FeO, CaO) and orthophosphoric acid (H_3PO_4) or its derivatives (for example, (di) hydrogenphosphates of metals (I) or ammonium) [7,8] are of particular interest.

Previously, we and other researchers demonstrated [9–16] that magnesium potassium phosphate (MKP) compound based on the $MgKPO_4 \times 6H_2O$ matrix obtained as a result of the reaction (1), which is an analog of the natural mineral K-struvite [17], is a promising low-temperature material for the immobilization of various RW types.

$$MgO + KH_2PO_4 + 5H_2O \rightarrow MgKPO_4 \times 6H_2O \qquad (1)$$

This method of RW management combines versatility, equipment simplicity and economic efficiency similar cementation, and the obtained MKP compound has a high physical and chemical stability.

The practical use possibility of MKP compound in RW management has to be explained in the context of reliability under the long storage of hazardous RW components mainly highly toxic plutonium and minor actinides, as well as REE, whose content is about half the content of all the metals in HLW. It should also be noted that, although uranium is maximally recovered from solutions during SNF reprocessing for its reuse in the fuel cycle, the residual uranium content (including isotopes U-232, 235, 236, 238) in HLW is about 3 g·L^{-1}. Thus, information on the behavior of uranium during immobilization in the MKP compound also has scientific interest.

The data on the phase composition, structure, and hydrolytic stability of synthesized MKP compounds containing uranium, plutonium, americium, and REE (on the example, lanthanum and neodymium) are presented in this article.

2. Materials and Methods

The experiments were performed in the glove box (PERERABOTKA, Dzerzhinsk, Russia). The chemicals used in the experiments were of no less than chemically pure grade. Samples of MKP compounds were prepared, according to the procedure previously given in reference [10]. For study, the forms of location and behavior during leaching of uranium and REE by the example of lanthanum in the MKP compound, concentrated aqueous solutions of their nitrates with a metal concentration 228.3 and 242.4 g·L^{-1}, respectively, were solidified.

The hydrolytic stability of MKP compound to the leaching of actinides and neodymium as a simulator of the REE group was carried out after the solidification of the HLW surrogate solution of 1000 MW water-water energetic reactor (WWER-1000). The HLW surrogate solution was prepared by dissolving the metal nitrates in an aqueous solution of nitric acid, molybdenum was added in the form of MoO_3 (Table 1). Preparation of the surrogate solution to solidification was carried out by neutralizing it to pH 8.0 ± 0.1 with sodium hydroxide solution at concentration 15.0 ± 0.1 mol·L^{-1}.

Table 1. Characteristics of high level waste (HLW) surrogate solution.

Specific Activity of Actinides (Bq·L^{-1})	Metal Content (g·L^{-1})	HNO$_3$ Content (mol·L^{-1})	Density (g·L^{-1})	Salt Content (g·L^{-1})
^{239}Pu – 3.8×10^8 ^{241}Am – 5.2×10^7	Na – 13.3; Sr – 3.9; Zr – 7.6; Mo – 0.9; Pd – 5.4; Cs – 9.3; Ba – 6.4; Nd – 28.8; Fe – 1.0; Cr – 2.8; Ni – 0.5; U – 3.1	3.2	1210	206.6

The phase composition of the prepared MKP compounds was determined by X-ray diffraction (XRD) (Ultima-IV, Rigaku, Tokyo, Japan). The X-ray diffraction data were interpreted using the specialized Jade 6.5 program package (MDI, Livermore, CA, USA) with PDF-2 powder database.

The structure of samples containing uranium and lanthanum was studied by the scanning electron microscopy (SEM) using microscopes Jeol JSM-6480LV (JEOL, Tokyo, Japan) and LEO Supra 50 VP (LEO Carl Zeiss SMT Ltd, Oberkochen, Germany), respectively. The electron probe microanalysis of the samples was performed using an energy-dispersive analyzer X-MAX 80 (Oxford Instruments plc, Abingdon, England).

The hydrolytic stability of compounds was determined using the semidynamic test, in accordance with GOST R 52126-2003 [18]. Conditions: Monolithic compound 2 cm × 2 cm × 2 cm; leaching agent—bidistilled water (pH 6.6 ± 0.1, volume 200 mL), temperature 23 ± 2 °C, periodic replacement of the leaching agent after 1, 3, 7, 10, 14, and 21 days, the total duration of the test was limited to 28 days. The content of lanthanum, neodymium, and uranium in solutions after leaching was determined by inductively coupled plasma atomic emission spectrometry (ICP-AES) (iCAP-6500 Duo, Thermo Scientific, Waltham, MA, USA), inductively coupled plasma mass spectrometry (ICP-MS) (X Series2, Thermo Scientific, Waltham, MA, USA), and by spectrophotometry (Cary 100 Scan, Varian, Palo Alto, CA, USA), and the content of ^{239}Pu and ^{241}Am—radiometric method with using of the α-spectrometer (Alpha Analyst, Canberra, Australia).

The mechanism of leaching of the compound components (lanthanum, neodymium, uranium, plutonium, and americium) from the samples was evaluated according to the model [19], described by the linear relationship of log (B_i) from log (t), where B_i is the total yield of the element from the compound during contact with water, mg·m^{-2}; t is the contact time, days. The calculate procedure of B_i is given in [9,20]. The following mechanisms of element leaching from the compound correspond to various values of the slope in this equation: >0.65—surface dissolution; 0.35–0.65—diffusion transport; <0.35—surface wash off (or a depletion if it is found in the middle or at the end of the test) [9,10,20–23].

3. Results

As a result of the performed experiments, the samples of MKP compounds with a density of 1.75 ± 0.07 g·cm^{-3} were prepared. The content of metals in compounds that were obtained under solidification of uranium and lanthanum nitrate solutions was 6.2 wt % uranium (hereinafter, compound #1) and 6.7 wt % lanthanum (hereinafter, compound #2), respectively. The salt content of the HLW surrogate solution that was obtained under neutralization was 369.5 g·L^{-1}, and filling of the compound by the salts of neutralized surrogate solution was 12.8 wt %, and the specific activity of ^{239}Pu and ^{241}Am was 1.8 × 10^5 and 2.4 × 10^4 Bq·g^{-1} (hereinafter, compound #3). The content of the components of the MKP compounds is presented in Table 2.

Table 2. Composition of magnesium potassium phosphate (MKP) compounds under study.

Compound	Liquid Waste (wt %)	Binders (wt %)		
		KH$_2$PO$_4$	H$_3$BO$_3$	MgO
#1	39.5	44.2	1.5	14.8
#2	43.4	41.3	1.5	13.8
#3	41.5	42.9	1.3	14.3

The obtained data on the study of the phase composition and structure of synthesized compounds #1–3 by XRD and SEM methods are shown in Figures 1 and 2, respectively.

In accordance with GOST R 52126-2003 [18], the differential leaching rates of actinides and REE from synthesized compound #1–3 during 28 days of contact with water (Figure 3a,c,e) were determined, and the mechanisms of their leaching (Figure 3b,d,f, as summarized in Table 3) were estimated.

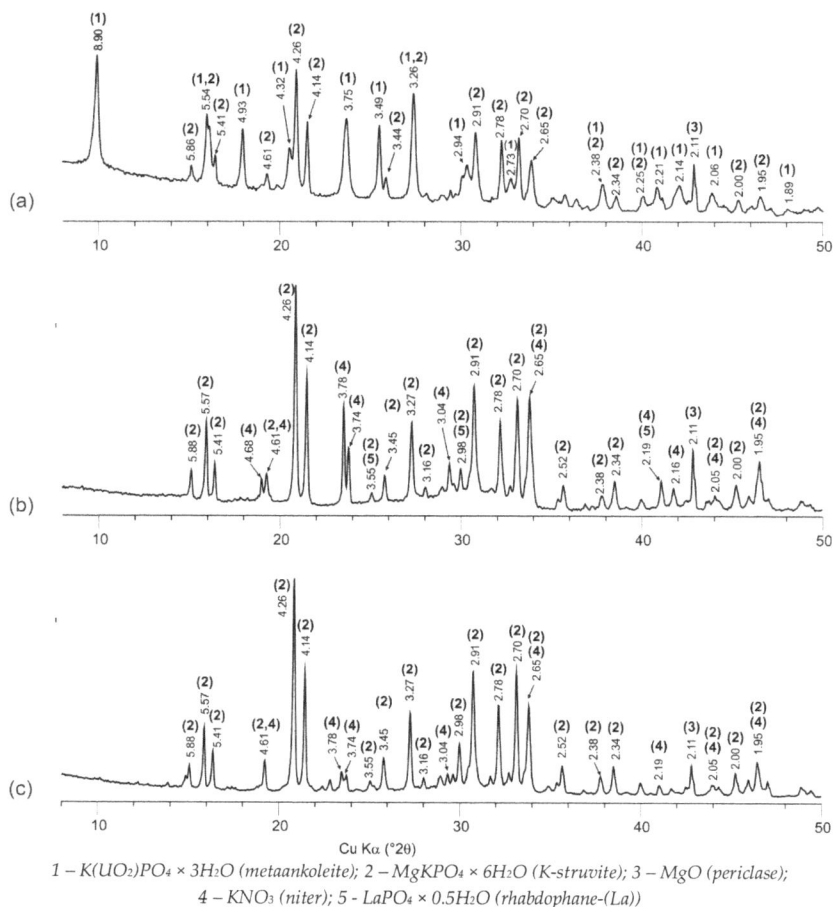

1 – K(UO₂)PO₄ × 3H₂O (metaankoleite); 2 – MgKPO₄ × 6H₂O (K-struvite); 3 – MgO (periclase);
4 – KNO₃ (niter); 5 - LaPO₄ × 0.5H₂O (rhabdophane-(La))

Figure 1. X-ray diffraction patterns of the compounds: #1 (**a**) and #2 (**b**), containing 6.2 wt % and 6.7 wt % uranium and lanthanum, respectively, and #3 (**c**), obtained by solidification of HLW surrogate solution.

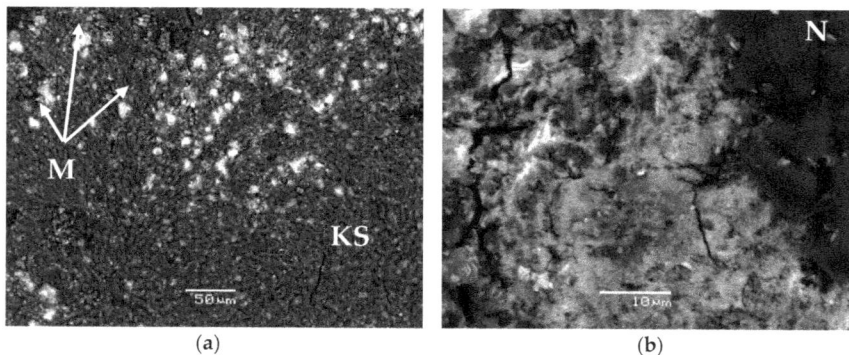

Figure 2. *Cont.*

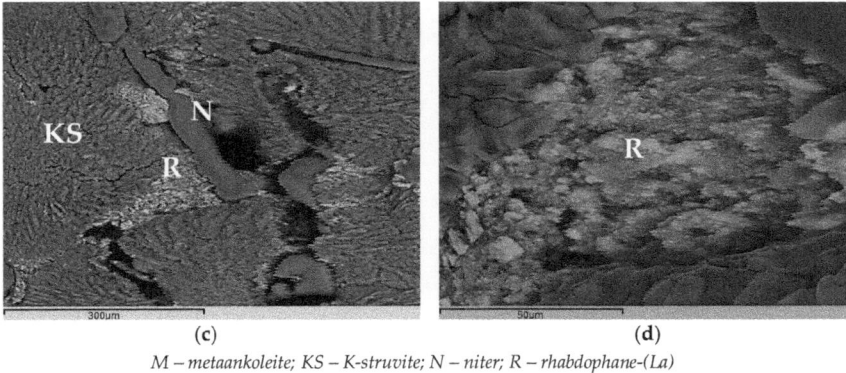

(c)

(d)

M – metaankoleite; KS – K-struvite; N – niter; R – rhabdophane-(La)

Figure 2. Scanning electron microscopy (SEM) images of the compounds #1 (**a,b**) and #2 (**c,d**), containing 6.2 wt % and 6.7 wt % uranium and lanthanum, respectively.

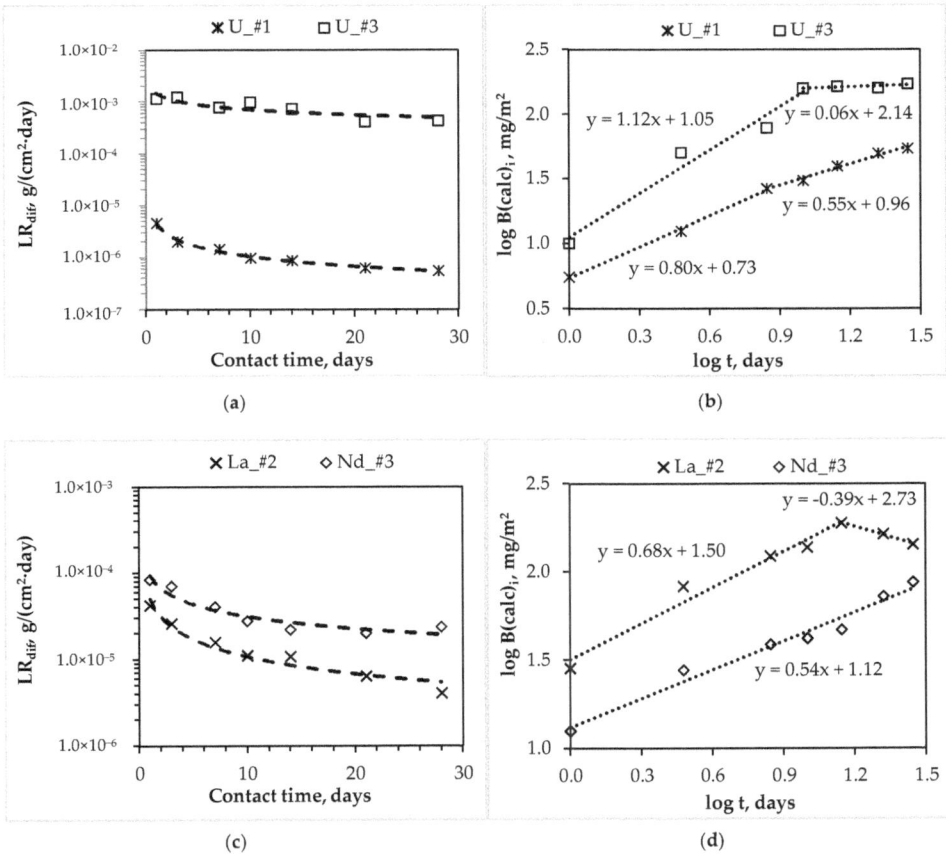

(a)

(b)

(c)

(d)

Figure 3. *Cont.*

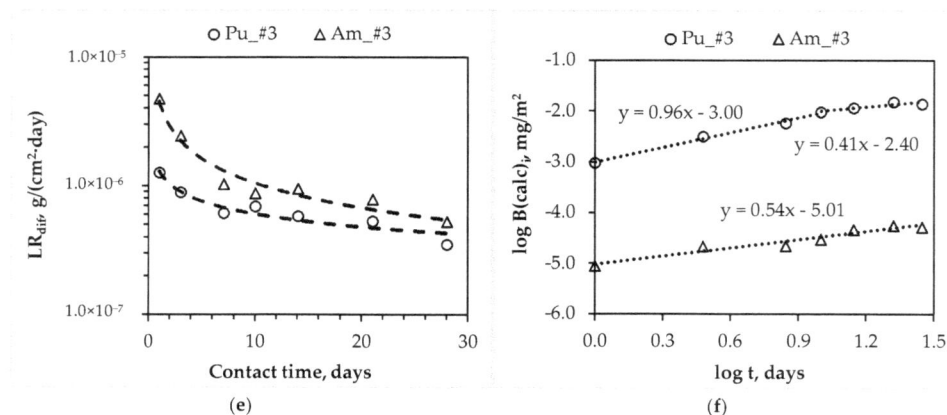

Figure 3. Dependence of the differential leaching rate (LR_{dif}) of actinides and rare earth elements (REE) (**a,c,e**) and the logarithmic dependence of their yield (log B (calc)$_i$) from compounds (**b,d,f**) on contact time with water (indices #1, 2, 3 correspond to compounds containing actinides and REE).

Table 3. The leaching mechanism of components of the MKP compounds (indices #1, 2, 3 correspond to the names of compounds containing actinides and REE).

Components of the MKP Compounds	Correspond Figure	Contact Time of the Samples with Water, Days	Slope of the Lines	Leaching Mechanism
U_#1	3b	1–7	0.80	dissolution
		7–28	0.55	diffusion
U_#3	3b	1–10	1.12	dissolution
		10–28	0.06	depletion
La_#2	3d	1–14	0.68	dissolution
		14–28	−0.39	depletion
Nd_#3	3d	1–28	0.54	diffusion
Am_#3	3f	1–28	0.54	diffusion
Pu_#3	3f	1–10	0.96	dissolution
		10–28	0.41	diffusion

4. Discussion

4.1. Phase Composition and Structure of MKP Compounds

The base of all the studied samples of compounds is the crystalline phosphate phase—the synthetic analogue of the K-struvite natural mineral $MgKPO_4 \times 6H_2O$ (the characteristic peaks at 4.26, 4.14, 2.91, 2.70 Å) (Figure 1a–c). In this case, compound #1 also contains a significant amount of the hydrated potassium uranyl orthophosphate phase, the X-ray diffraction parameters, of which correspond to the metaankoleite mineral $K(UO_2)PO_4 \times 3.0H_2O$ (the characteristic peaks at 8.90, 3.75, 3.49 Å) (Figure 1a). This is confirmed by the results of calculating the compound #1 phase composition, according to the microanalysis data: the uranium-enriched particles (denoted by M in Figure 2a) contain up to 45 wt % of uranium and have an average composition $Mg_{0.33}K(UO_2)_{0.67}PO_4 \times 4.0H_2O$, which corresponds to a mixture of metaankoleite $K(UO_2)PO_4 \times 3.0H_2O$ and the K-struvite $MgKPO_4 \times 6.0H_2O$ in a molar ratio of 2/1. In this case, the main phase of compound #1 (phase KS in Figure 2a) contains up to 3 wt % of uranium.

As a result of potassium replacement with metals of nitric acid solutions, the KNO_3 (niter) phase (the characteristic peaks at 3.78, 3.74, 3.04 Å) is formed in the obtained compounds (Figure 1), which were also shown previously in [9,10]. However, it was not possible to identify this phase in compound #1 (Figure 1a) by XRD method, probably because of its small content (theoretically not more than 5 wt %), but its presence is clearly confirmed by microanalysis data (phase N in Figure 2b). Impurities of magnesium, phosphorus, and uranium in phase N (Figure 2b) do not exceed 0.6, 1.4, and 1.3 wt %, respectively.

According to XRD (Figure 1b) and SEM (R in Figure 2c,d), it was established that lanthanum as REE in compound #2 is present as a phosphate compound of the analogue of the natural mineral rhabdophane—(La) $LaPO_4 \times 0.5H_2O$. In this case, according to the microanalysis data, the main phase of compound #2 (phase KS Figure 2c) is a phosphate compound of the composition $Mg_{0.60}K_{0.68}La_{0.36}PO_4 \times 6.3H_2O$, similar to K-struvite. In compound #2, the presence of the KNO_3 phase (N in Figure 2c) is also clearly confirmed by SEM data.

The MgO (periclase) phase (the characteristic peak at 2.11 Å) is present in all of the studied compounds (Figure 1), and it is associated with an excess of the 10 wt % used MgO relative to the stoichiometry of the reaction (1) in accordance with the technique [10].

4.2. The Leaching Rate and Mechanism of Actinides and REE from MKP Compounds

The leaching rate of all metals decreases depending on the contact time of the studied compounds with water (Figure 3a,c,e). However, a significant difference in the rate of uranium leaching from compounds #1 and #3 was established: At the 28th day, the differential uranium leaching rate is 5.5×10^{-7} and 4.4×10^{-4} g/(cm^2·day), respectively (Figure 3a). It was determined by XRD and SEM (Figures 1a and 2a) that uranium in compound #1 is bound in a slightly soluble phosphate, which is analog of the natural mineral metaankoleite [24,25], which provides a high resistance of compound #1 to uranium leaching. Obviously, the formation of such stable phase did not occur under the high-salt HLW surrogate solution solidification in compound #3, due to the presence of a large amount of nitrates of various metals (salt background—369.5 g·L^{-1}).

Data on the uranium leaching mechanism from compounds #1 and #3 (Figure 3b, Table 3) confirm the difference in leaching behavior. The logarithmic dependence of the uranium yield from compound #1 on the time of contact with water can be divided into two sections, which are described by linear equations with the slopes are 0.80 and 0.55 for seven days from the beginning of the test and the next 21 days, respectively. Thus, the uranium leaching mechanism varies depending on the duration of contact of the compound with water. So, during the first seven days, uranium leaching occurs due to surface dissolution of the compound, where individual particles of hydrated uranyl nitrate were localized. In the next 21 days, the uranium leaching is precisely determined by the diffusion transport from the inner layers of the compound. The uranium leaching from compound #3 in the first 10 days, is determined by the intensive surface dissolution, which is probably enriched by uranyl nitrate, and in the next 18 days—due to the gradual surface depletion (the slopes are 1.12 and 0.06, respectively).

The leaching rate of trivalent REE from the MKP compound increases with the increasing of content of various salts in the compound (for example, lanthanum and neodymium, Figure 3c). So, for compounds #2 and #3 the differential leaching rate of REE on the 28th day is 4.1×10^{-6} g/(cm^2·day) for lanthanum from compound #2, and 2.4×10^{-5} g/(cm^2·day) for neodymium from compound #3. It is obvious (Figure 3d) that, under the contact of compound #2 with water, the lanthanum leaching rate will decrease, since lanthanum leaching for 14 days is probably due to the surface dissolution (slope = 0.68) of soluble lanthanum forms of compound #2 in consequence of significant its content in the compound (6.2 wt %), and after 14 days by surface depletion (slope = −0.39). It is important to note that the neodymium leaching from compound #3 is uniquely determined by diffusion from the inner layers of the compound (slope = 0.54), which probably contains a uniformly distributed phase of hydrated neodymium nitrate that is unbound in slow-soluble phosphate forms.

The ^{239}Pu leaching rate is the main criterion of matrix quality evaluation for HLW immobilization. It has been established that compound #3 reliably kept both plutonium and americium: the differential leaching rate of ^{239}Pu and ^{241}Am on the 28th day is 3.5×10^{-7} and 5.3×10^{-7} g/(cm^2·day), respectively (Figure 3e). The plutonium yield from compound #3 in the leaching agent in the first 10 days occurred under the surface dissolution of the compound (slope = 0.96), and then by diffusion transport (slope = 0.41, Figure 3f). Americium leaching is also determined by diffusion transport (slope = 0.54), which is probably from the slow-soluble mixed orthophosphate (Am, REE)PO$_4$, which is an analogue of the natural mineral monazite.

The established low value of the ^{239}Pu leaching rate from MKP compound is close to the standard requirements for the glass-like compound for HLW immobilization (1×10^{-7} g/(cm^2·day)). However, it is important to note that MKP compound is synthesized at room temperature, whereas vitrification requires the use of expensive high-temperature electric furnaces or special melters, the liquidation of which after the end of the service life is a great radioecological problem, which is yet unsolved. Thus, the MKP compound approbation for immobilization of real wastes samples that were obtained by radiochemical plants during reprocessing of SNF, and a systematic comparison of MKP and glass-like compound quality indicators, including taking into account the technical and the economic evaluation of these technologies, are of scientific interest.

5. Conclusions

As a result of the research, it was established that MKP compounds that were synthesized at room temperature under solidification of nitric acid solutions, which are the surrogate solution of LRW, having complex chemical composition and containing actinides and REE, consisting of crystalline phases—analogues of natural phosphate minerals and possess high hydrolytic stability. Thus, the MKP compound is promising for the immobilization of LRW and it can be an alternative material combining the advantages of technology implementation simplicity that is similar to cementation and high physical and chemical stability corresponding to the glass-like compound.

Author Contributions: S.E. Vinokurov and B.F. Myasoedov conceived and designed the experiments; S.E. Vinokurov and S.A. Kulikova performed the experiments; S.E. Vinokurov, S.A. Kulikova and B.F. Myasoedov wrote the paper.

Funding: The research was funded by the Russian Science Foundation (No 16-13-10539).

Acknowledgments: Determination of element concentrations in solutions by ICP-MS and ICP-AES was performed at the Laboratory of Methods for Investigation and Analysis of Substances and Materials, Vernadsky Institute of Geochemistry and Analytical Chemistry, RAS (A.V. Zhilkina, I.N. Gromyak).

Conflicts of Interest: The authors declare no conflict of interest. The founding sponsors had no role in the design of the study; in the collection, analyses, or interpretation of data; in the writing of the manuscript, and in the decision to publish the results.

References

1. Ojovan, M.I.; Lee, W.E. *An Introduction to Nuclear Waste Immobilisation*, 2nd ed.; Elsevier: Amsterdam, The Netherlands, 2014; pp. 1–362, ISBN 978-0-08-099392-8.
2. Stefanovsky, S.V.; Yudintsev, S.V.; Vinokurov, S.E.; Myasoedov, B.F. Chemical-technological and mineralogical-geochemical aspects of the radioactive waste management. *Geochem. Int.* **2016**, *54*, 1136–1156. [CrossRef]
3. Stefanovsky, S.V.; Stefanovskaya, O.I.; Vinokurov, S.E.; Danilov, S.S.; Myasoedov, B.F. Phase composition, structure, and hydrolytic durability of glasses in the Na$_2$O-Al$_2$O$_3$-(Fe$_2$O$_3$)-P$_2$O$_5$ system at replacement of Al$_2$O$_3$ by Fe$_2$O$_3$. *Radiochemistry* **2015**, *57*, 348–355. [CrossRef]
4. Ewing, R.C.; Lutze, W.F. High-level nuclear waste immobilization with ceramics. *Ceramics Int.* **1991**, *17*, 287–293. [CrossRef]
5. Schlenz, H.; Neumeier, S.; Hirsch, A.; Peters, L.; Roth, G. Phosphates as safe containers for radionuclides. In *Highlights in Applied Mineralogy*; Heuss-Aßbichler, S., Amthauer, G., John, M., Eds.; De Gruyter: Munich, Germany, 2017; pp. 171–196, ISBN 9783110497342.
6. Schlenz, H.; Heuser, J.; Neumann, A.; Schmitzl, S.; Bosbach, D. Monazite as a suitable actinide waste form. *Z. Kristallogr. Cryst. Mater.* **2013**, *228*, 113–123. [CrossRef]

7. Wagh, A.S. *Chemically Bonded Phosphate Ceramics: Twenty-First Century Materials with Diverse Applications,* 2nd ed.; Elsevier: Amsterdam, Netherlands, 2016; pp. 1–422, ISBN 978-0-08-100380-0.

8. Roy, D.M. New Strong Cement Materials: Chemically Bonded Ceramics. *Science* **1987**, *235*, 651–658. [CrossRef] [PubMed]

9. Vinokurov, S.E.; Kulikova, S.A.; Krupskaya, V.V.; Danilov, S.S.; Gromyak, I.N.; Myasoedov, B.F. Investigation of the leaching behavior of components of the magnesium potassium phosphate matrix after high salt radioactive waste immobilization. *J. Radioanal. Nucl. Chem.* **2018**, *315*, 481–486. [CrossRef]

10. Vinokurov, S.E.; Kulikova, S.A.; Krupskaya, V.V.; Myasoedov, B.F. Magnesium Potassium Phosphate Compound for Radioactive Waste Immobilization: Phase Composition, Structure, and Physicochemical and Hydrolytic Durability. *Radiochemistry* **2018**, *60*, 70–78. [CrossRef]

11. Myasoedov, B.F.; Kalmykov, S.N.; Kulyako, Y.M.; Vinokurov, S.E. Nuclear fuel cycle and its impact on the environment. *Geochem. Int.* **2016**, *54*, 1156–1167. [CrossRef]

12. Vinokurov, S.E.; Kulyako, Y.M.; Slyunchev, O.M.; Rovny, S.I.; Myasoedov, B.F. Low-temperature immobilization of actinides and other components of high-level waste in magnesium potassium phosphate matrices. *J. Nucl. Mater.* **2009**, *385*, 189–192. [CrossRef]

13. Vinokurov, S.E.; Kulyako, Y.M.; Slyunchev, O.M.; Rovnyi, S.I.; Wagh, A.S.; Maloney, M.D.; Myasoedov, B.F. Magnesium potassium phosphate matrices for immobilization of high-level liquid wastes. *Radiochemistry* **2009**, *51*, 65–72. [CrossRef]

14. Wagh, A.S.; Sayenko, S.Y.; Shkuropatenko, V.A.; Tarasov, R.V.; Dykiy, M.P.; Svitlychniy, Y.O.; Virych, V.D.; Ulybkina, E.A. Experimental study on cesium immobilization in struvite structures. *J. Hazard. Mater.* **2016**, *302*, 241–249. [CrossRef] [PubMed]

15. Wagh, A.S.; Strain, R.; Jeong, S.Y.; Reed, D.; Kraus, T.; Singh, D. Stabilization of Rocky Flats Pu-contaminated ash within chemically bonded phosphate ceramics. *J. Nucl. Mater.* **1999**, *265*, 295–307. [CrossRef]

16. Singh, D.; Mandalika, V.R.; Parulekar, S.J.; Wagh, A.S. Magnesium potassium phosphate ceramic for ^{99}Tc immobilization. *J. Nucl. Mater.* **2006**, *348*, 272–282. [CrossRef]

17. Graeser, S.; Postl, W.; Bojar, H.-P.; Berlepsch, P.; Armbruster, T.; Raber, T.; Ettinger, K.; Walter, F. Struvite-(K), $KMgPO_4 \cdot 6H_2O$, the potassium equivalent of struvite – a new mineral. *Eur. J. Miner.* **2008**, *20*, 629–633. [CrossRef]

18. GOST R 52126-2003. *Long Time Leach Testing of Solidified Radioactive Waste Forms*; Gosstandart of Russia: Moscow, Russia, 2003; pp. 1–8.

19. De Groot, G.J.; van der Sloot, H.A. Determination of leaching characteristics of waste materials leading to environmental product certification. In *Stabilization and Solidification of Hazardous, Radioactive and Mixed Wastes: 2nd Volume*; Gilliam, T.M., Wiles, C.C., Eds.; ASTM International: West Conshohocken, PA, USA, 1992; Volume 2, pp. 149–170. [CrossRef]

20. Torras, J.; Buj, I.; Rovira, M.; de Pablo, J. Semi-dynamic leaching tests of nickel containing wastes stabilized/solidified with magnesium potassium phosphate cements. *J. Hazard. Mater.* **2011**, *186*, 1954–1960. [CrossRef] [PubMed]

21. Al-Abed, S.R.; Hageman, P.L.; Jegadeesan, G.; Madhavan, N.; Allen, D. Comparative evaluation of short-term leach tests for heavy metal release from mineral processing waste. *Sci. Total Environ.* **2006**, *364*, 14–23. [CrossRef] [PubMed]

22. Moon, D.H.; Dermatas, D. An evaluation of lead leachability from stabilized/solidified soils under modified semi-dynamic leaching conditions. *Eng. Geol.* **2006**, *85*, 67–74. [CrossRef]

23. Xue, Q.; Wang, P.; Li, J.-S.; Zhang, T.-T.; Wang, S.-Y. Investigation of the leaching behavior of lead in stabilized/solidified waste using a two-year semi-dynamic leaching test. *Chemosphere* **2017**, *166*, 1–7. [CrossRef] [PubMed]

24. Gallagher, M.J.; Atkin, D. V-Meta-ankoleïte, hydrated potassium uranyl phosphate. *Bull. Geol. Soc. Great Britain.* **1966**, *25*, 49–54.

25. Fleischer, M. New mineral names. *Am. Miner.* **1967**, *52*, 559–564.

materials

MDPI

Article

Analysis of the Secondary Phases Formed by Corrosion of U_3Si_2-Al Research Reactor Fuel Elements in the Presence of Chloride Rich Brines

Andreas Neumann [1,*], Martina Klinkenberg [2] and Hildegard Curtius [3]

[1] Institute of Geoscience and Geography, Mineralogy & Geochemistry, Martin-Luther University Halle-Wittenberg, Von-Seckendorff-Platz 3, D-06120 Halle (Saale), Germany

[2] Institute of Energy and Climate Research, IEK-6 Nuclear Waste Management, Forschungszentrum Jülich GmbH, Wilhelm-Johnen-Strasse, D-52425 Jülich, Germany; m.klinkenberg@fz-juelich.de

[3] Building and Property Management, Forschungszentrum Jülich GmbH, Wilhelm-Johnen-Strasse, D-52425 Jülich, Germany; h.curtius@fz-juelich.de

* Correspondence: andreas.neumann@geo.uni-halle.de; Tel.: +49-345-55-26076

Received: 25 May 2018; Accepted: 27 June 2018; Published: 30 June 2018

Abstract: Corrosion experiments with non-irradiated U_3Si_2-Al research reactor fuel samples were carried out in synthetic $MgCl_2$-rich brine to identify and quantify the secondary phases because depending on their composition and on their amount, such compounds can act as a sink for the radionuclide release in final repositories. Within the experimental period of 100 days at 90 °C and anoxic conditions the U_3Si_2-Al fuel sample was completely disintegrated. The obtained solids were subdivided into different grain size fractions and non-ambient X-ray diffraction (XRD) was applied for their qualitative and quantitative phase analysis. The secondary phases consist of lesukite (aluminum chloro hydrate) and layered double hydroxides (LDH) with varying chemical compositions. Furthermore, iron, residues of non-corroded nuclear fuel (U_3Si_2), iron oxy hydroxides and chlorides were also observed. In addition to high amorphous contents (>45 wt %) hosting the uranium, the quantitative phase analysis showed, that LDH compounds and lesukite were the major crystalline phases. Scanning electron microscopy (SEM) and energy dispersive -Xray spectroscopy (EDS) confirmed the results of the XRD analysis. Elemental analysis revealed that U and Al were concentrated in the solids. However, most of the iron, added as Fe(II) aqueous species, remained in solution.

Keywords: research reactor fuel element U_3Si_2-Al; spent nuclear fuel; corrosion; secondary phases; layered double hydroxides LDH; lesukite

1. Introduction

Due to considerable long-term impacts on the environment and society the waste management of spent nuclear fuel (SNF) is one of the most challenging issues for which sustainable disposal solutions must be found [1–3]. Yet, SNF arises not only from nuclear power plants it is also accumulated in research reactors of which currently around 250 are globally in operation. The composition of the fuel elements of such facilities differs from the oxide fuel types (UO_2) of power plants and often consists of uranium bearing metallic alloys which are dispersed in an aluminum matrix. However, radioactivity is produced likewise by fission of uranium and therefore minor actinides, plutonium, and fission products are also a critical feature of spent research reactor fuel elements. Nuclear waste management has thus to cover this type of high-level waste (HLW) on a scientific (e.g., [4–15]) and regulative basis (e.g., [16,17]) as well. In some cases, existing contracts with manufactures of research reactor fuel elements regulate the return shipment of medium enriched fuel types (research reactors FRG-1 in Geesthacht and BER II in Berlin).

However, spent fuel (highly enriched U_3Si_2-Al with U-235 > 90 wt %) from the research reactor FRM II in Munich is considered to be disposed of in a final repository for HLW waste. It is therefore necessary to store the irradiated U_3Si_2-Al fuel elements in massive iron bearing containers (CASTOR® MTR2 casks or alternatively in modified BSK-3 spent fuel coquilles) [18,19].

The long-term performance of the waste package, which has hence to be evaluated, will not only be governed by its design, it also depends on the chemical and physical situation in the deep geologic repository. Considering the long term safety of approx. 10^6 years, storage conditions may change over time. The safety assessment of final repositories has to consider processes, which will lead to an alteration of the disposed SNF.

Among those, the formation of secondary phases is relevant and has to be investigated, because the corrosion products constitute a sink for the radionuclide release and define thus parameters for the source term. This study focuses on the identification and quantification of secondary phases which were retrieved by corrosion of non-irradiated U_3Si_2-Al research reactor fuel elements in $MgCl_2$ rich brine (which accounts for a repository in salt formations). Such investigations are also important due to the fact that corrosion of aluminum dispersed fuels exhibit higher degradations rates than those being determined for UO_2 fuels of commercial nuclear power plants.

Wiersma [9] investigated the corrosion of different (non- and irradiated) fuel types (UAl, UAlx, U_3O_8, and U_3Si_2 alloys and chemical compounds). The microstructural investigations assumed the formation of gibbsite, hydragillite, or bayerite as corrosion products. Surface analysis revealed also the formation of boehmite (cf. also [15]).

Corrosion rates presented and reviewed by Hilton [20] refer, among other fuel types, to the interaction of UAl$_x$- and U$_x$Si$_y$-based fuels only with pure water. Therein, it was reported that at 80 °C the corrosion of the Al component of this fuel type leads to the formation of boehmite (AlO(OH)). The reaction rates of UAl$_x$-Al dispersion fuel in water were essentially the same as those of the aluminum alloys. The Arrhenius expression for the aluminum alloy-H_2O reaction was determined as k_{linear} = 4.29exp(32.8 ± 1.8 kJ/(molRT)) (where k has units of mg metal/cm^{2}h) for temperatures ranging from 25 to 360 °C. At 80 °C the rate constant could be estimated as 0.11 g/(m^{2}d). The reported behavior of U$_x$Si$_y$-Al based fuels differs considerably. Most of the defected fuel plates exhibited a reaction rate ~10^5 times faster than the rate of aluminum alloys and 100–1000 times faster than uranium silicide inter-metallics. This observation may be due to freshly exposed metal (machined holes as defect source) and/or an accelerated crevice corrosion.

Kaminski et al. [12] determined a corrosion rate of 9.7×10^{-2} g/(m^{2}d) for a UAl$_x$-Al sample for repository relevant conditions; the experiments were carried out at 90 °C by dripping permanently EJ-13 modified well water (Yucca Mountain site) up to 183 d on the specimen.

Although test conditions were different (accounting for different national disposal regulations/strategies) corrosion rates for research reactor fuel elements determined by Curtius et al. [21] are of similar magnitude. The experiments were carried out at 90 °C under anoxic conditions in the presence of Fe(II) with irradiated and non-irradiated samples in $MgCl_2$ rich brine (U_3Si_2-Al$_{irr}$: 4.24×10^{-2} g/(m^{2}d), U_3Si_2-Al: 5.74×10^{-2} g/(m^{2}d), UAl$_x$-Al$_{irr}$: 7.69×10^{-3} g/(m^{2}d), UAl$_x$-Al: 1.02×10^{-3} g/(m^{2}d)) and in clay pore water (U_3Si_2-Al$_{irr}$: 6.93×10^{-2} g/(m^{2}d), U_3Si_2-Al: 2.36×10^{-2} g/(m^{2}d), UAl$_x$-Al$_{irr}$: 1.05×10^{-3} g/(m^{2}d), UAl$_x$-Al: 2.68×10^{-3} g/(m^{2}d)). This study showed that corrosion in $MgCl_2$ rich brine is somewhat faster than it was observed in clay pore water. However, comparing irradiated U_3Si_2-Al/UAl$_x$-Al fuels [21] with irradiated UO_2 (2.36×10^{-6} g/(m^{2}d) [22], the corrosion rate of the alloys was increased up to ~3–4 orders of magnitude. Both experiments considered corrosion in chloride rich solution, but the setup of Loida [22] differs in some aspects, considering fuel sample specifications, temperature (25 °C) and composition of the brine (NaCl$_{sat}$) as well as the iron supply simulating the waste package. After an experimental period of 3.5 years the irradiated U_3Si_2-Al and UAl$_x$-Al fuel samples were fully decomposed. This implies a very fast release of the radioactive inventory. Yet, the radio analytical investigations of the secondary phases of the U_3Si_2-Al$_{irr}$/UAl$_x$-Al$_{irr}$ fuel sample corrosion showed that the long-lived ^{234}U, ^{238,239,240}Pu and ^{241}Am isotopes were immobilized by the solids [21]. This is observed for MgCl$_2$ rich brine and for Mont Terri clay pore water as well.

For the phase specific evaluation, considering the retention capacity of the solids, complementary tests with non-irradiated fuel were performed to identify and quantify the secondary phases. Kaminski et al. [12,14] observed the formation of a silica-substituted hydrous aluminum gel layer on the sample surface. Additionally, dehydrated uranyl oxyhydroxides, schoepite ([(UO$_2$)$_4$ | O | (OH)$_6$]·6H$_2$O), becquerelite (Ca(UO$_2$)$_6$O$_4$(OH)6·8(H$_2$O)) and colloids, prevailingly silica rich, were also formed. More than 99 wt % of the dissolved uranium was bound to the colloids which exhibit a diameter of some hundred nanometers. Mazeina et al. [23,24] carried out experiments with UAl$_x$-Al fuels under reducing conditions (identically to Curtius et al. [21]) and observed the formation of the crystalline phases hydrotalcite, i.e., LDH (layered double hydroxides—common composition: [M^{2+}$_{1-x}$ M^{3+}$_x$(OH)$_2$]$^{q+}$(X^{n-})$_{q/n}$·yH$_2$O) and bischofite (MgCl$_2$·6H$_2$O). In a more recent study with UAl$_x$-Al by Klinkenberg et al. [25] LDH was again observed, yet with varying chemical compositions. Additionally, lesukite, an aluminum hydroxy chloride hydrate, being described by Vergasova et al. [26] and Witzke [27], was also identified as a major phase. Small amounts of iron (III) oxy hydroxides and iron (II) chlorides were observed as well. Further accessories like metallic iron and residues of nuclear fuel were also present. The amorphous content mounts up to ~20 wt % for UAl$_x$-Al. In these studies [25,28], it was shown that the stability of observed phases strongly depends on the post treatment, i.e., on the chosen liquid (water or isopropanol) for the retrieval of the secondary phases. After the corrosion experiment of the non-irradiated fuel sample was finished the residues which were treated with water showed different secondary phases compared to those being treated with isopropanol. This is especially valid for lesukite which was not observed as a secondary phase considering the water treatment. Instead, different aluminum hydroxides (boehmite, nordstrandite, gibbsite) were observed [21,28] and hence indicating that isopropanol is more beneficial for a post treatment of corrosion solids.

Further studies [29] with UAl$_x$-Al in the presence of standardized clay pore water (Mont Terri type [30]) were carried out. The secondary crystalline phases gypsum, bassanite, goethite, and boehmite were identified. Non-corroded leftovers of UAl$_4$ were also observed. The amorphous content exceeded 80 wt % for the system UAl$_x$-Al in clay pore solution.

In this study non-irradiated U$_3$Si$_2$-Al fuel elements were corroded in MgCl$_2$ rich solution (salt host rock). Fe(II)$_{aq.}$ was also added to simulate the decomposition of the waste package. The experiments focus on the non-ambient laboratory XRD phase analysis of the secondary phases. Efforts were taken and unique experimental equipment was applied to prevent the secondary phases from alteration by oxidation during retrieving, treatment, and analysis.

2. Materials and Methods

2.1. Setup of the Corrosion Experiments and Sample Pre-Treatment

The corrosion experiment (static batch, 90 °C) with an U$_3$Si$_2$-Al sample was carried out under anoxic condition in standardized solution (brine 2, cf. [6]). The small cut non-irradiated fuel platelet (40.2 × 20.0 × 1.4 mm^3, S/V = 15.8 cm^{-1}) weighed totally 4.40 g (m(U) = 1.6 g, m(Al) = 2.30 g, m(Si) = 0.28 g). The U$_3$Si$_2$-Al fuel matrix was two-sided covered with an aluminum cladding. The sample was put into a glass autoclave with 400 mL of magnesium chloride rich brine. 10 g of FeCl$_2$·H$_2$O were added to the solution to simulate the corrosion of the iron bearing waste package. The vessel was tightly closed, put into a drying oven and was heated to 90 °C. The corrosion progress was monitored by a probe measuring the hydrogen pressure built up. pH was also recorded; at the beginning of the experiment pH was little more than 1 and reached after ~75 days a constant value between 4 and 5. More specific details concerning setup, data monitoring, and the fuel sample are described by Curtius et al. [21].

The hydrogen pressure built up in the autoclave due to reducing conditions was monitored to observe indirectly the corrosion progress. After 100 days no further increase of the pressure was observed and the secondary phases were retrieved out of the vessel. Efforts were taken that every work step considering sample retrieval, pre-treatment, and drying was carried out under argon atmosphere. Inert conditions were necessary to prevent the secondary phases from alteration by oxidation. The suspension has been retrieved and separated for the pre-treatment. One part was used for the grain size classification into the fractions >63 μm, 2–63 μm, and <2 μm. This was achieved by wet sieving with isopropanol to obtain the fraction >63 μm. The subdivision of the smaller fraction was carried out also in isopropanol by a sedimentation procedure according to Atterberg [31]. Additionally, an analogous treatment for retrieving the secondary phases carried out again according to the protocol as described by Klinkenberg et al. [25].

The remaining part of the retrieved suspension was used to determine the amount of U, Al, Ca, Si, and Fe. After centrifugation the supernatant was used for elemental analysis. To determine the U content a Liquid Scintillation Counter (LSC) TRI-CARB 2020 (PerkinElmer, Waltham, MA, USA) and an α-analyzer for α-spectrometry (Canberra-Packard GmbH, Schwadorf, Austria) were used. Al, Ca, Si, and Fe were analyzed by ICP-OES (inductively coupled plasma optical emission spectroscopy ELAN 6100 DRC (PerkinElmer, Waltham, MA, USA) with a TJA-IRIS instrument. A full description of these analytical procedures is given elsewhere [21,25]. The estimation of the water content of the untreated sample was determined by drying for one week at 105 °C in argon atmosphere in which 82.57% of the weight accounted for water.

2.2. X-ray Diffraction Analysis

The phase identification was evaluated with the DiffracPlus software from Bruker-AXS (Karlsruhe, Germany) by retrieving the powder diffraction file PDF-2 (ICDD Release 2007). The amount of each crystalline phase and the amorphous content was determined by the Rietveld method [32,33]. Therefore, an internal non-certified zincite (ZnO) standard (Merck, Darmstadt, Germany) of known weight has been added for the quantitative X-ray phase analysis (QPA).

The applied structures, i.e., the retrieved CIF-Files of the ICSD (Inorganic Crystal Structure Database) of the identified phases are summarized in Table 1. Exceptionally for lesukite a structure did not exist. Therefore, a model was derived [28].

Table 1. Phase quantities (crystalline, amorphous, and total) in dependence of the different grain size fractions and R_{wp} values of the Rietveld refinements. The left column also features the phase composition, phase name, and database (PDF-2 and ICSD) reference numbers of the identified crystalline phases.

Phase (PDF-2 No./ICSD No.)	Weight/%		
	<2 μm	2–63 μm	>63 μm
$Al_2Cl(OH)_5 \cdot 2H_2O$ lesukite (00-031-0006/-)	95.37 ± 0.33	51.00 ± 2.75	13.87 ± 1.26
$(Mg_{0,67}Al_{0,33}(OH)_2) \cdot (CO_3)_{0.165} \cdot (H_2O)_{0.48}$ LDH 3R (01-089-0460/86655)	1.43 ± 0.28	17.00 ± 4.08	7.37 ± 0.66
$Al_2Mg_4(OH)_{12}(CO_3)(H_2O)_3$ LDH 2H 00-020-0658/82874		12.35 ± 1.33	10.21 ± 0.96
$((Zn_{0,625}Al_{0,375})(OH)_2)(SO_4)_{0.188}$ LDH sulphate (01-070-6422/91859)	0.25 ± 0.08	3.94 ± 0.28	1.26 ± 0.19
$(Fe(OH)_2)((OH)_{0.25}(H_2O)_{0.5})$ green rust (00-040-0127/159700)		0.20 ± 0.06	2.15 ± 0.21
$Fe_8O_8(OH)_8Cl_{1.35}$ akaganeite (00-034-1266/69606)	2.05 ± 0.16	1.55 ± 0.15	0.65 ± 0.17
$FeO(OH)$ goethite (01-081-0462/245057)		0.53 ± 0.08	

Table 1. *Cont.*

Phase (PDF-2 No./ICSD No.)	Weight/%		
	<2 µm	2–63 µm	>63 µm
FeO(OH) lepidocrocite (01-070-8045/93948)		0.73 ± 0.11	
FeCl$_2$ lawrencite (01-070-1634/64830)			0.68 ± 0.07
U$_3$Si$_2$ uranium silizide (00-005-0628/73695)			0.78 ± 0.07
Fe iron (00-006-0696/84483)			0.66 ± 0.10
Amorphous	0.90 ± 2.00	12.70 ± 5.30	62.36 ± 2.74
Total	100.00	100.00	100.00
Relative fraction amount (%)	20.30	8.20	71.50
R$_{wp}$ (%)	12.21	8.20	0.95 *

* The very low weighted profile R-factor (R$_{wp}$) (%) of 0.95% for the fraction >63 µm results from a manipulation of the background by adding stepwise intensity to the diffractograms in order to improve the description of the background by polynomials. The given R$_{wp}$ is the mean value obtained by adding 100, 200, 300, 400, and 500 counts.

The diffractograms were recorded with a D8 diffractometer from Bruker-AXS (Karlsruhe, Germany). The space group and lattice parameters of lesukite and the secondary phase quantification have been computed with TOPAS [34,35] and BGMN [36]. Both programs use the fundamental parameter approach (FPA), i.e., the full diffractometer device function is thereby defined by the emission spectra of the X ray tube [37] and by the geometry of the beam path. The goniometer of the diffractometer features a θ–θ geometry. For the XRD measurements CuK$_\alpha$ radiation (λ_1 = 1.54059 Å) at 40 kV and 40 mA was applied. Further details about the diffractometer setup are given elsewhere (cf. [28,37]).

For the analyses it was crucial to avoid a sample alteration due to oxidation during the measurements. Therefore, the samples were put into a climate chamber from MRI (Materials research instruments). This device has been purged permanently with nitrogen while the diffractograms were recorded at room temperature.

2.3. SEM/EDS Analysis

The morphology and the chemical composition of the secondary phases were investigated with a FEI Quanta 200 ESEM FEG (Hillsboro, OR, USA). The instrument was equipped with an Apollo X silicon drift detector (EDAX, Mahwah, NJ, USA) for energy-dispersive X-ray spectroscopic (EDS) measurements. The particles of the different grain size fraction were prepared on adhesive carbon tabs without any previous sputtering. The samples were analyzed in low vacuum mode (0.6 mbar) at 20 kV with spot size 4, and 10 mm working distance. The investigations were carried out with the large field low vacuum detector LFD for secondary electrons and BSED (backscattered electrons) detectors of the SEM device.

To account for micro-absorption [38] the particle dimensions of the different grain size fractions were also measured. Some micrograms of each fraction were thus suspended in isopropanol, sonicated for several minutes, and then prepared on non-adhesive carbon tabs. Based on the assumption of spherical particle shape, the average diameter was determined with the image analysis software EDAX Genesis V 6.2.

3. Results and Discussion

3.1. X-ray Analysis of the Secondary Phases

Figure 1 shows the results of the qualitative and the quantitative phase analyses in dependence of the different grain size fraction <2 µm, 2–63 µm, and >63 µm. The qualitative phase analysis is given by the diffractograms shown left (Figure 1a,c,e). Results of quantitative phase analysis are given by the related Rietveld plots on the right column (Figure 1b,d,f).

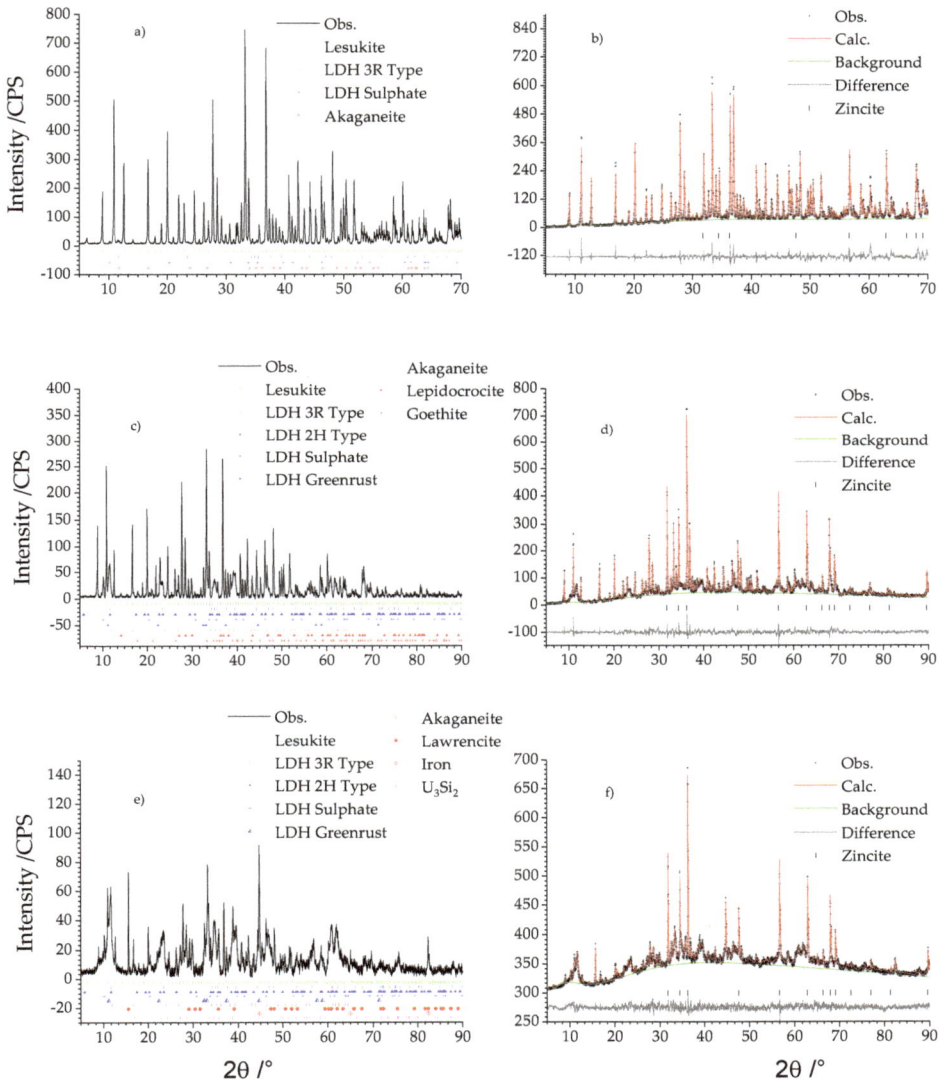

Figure 1. Qualitative (**left**) and quantitative (**right**) phase analyses of the secondary phases (**a**) Qualitative phase analysis of the fraction <2 μm, (**c**) qualitative phase analysis of the fraction 2–63 μm and (**e**) qualitative phase analysis of the fraction >63 μm. (**b**) QPA of the fraction <2 μm, (**d**) QPA of the fraction 2–63 μm, and (**f**) QPA of the fraction >63 μm.

The fraction <2 μm exhibited only three different compounds (cf. Table 1): akaganeite, lesukite, and two types of LDH of which one incorporated sulphate and the other chloride in the interlayer. This was inferred by analysing the respective (001) basal reflections, which showed for sulphate intercalation an increase of the *d* spacing of adjacent layers normal to the *c* axis. The *d* spacing for the chloride type LDH is approx. 8 Å and approx. 8.6 Å for the sulphate type.

Lesukite was quantitatively the major phase and mounted up to 95.37 ± 0.31 wt %. Chloride and sulphate type LDH exhibit accessory amounts of 1.43 ± 0.26 and 0.25 ± 0.08 wt %. A further accessory mineral is akageneite with 2.05 ± 0.15 wt %. The amorphous content was practically not existent due to the determined uncertainties.

The medium sized fraction from 2 to 63 µm featured additionally three different Fe(III) oxy hydroxides (akaganeite, goethite, lepidocrocite) and two new LDH compounds manasseite (2H LDH type) and greenrust (cf. Table 1). Quantitatively the grain size fraction 2–63 µm was still dominated by lesukite although its relative amount has nearly been halved to 51.00 ± 2.90 wt %. Approx. 30 wt % could be ascribed to the 3R (17.00 ± 4.30 wt %) and 2H (12.35 ± 1.40 wt %) LDH types. 3.94 ± 0.30 wt % was calculated for the sulphate LDH. The content of greenrust was very low (0.20 ± 0.06 wt %). Likewise, the Fe(III) oxy hydroxides: akaganeite (1.55 ± 0.16 wt %), goethite (0.53 ± 0.12 wt %), and lepidocrocite (0.73 ± 0.09 wt %). The amorphous content increased to 12.7 ± 5.30 wt %.

In the fraction >63 µm (cf. Table 1) residues of non-corroded fuel with the composition U_3Si_2 could be observed. Lesukite and LDH compounds were still present. This was also valid for akaganeite. Moreover, Fe(0) and lawrencite (Fe(OH,Cl)$_2$), were also observed. Compared to the fractions <63 µm the quantity of lesukite was again reduced and exhibited 13.87 ± 3.02 wt %. This trend is also valid for the LDH compounds: 3R-type (7.37 ± 1.58 wt %), 2H-type (10.21 ± 3.74 wt %), sulphate type (1.26 ± 0.45 wt %). However, the greenrust increased to 2.15 ± 0.51 wt %. Considering the iron oxy hydroxides, akaganeite was reduced to 0.65 ± 0.40 wt.%. Lepidocrocite and goethite were not present anymore. Iron (0.66 ± 0.23 wt %), lawrencite (0.68 ± 0.16 wt %), and U_3Si_2 were of equal but just of minor content. The amorphous part increased to 62.36 ± 2.74 wt % constituting the most abundant phase in the fraction >63 µm. This increase could be explained by the sample preparation. This fraction was obtained just by sieving whereas the smaller fraction was additionally subjected to the Atterberg procedure for the further grain size subdivision. Thereby, it could not be ruled out that amorphous parts which may have been present after sieving in the smaller fraction were dissolved during this application. This assumption is also valid for the UAl$_x$-Al being subjected to MgCl$_2$ rich solution [25] solution and clay pore water (Mont Terri type) [29].

Especially with respect to the study of UAl$_x$-Al fuel in brine 2 [25,28] the system U_3Si_2-Al in brine 2 behaved similarly in many aspects. In both systems lesukite and LDH compounds are the major phases. Trace amounts of non-corroded fuel were also present and considerable amounts of amorphous phases were observed as well.

The occurrence of the different observed phases was also dependent on the grain size and showed a similar distribution (cf. [25,28]). Disregarding the observed residues of non-corroded nuclear fuel, other crystalline uranium bearing phases could neither be observed for the U_3Si_2-Al nor for the UAl$_x$-Al system [25,28] in chloride rich solution. However, the disintegration of UAl$_x$-Al fuel element in Mont Terri solution resulted in different corrosion behavior. Observed crystalline phases were goethite, calcium sulfates and residues of UAl$_4$, yet most of the solids were amorphous compounds which represented the greatest solid part [29]. Therefore, composition and specific surface are critical parameters which will have an impact on the source term. The results of this study and the investigations of [21,25,28,29] generally support the assumption of a similar corrosion behavior of Al dispersed UAl$_x$ and U_3Si_2 fuels (cf. [5]) and a faster corrosion of the aluminum component of Al based fuels was also affirmed because pure Al metal was not observed in the corrosion residues.

Figure 2 shows an overview of the quantities being normalized for each fraction to the total amount of all obtained secondary phases. From the magenta colored columns in the last row, representing the total of all fractions could be seen that the amorphous part with more than 45 wt % is the most abundant phase. Second and third ranked in quantity were lesukite and various LDH types. All other phases were just present as accessories. It is expected that the amorphous part contains the uranium because—with exception of the residues of U_3Si_2—no further crystalline uranium phases were observed. The observed content of uranium and silicon (as U_3Si_2) was very low compared to the originally supplied quantities. With respect to the findings of [39] and taking into account that uranium was not found in the liquid part of the suspension of brine 2 it could be assumed that uranium is thus quantitatively constituents of the amorphous phases. This result constitutes an important finding as the analogue experiments of Curtius et al. [39] with irradiated U_3Si_2-Al showed that not only uranium but also americium, plutonium, and europium, as well were mostly immobilized in the solid phase. Therefore, possible implications for the source term must be evaluated whether the radionuclides in the SNF are also part of the amorphous phases, because the solubility is a critical parameter for their immobilization.

The ICP-OES results for aluminum and iron of the liquid phase of the corrosion products showed that iron is found in solution with 61.0 wt % ($\pm$0.5 wt %) whereas aluminum was totally part of the solid phase. Results indicated that the latter was part of crystalline and amorphous phase as well. Most of the magnesium was detected in the liquid phase due to the high solubility of $MgCl_2 \cdot 6H_2O$ which has been used for the preparation of the chloride rich brine. Yet, LDH phases contain considerable amounts of magnesium and constitute a major secondary phase.

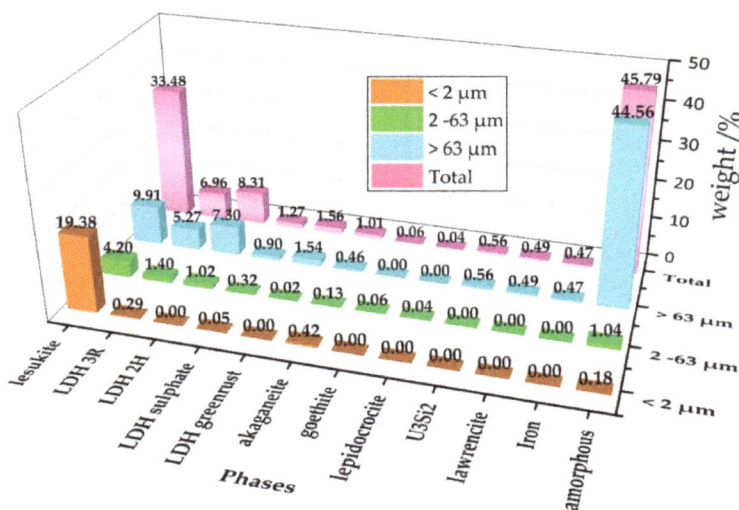

Figure 2. Quantitative phase distribution in dependence of the grain size fraction (Orange columns: <2 µm, green columns: 2–63 µm, light blue: >63 µm, and magenta: Total of all fractions).

Contrary, silicon is mainly found the solid part. Most of silicon is probably part of the amorphous phase because the only crystalline Si bearing phases were remnants of U_3Si_2 which only host minor amounts of silicon by comparing its amount with original quantities of the non-corroded fuel sample. The elemental analysis of calcium indicated to be dissolved in the liquid phase. Neither it was observed in a crystalline secondary phase nor could calcium be detected in the solid phase. The proportion of sulphur (122.46 wt %) was slightly overestimated compared to its initially supplied amounts (100 wt %). Yet, this could be due to the uncertainties given by the very low originally supplied amounts (<0.007 g), by the sample quartering and preparation [21], and by the quantification (cf. Table 1). Nevertheless, from this finding it could be concluded that sulphur neither was a part of the amorphous phases nor has it been dissolved. It was totally fixed in the sulphate LDH.

With respect to the safety assessment it is important to consider the phase stability of selected phases. The observed iron bearing phases exhibited valence states of 0, 2+, and 3+. One may interpret such a condition as a non-equilibrated system, yet artifacts due to preparation may also have an impact. Taking into account the quantitative development of the amorphous content of the different fractions it could not be ruled out that the Atterberg procedure gave reason for the observations made considering the iron valence state. Fe(0) and Fe(II) compounds (lawrencite, greenrust) were basically only observed in the fraction >63 µm whereas the fractions <63 µm are dominated by the Fe(III) compounds (akaganeite, goethite, and lepidocrocite). Although greenrust can accommodate Fe(II) as well, in the fraction 2–63 µm it was only of minor content. Therefore it could not be ruled out that during the Atterberg procedure Fe(0) and Fe(II) were oxidized although special care was taken to prevent the iron bearing secondary phases being altered. The impact of oxidation of Fe(0) and lawrencite (Fe(II)) of the fraction >63 µm is shown in Figure 3.

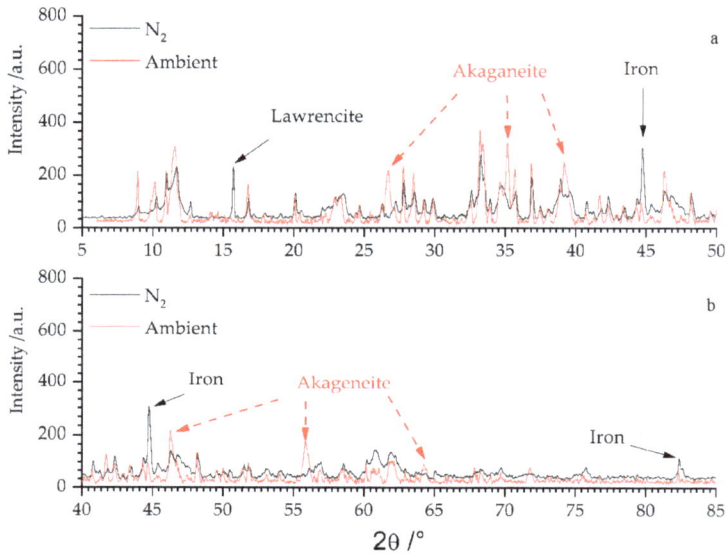

Figure 3. Comparison of diffractograms (divided for clarity's sake in two ranges from 5–50° 2θ (**a**) and from 40–85° 2θ (**b**); overlap region between 40–50° 2θ) of fraction >63 µm being analyzed under inert conditions (N$_2$—black line) and being subjected to ambient conditions (air: red line).

The black diffractogram, being recorded in N$_2$ atmosphere, shows the fraction >63 µm (cf. Figures 1 and 3). The peak positions of lawrencite and iron have been marked with black arrows. These reflections of the Fe$^{0/2+}$ species vanished and new peaks (red arrows) of akaganeite (Fe^{3+}) could be observed, when the sample was subjected to air for several weeks.

3.2. SEM and EDS Analysis of the Secondary Phases

Figure 4a–f shows the SEM/EDS analyses of the pre-treated solid secondary phases obtained from the autoclave. In Figure 4a the observed compound exhibited perfectly cubic shaped lesukite crystals with an edge of several hundred nanometers. These crystals were observed in each grain size fractions. The related EDS spectrum showed typical lines of aluminum, chloride, and oxygen. This observation is in good agreement with the results reported by Vergasova et al. [26] and Witzke [27]. In Figure 4b lawrencite (FeCl$_2$) is shown. This phase exhibited also platelet morphology. However, the crystals are ~10 times larger than the observed LDH compounds. The related EDS spectrum featured distinct iron and chlorine lines. This mineral belongs to the trigonal system and exhibited a layered structure which, contrary to LDH compounds (Figure 4d), did not feature any interlayer constituents (e.g., H$_2$O, Cl$^-$, SO$_2^{4-}$). The observed oxygen line could indicate quantitative exchange of chloride anions

($r_{shannon}$ = 1.7 Å) by hydroxide anions ($r_{shannon}$ = 1.4 Å). This assumption is based on the observation that the determined c lattice parameter—distance d of the layer spacing—is gradually smaller (5.62 Å) than the theoretical one (5.83 Å, cf. Table 1—ICSD code FeCl$_2$: 64830 [40]). Figure 4c showed a very bright phase. This could be attributed to non-corroded leftovers of U$_3$Si$_2$-fuel. The respective EDS showed the expected signals for U, Si, and Al.

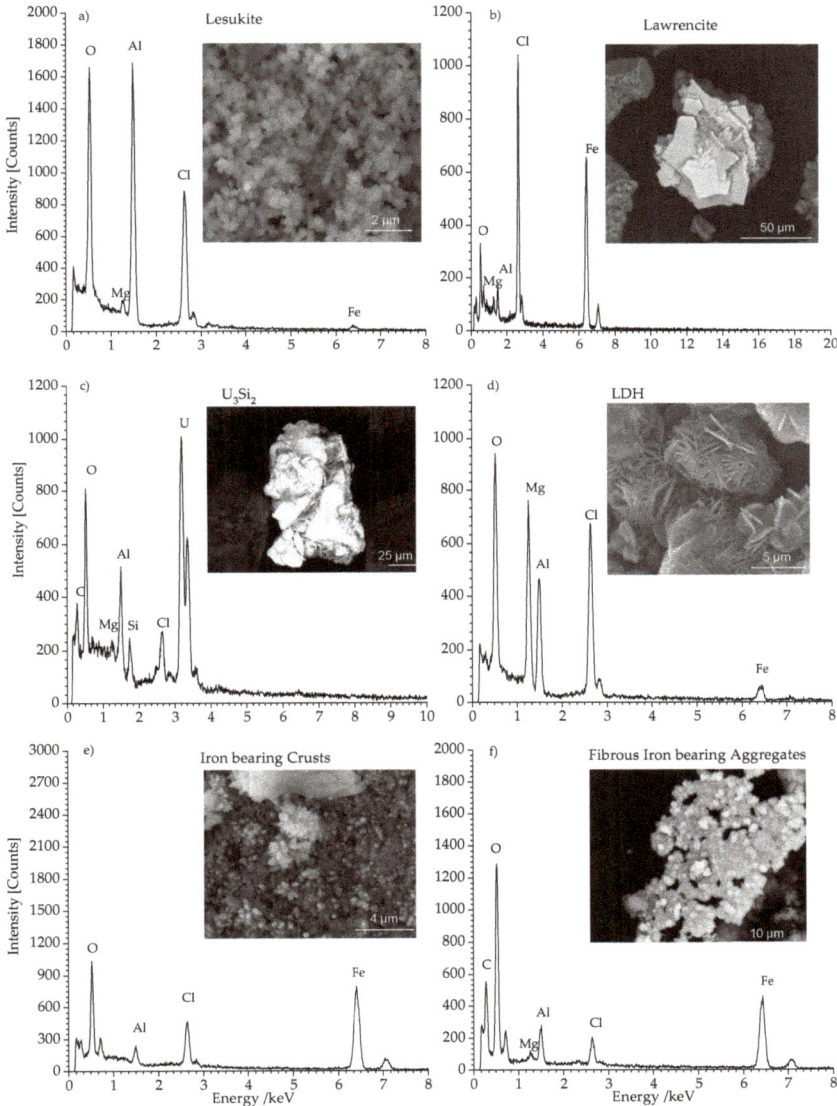

Figure 4. SEM (BSE) micrographs and EDS analyses of selected secondary phases: (**a**) shows cubic shaped lesukite, (**b**) shows platelets of Lawrencite FeCl$_2$. (**c**) shows remnants of non-corroded U$_3$Si$_2$. (**d**) shows the typical sand rose like appearance of LDH phase. (**e,f**) show Iron bearing phases.

This phase was prevailingly located in the largest fraction >63 µm. Its presence was already evidenced in the related diffractogram (cf. Figure 1c). Traces of U_3Si_2 could also be observed in the fraction 2–63 µm.

Figure 4d shows sand rose shaped aggregations of laminar crystals. These platelet crystals had an in plane dimension of ~5 µm. Normal to these planes the thickness was clearly less than 0.5 µm. This phase was identified as a LDH-type compound. The EDS spectrum in Figure 4d exhibited the expected lines of magnesium, aluminum, chlorine, and oxygen.

The LDH crystals are commonly observed in each grain size fraction (cf. Figure 1). The sulphate bearing LDH being clearly identified in the XRD analysis has not been observed by SEM/EDS analysis possibly due to the very low content of sulphate (cf. Table 1).

In Figure 4e the micrographs show iron compounds which formed coarse crusts. The crust formation was due to the sample desiccation especially in the fraction >63 µm. Fibrous aggregates were typically observed in fraction 2–63 µm (Figure 4f). Although Fe(III) oxy hydroxides (akaganeite, lepidocrocite, goethite), green rust, and Fe(0) have been observed in the diffractograms (cf. Figure 1), this phase did not show up in the related SEM/EDS analysis. This observation could be attributed to the poorly developed morphology and the small size of these iron bearing phases.

4. Conclusions

Within a short period (~100 days) the U_3Si_2-Al fuel sample corroded completely in the $MgCl_2$ brine in the presence of Fe(II)$_{aq}$. Elemental analysis (ICP-OES and LSC) showed that aluminum and uranium were quantitatively found in the secondary phases. Yet approx. 65 wt % of iron is remained in solution.

Special treatment was necessary for the characterization of the corrosion products, i.e., secondary phases were subdivided by sieving and by the Atterberg method in an inert atmosphere to prevent oxidative alteration, because XRD and SEM analyses revealed the presence of phases being sensitive to oxidation (iron, greenrust, and lawrencite, cf. Table 1).

As summarized in Table 1 the fraction <2 µm mainly consisted of cubic shaped lesukite. LDH, akaganeite, and the amorphous phases were of minor content. In the fractions >2 µm the LDH compounds became besides lesukite also major phases. The amount of the other crystalline phases still remained less than 5 wt %. Residues of non-corroded fuel U_3Si_2, Fe (0), and lawrencite (Fe(II) compound) were exclusively present in the fraction >63 µm.

Depending on grain sizes fraction the content of amorphous phases varied and iron compounds with different valance states were observed. Although efforts were taken oxidation during the pre-treatment of the samples could not be ruled out and may hence explain the presence of Fe^{3+} in some compounds (cf. Table 1). This seemed notably true for the smaller grain sizes as within the treatment procedure this fraction was the more sensitive to alteration considering specific surface of the samples and the treatment duration.

The amount of the amorphous phase could also be underestimated as during wet sieving some of the amorphous phases could be dissolved. This effect may even be increased for the fractions <63 µm because this material was additionally subdivided by the Atterberg method where further amorphous solids could be dissolved.

In addition to corrosion rates future prospects of safety related issues of research reactor fuel elements must thus focus on the characterization of uranium with respect to its physicochemical properties in the amorphous part. This is an important issue considering the release and sorption of radionuclides within this uranium bearing solid.

Furthermore, with respect to the corrosion rates the stability of each observed crystalline phase has to be determined individually. Consequently, it is important to get more insights into the physiochemical properties of lesukite in order to predict the sorption behavior of this compound for radionuclides under repository relevant conditions. First results with Eu^{3+} and SeO_4^{2-} indicated a potential of retardation of anionic species (selenate) whereas europium interacts only weakly with lesukite [28]. The interaction of LDH phases with nuclear relevant compounds are also under investigation [41–43], because the incorporation of radioactive cations (e.g., cobalt, europium etc.) in the main layer and radioactive anions in the interlayer could lead to an immobilization these compounds.

Author Contributions: The authors worked together in a research project. The project was supervised and conceived by H.C., A.N. and M.K. carried out material analysis as reported in this study. The obtained data of this study were analyzed by A.N. and M.K. Results were discussed by A.N., M.K., and H.C., A.N. wrote the paper.

Funding: This research was funded by the Federal Ministry Education and Research BMBF (grant number 02 E 10357). The German Research Foundation (DFG) is gratefully acknowledged for the financial support within the funding program Open Access Publishing.

Acknowledgments: The authors want to thank Reinhard Odoj, Dirk Bosbach, Georg Roth, and Bender Willich for their support and fruitful discussions.

Conflicts of Interest: The authors declare no conflict of interest. The founding sponsors had no role in the design of the study; in the collection, analyses, or interpretation of data; in the writing of the manuscript, and in the decision to publish the results.

References

1. Johnson, L.H.; Shoesmith, D.W. Spent fuel. In *Radioactive Waste Forms for the Future*; Lutze, W., Ewing, R.C., Eds.; Elsevier: Amsterdam, The Netherlands, 1988; p. 635, ISBN 0444871047.
2. Kienzler, B.; Loida, A. *Endlagerrelevante Eigenschaften von Hochradioaktiven Abfallprodukten—Charakterisierung und Bewertung—Empfehlungen des Arbeitskreises HAW-Produkte*; Karlsruhe Institute of Technology: Karlsruhe, Germany, 2001.
3. Fein, E.; Müller-Lyda, I.; Rübel, A. Anhang Langzeitsicherheitsanalyse. In *GRS Bericht 247/7—Endlagerung Wärmeentwickelnder Radioaktiver Abfälle in Deutschland*; Gesellschaft für Anlagen- und Reaktorsicherheit GRS: Cologne, Germany, 2008; ISBN 978-3-939355-22-9.
4. Brücher, H.; Curtius, H.; Kaiser, G.; Mazeina, L.; Fachinger, J. *Untersuchungen zur Radionuklidfreisetzung und zum Korrosionsverhalten von Bestrahltem Kernbrennstoff aus Forschungsreaktoren unter Endlagerbedingungen*; Berichte des Forschungszentrum Jülich 4104: Jülich, Germany, 2001.
5. Cloke, P.L.; Gottlieb, P.; Lester, D.H.; Fuenties, E.; Benton, H.A. *Geochemical Analysis of Degradation Modes of HEU SNF in a Codisposal Waste Package with HLW Canisters*; BBA000000-0171740200-0059 REV 01; CRWMS/M&O: Las Vegas, NV, USA, 1998.
6. Fachinger, J.; Brücher, H.; Rainer, H.; Kaiser, G.; Syuhada, I.; Zschunke, S.; Nau, K. *Untersuchungen zur Radionuklidfreisetzung durch Einwirkung konzentrierter Salzlaugen auf Alu-MTR-Brennelemente*; Berichte des Forschungszentrums Jülich 4104: Jülich, Germany, 1998.
7. Curtius, H.; Kaiser, G.; Paparigas, Z.; Ufer, K.; Müller, E.; Enge, E.; Brücher, H. *Untersuchungen zum Verhalten von FR-BE in Wirtsgesteinswässern Möglicher Endlager*; Berichte des Forschungszentrums Jülich 4237: Jülich, Germany, 2006.
8. Curtius, H.; Paparigas, Z.; Kaiser, G. Sorption of selenium on Mg-Al and Mg-Al-Eu layered double hydroxides. *Radiochim. Acta* **2008**, *96*, 651–655. [CrossRef]
9. Wiersma, B.J.; Mickalonis, J.I. *Preliminary Report on the Dissolution Rate and Degradation of Aluminum Spent Nuclear Fuels in Repository Environments*; WSRC-TR-98-00290 (U) Report; U.S. Department of Energy: Washington, DC, USA, 1998.
10. McClure, J.A.; Davis, J.W.; Gottlieb, P.; Cloke, P.L.; Nitti, D.A.; Benton, H.A. *Evaluation of Codisposal Viability for Aluminum-Clad DOE-Owned Spent Fuel: Phase II—Degraded Codisposal Waste; Package Internal Criticality*; BBA000000-01717-5705-00017 REV 01; CRWMS/M&O: Las Vegas, NV, USA, 1998.
11. Kaminski, M.D.; Goldberg, M.M. Corrosion of breached Aluminide Fuel under Potential Repository Conditions. In Proceedings of the International High Level Radioactive Waste Management Conference, Las Vegas, NV, USA, 29 April–3 May 2001.
12. Kaminski, M.D.; Goldberg, M.M. Aqueous corrosion of aluminum-based nuclear fuel. *J. Nucl. Mater.* **2002**, *304*, 182–188. [CrossRef]
13. Kaminski, M.D. *Aqueous Corrosion of Aluminum-Based Nuclear Fuel*; DOE: ANL-CMT-03/1; Argonne National Laboratory: Argonne, IL, USA, 2003.

14. Kaminski, M.D.; Goldberg, M.M.; Mertz, C.J. Colloids from the aqueous corrosion of aluminium-based nuclear fuel. *J. Nucl. Mater.* **2005**, *347*, 88–93. [CrossRef]

15. Shelton-Davis, C.; Loo, H.; Mackay, N.; Wheatley, P. *Review of DOE Spent Nuclear Fuel Release Rate Test Results*; DOE/SNF/REP-073; Idaho National Laboratory: Idaho Falls, ID, USA, 2003.

16. *Standortauswahlgesetz—StandAG: Gesetz zur Suche und Auswahl Eines Standortes für ein Endlager für Wärme Entwickelnde Radioaktive Abfälle*; Bundesgesetzblatt Teil I Nr. 26, S. 1074-1102: Bonn, Germany, 2017.

17. *Code of Federal Regulations (Annual Edition)—Titel 10: Energy*; Office of the Federal Register: Washington, DC, USA, 2017.

18. Dörr, S.; Bollingerfehr, W.; Filbert, W.; Tholen, M. *Status quo der Lagerung ausgedienter Brennelemente aus stillgelegten/rückgebauten deutschen Forschungsreaktoren und Strategie (Lösungsansatz) zu deren künftigen Behandlung/Lagerung—LABRADOR*; Abschlussbericht FKZ 02 S 8679; DBE Technology GmbH: Peine, Germany, 2011.

19. Dörr, S.; Bollingerfehr, W.; Filbert, W.; Tholen, M. Quantity and Management of Spent Fuel of Prototype and Research Reactors in Germany. In Proceedings of the German Annual Meeting on Nuclear Technology 2012, Stuttgart, Germany, 22–24 May 2012.

20. Hilton, B.A. *Review of Oxidation Rates of DOE Spent Nuclear Fuel, Part 1: Metallic Fuel*; W-31109-ENG-38, ANL-00/24 Report; Argonne National Laboratory: Argonne, IL, USA, 2000.

21. Curtius, H.; Kaiser, G.; Paparigas, Z.; Hansen, B.; Neumann, A.; Klinkenberg, M.; Müller, E.; Brücher, H.; Bosbach, D. *Wechselwirkung Mobilisierter Radionuklide mit Sekundären Phasen in Endlagerrelevanten Formationswässern*; Berichte des Forschungszentrums Jülich 4333: Jülich, Germany, 2010.

22. Loida, A.; Grambow, B.; Geckeis, H. Anoxic corrosion of various high burnup spent fuel samples. *J. Nucl. Mater.* **1996**, *238*, 11–22. [CrossRef]

23. Mazeina, L.; Curtius, H.; Fachinger, J.; Odoj, R. Characterisation of secondary products of uranium-aluminium material test reactor fuel element corrosion in repository-relevant brine. *J. Nucl. Mater.* **2003**, *323*, 1–7. [CrossRef]

24. Mazeina, L. *Investigation of the Corrosion Behaviour of U-Al Material Test Reactor Fuel Elements in Repository-Relevant Solutions and Characterisation of the Secondary Phases Formed*; Berichte des Forschungszemtrums Jülich 4063: Jülich, Germany, 2003.

25. Klinkenberg, M.; Neumann, A.; Curtius, H.; Kaiser, G.; Bosbach, D. Research reactor fuel element corrosion under repository relevant conditions: Seperation, identification, and quantification of secondary alteration phases of UAl_x-Al in $MgCl_2$ rich brine. *Radiochim. Acta* **2014**, *102*, 311–324. [CrossRef]

26. Vergasova, L.P.; Stepanova, E.L.; Serafimova, E.K.; Filatov, S.K. Lesukite $Al_2(OH)_5Cl \cdot 2H_2O$—A new mineral from volcanic exhalation. *Proc. Russ. Miner. Soc.* **1997**, *126*, 104–110.

27. Witzke, T. A new aluminium chloride mineral from Oelsnitz near Zwickau, Saxony, Germany. *Neues Jahrb. Mineral. Monatshefte* **1997**, *7*, 301–308.

28. Neumann, A. *Bildung von Sekundären Phasen bei Tiefengeologischer Endlagerung von Forschungsreaktor-Brennelementen—Struktur-und Phasenanalyse*; Schriften des Forschungszentrums Jülich, Reihe Energie & Umwelt, Band 153: Jülich, Germany, 2012.

29. Neumann, A.; Klinkenberg, M.; Curtius, H. Corrosion of non-irradiated UAl_x–Al fuel in the presence of clay pore solution: A quantitative XRD secondary phase analysis applying the DDM method. *Radiochim. Acta* **2017**, *105*, 85–94. [CrossRef]

30. Pearson, F.J. *Opalinus Clay Experimental Water: A1 Type*, version 980318; PSI Internal Report TM-44-98-07; Paul Scherrer Institut: Villigen, Switzerland, 1998.

31. Atterberg, A. Die rationelle Klassifikation der Sande und Kiese. *Chem.-Zeitung* **1905**, *29*, 195–198.

32. Rietveld, H.M. Line profiles of neutron powder-diffraction peaks for structure refinement. *Acta Cryst.* **1967**, *22*, 151–152. [CrossRef]

33. Rietveld, H.M. A profile refinement method for nuclear and magnetic structures. *J. Appl. Crystallogr.* **1969**, *2*, 65–71. [CrossRef]

34. Cheary, R.W.; Coelho, A.A. A fundamental parameters approach to X-ray line-profile fitting. *J. Appl. Crystallogr.* **1992**, *25*, 109–121. [CrossRef]

35. Coelho, A.A. Indexing of powder diffraction patterns by iterative use of singular value decomposition. *J. Appl. Crystallogr.* **2003**, *36*, 86–95. [CrossRef]

36. Bergmann, J.; Friedel, P.; Kleeberg, R. BGMN—A new fundamental parameters based Rietveld program for laboratory X-ray sources, its use in quantitative analysis and structure investigations. *CPD Newslett.* **1998**, *20*, 5–8.

37. Hölzer, G.; Fritsch, M.; Deutsch, M.; Härtwig, J.; Förster, E. $K_{\alpha 1,2}$ and $K_{\beta 1,3}$ X-ray emission lines of the 3d transition metals. *Phys. Rev. A* **1997**, *56*, 4554–4568. [CrossRef]

38. Brindley, G.W. The effect of grain or particle size on X-ray reflections from mixed powder and alloys considered in relation to the quantitative determination of crystalline substances by X-ray methods. *Philos. Mag.* **1945**, *36*, 347–369. [CrossRef]

39. Curtius, H.; Kaiser, G.; Müller, E.; Bosbach, D. Radionuclide release from research reactor spent fuel. *J. Nucl. Mater.* **2011**, *416*, 211–215. [CrossRef]

40. Vettier, C.; Yelon, W.B. The structure of $FeCl_2$ at high pressures. *J. Phys. Chem. Solids* **1975**, *36*, 401–405. [CrossRef]

41. Curtius, H.; Kaiser, G.; Rozov, K.; Neumann, A.; Dardenne, K.; Bosbach, D. Preparation and characterization of Fe-, Co-, and Ni-containing Mg-Al-layered double hydroxides. *Clays Clay Miner.* **2013**, *61*, 424–439. [CrossRef]

42. Rozov, K.; Curtius, H.; Bosbach, D. Preparation, characterization and thermodynamic properties of Zr-containing Cl-bearing layered double hydroxides (LDHs). *Radiochim. Acta* **2015**, *103*, 369–378. [CrossRef]

43. Poonoosamy, J.; Brandt, F.; Stekiel, M.; Kegler, P.; Klinkenberg, M.; Winkler, B.; Vinograd, V.; Bosbach, D.; Deissmann, G. Zr-containing layered double hydroxides: Synthesis, characterization, and evaluation of thermodynamic properties. *Appl. Clay Sci.* **2018**, *151*, 54–65. [CrossRef]

materials

MDPI

Article

Synthesis and Physical Property Characterisation of Spheroidal and Cuboidal Nuclear Waste Simulant Dispersions

Jessica Shiels *, David Harbottle and Timothy N. Hunter *

School of Chemical and Process Engineering, University of Leeds, Leeds LS2 9JT, UK; d.harbottle@leeds.ac.uk
* Correspondence: pmjas@leeds.ac.uk (J.S.); t.n.hunter@leeds.ac.uk (T.N.H.)

Received: 30 May 2018; Accepted: 15 July 2018; Published: 18 July 2018

Abstract: This study investigated dispersions analogous to highly active nuclear waste, formed from the reprocessing of Spent Nuclear Fuel (SNF). Non-radioactive simulants of spheroidal caesium phosphomolybdate (CPM) and cuboidal zirconium molybdate (ZM-a) were successfully synthesised; confirmed via Scanning Electron Microscopy (SEM), powder X-ray diffraction (PXRD) and Fourier transform infrared (FTIR) spectroscopy. In addition, a supplied ZM (ZM-b) with a rod-like/wheatsheaf morphology was also analysed along with titanium dioxide (TiO_2). The simulants underwent thermal gravimetric analysis (TGA) and size analysis, where CPM was found to have a D50 value of 300 nm and a chemical formula of $Cs_3PMo_{12}O_{40} \cdot 13H_2O$, ZM-a a D50 value of 10 µm and a chemical formula of $ZrMo_2O_7(OH)_2 \cdot 3H_2O$ and ZM-b to have a D50 value of 14 µm and a chemical formula of $ZrMo_2O_7(OH)_2 \cdot 4H_2O$. The synthesis of CPM was tracked via Ultraviolet-visible (UV-Vis) spectroscopy at both 25 °C and 50 °C, where the reaction was found to be first order with the rate constant highly temperature dependent. The morphology change from spheroidal CPM to cuboidal ZM-a was tracked via SEM, reporting to take 10 days. For the onward processing and immobilisation of these waste dispersions, centrifugal analysis was utilised to understand their settling behaviours, in both aqueous and 2 M nitric acid environments (mimicking current storage conditions). Spheroidal CPM was present in both conditions as agglomerated clusters, with relatively high settling rates. Conversely, the ZM were found to be stable in water, where their settling rate exponents were related to the morphology. In acid, the high effective electrolyte resulted in agglomeration and faster sedimentation.

Keywords: inorganic synthesis; nuclear waste; caesium phosphomolybdate; zirconium molybdate; sedimentation

1. Introduction

The Highly Active Liquor Evaporation and Storage (HALES) plant at Sellafield, UK, consolidates waste raffinates from the reprocessing of Spent Nuclear Fuel (SNF) by dissolution of the waste fission products in nitric acid, before concentrating them via evaporation [1]. The waste Highly Active Liquor (HAL) is made up of several fission products, including caesium phosphomolybdate ($Cs_3PMo_{12}O_{40} \cdot xH_2O$, CPM) and zirconium molybdate ($[ZrMo_2O_7(OH)_2] \cdot xH_2O$, ZM) [2], which precipitate out during temporary storage within the Highly Active Storage Tanks (HASTs) before eventual vitrification [3]. Consequently, critical research is required on non-radioactive simulants to aid behavioural understanding of the HAL precipitated dispersions. In addition, more knowledge is needed on to how the physical properties of HAL may change once processing moves to a Post Operational Clean Out (POCO) stage, where the relative concentrations of nitric acid may be diluted, potentially altering dispersion stability and other properties.

ZM has been the focus of several studies within literature, where it has been synthesised for various applications, such as a Technetium-99m generator [4–7], a precursor for the formation of the negative thermal expansion material $ZrMo_2O_8$ [8,9] and predominately for nuclear waste based studies [10–23]. Clearfield et al. [24] were the first to publish a synthesis route for ZM to characterise its ion exchange properties. Paul et al. [21] further investigated the synthesis method, including using the addition of citric acid to alter the crystal morphology. ZM is known to cause issues within the nuclear industry, due to its mobility properties leading to potential problems with pipe blockages, for example [1].

There have been fewer studies investigating CPM, although there are a number of analogues that have been studied, such as ammonium phosphomolybdate, which is a potential cation-exchange material for selective recovery of Cs [25,26]. Indeed, CPM has also been synthesised itself for the same purpose, and as a photocatalyst for the photodegradation of dye pollutant [27–30]. The physical behaviour of CPM in nuclear waste HAL is of concern due to the presence of the radioactive isotopes ^{134}Cs and ^{137}Cs, which, if concentrated, could form potential hotspots within the HASTs [1]. Paul et al. [21] also published a synthesis route to CPM, in order to study its morphology in nuclear HAL systems, and this method was also used in the current research.

Several studies have looked into both CPM and ZM, specifically relating to the issues they cause within the HASTs, the challenge they pose to the Waste Vitrification Plant (WVP) and the in situ conversion of CPM to ZM [1,2,21–23,31–33]. Given the complexity of CPM and ZM dispersion behaviour, there is a critical need to study their synthesis and physical behaviour under a wide range of conditions. For example, there is no current information on the kinetics of CPM formation, or what impact storage temperature changes may have on growth rates and final morphology. Additionally, the main route for formation of ZM in nuclear operations is from metal substitution reactions with precipitated CPM, in current holding tanks. While it is known that these conversion reactions are very slow kinetically [21], exact time scales for ZM precipitation by this route are not known, although it has been reported that an increase in temperature and a decrease in acidity promote the conversion [31]. Different wash regents for both compounds have also been investigated with a suggestion that doping could be used to change the morphology of ZM, which has potential to be advantageous for transport or separation, depending on properties, such as sedimentation rate [1,32]. Consequently, a fuller understanding of the impact of ZM morphology on its dispersion behaviour is required.

Therefore, this study investigated a number of physical and chemical properties of synthesised non-active ZM of various morphologies and CPM, in different conditions. Additionally, the settling behaviour of these simulants was explored to understand the influence of acid concentration, which is of particular importance in the case of POCO, where the washout fluid pH and electrolyte concentration will be critical considerations. Titanium dioxide was also used, as a cheap and more easily obtainable comparison simulant, as it is a material that has been widely studied and has been previously used as a nuclear waste simulant in several studies [23,34].

2. Results and Discussion

2.1. Synthesis and Formation Tracking

The Scanning Electron Micrograph (SEM) images in Figure 1 show the CPM, TiO_2 and two morphologies of ZM (cuboidal and wheatsheaf) referred to as ZM-a and ZM-b, respectfully. The images for CPM and ZM-a are in agreement with those published by Paul et al. and Dunnett et al. [23,33]. CPM is known to generally exhibit a roughly spheroidal shape, consisting of agglomerated nanoclusters, whereas ZM is most commonly known to be discrete cuboidal in shape. TiO_2 is also spheroidal in shape, consisting of bound agglomerated clusters of nanocrystallites, which appear comparable to CPM, and an appropriate comparison material, from a morphological perspective. The SEM image of ZM-b shows a mixture of both rod-like particles and a shape somewhat resembling a sheaf of wheat, hence commonly named "wheatsheaf". The addition of citric acid in the ZM synthesis alters the cubic morphology of the particles as it binds to certain faces of the ZM that reduces their growth, and therefore the particle aspect ratio is increased [21]. What is unclear is why there is a mix of both wheatsheaf and rods formed. It is noted that the ZM-b particles used were produced at an industrial scale with different reagent conditions to the laboratory synthesised ZM-a, and it is likely that the final morphology of the particles is sensitive to various factors, such as concentration of the citric acid reagent added, potential trace contaminants or even different precursor materials. As the chemistry of the HAL and the conditions of the HASTs are largely unknown, it is not unreasonable to suggest a mix of ZM morphologies could be present, and therefore ZM-b may represent a more realistic simulant than the well-defined cuboidal ZM-a. Additionally, depending on the morphology of the ZM and the properties it exhibits, doping the HAL to promote a morphology change before POCO could be advantageous [1]. This complexity re-enforces the need to characterise various morphologies of ZM to investigate the potential differences in physical and chemical properties.

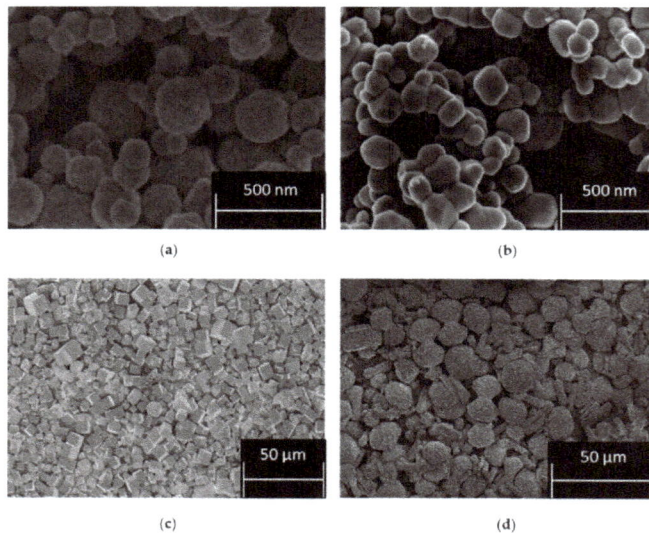

Figure 1. Scanning electron micrograph of: (**a**) caesium phosphomolybdate (CPM) formed at 50 °C; (**b**) titanium dioxide (TiO_2); (**c**) zirconium molybdate (ZM-a); and (**d**) zirconium molybdate (ZM-b).

The synthesis of CPM involves a double replacement reaction with two reactants, namely phosphomolybdic acid ($H_3PMo_{12}O_{40} \cdot xH_2O$) and caesium nitrate ($CsNO_3$), as discussed in detail by Paul et al. [21]. Phosphomolybdic acid is Ultraviolet–Visible (UV–Vis) active, which allowed the precipitation kinetics of CPM to be tracked via measuring the phosphomolybdic acid decrease in concentration over time. Figure 2 shows the rate of reaction curves determined for the CPM synthesis conducted at both 25 °C and 50 °C, through tracking of the phosphomolybdic acid concentration. It was found to be a first-order reaction in respect to phosphomolybdic acid, which is in excess to the caesium nitrate, giving a rate constant of 0.04 min^{-1} for the reaction at 25 °C and of 0.09 min^{-1} for the reaction at 50 °C. This result demonstrates the fast kinetics at which CPM is formed within laboratory conditions, and, while published synthesis routes often quote total reaction times of ~48 h [21], it is clear that the actual reaction is almost at completion within 1 h at the higher temperature. While the reaction environment will differ within the HASTs, average temperatures are kept within 50–60 °C, suggesting a similar reaction rate to that found at 50 °C could be feasible. It is also noted that, due to the high relative proportions of caesium within the fission products, it is expected that CPM will form easily and in high amounts.

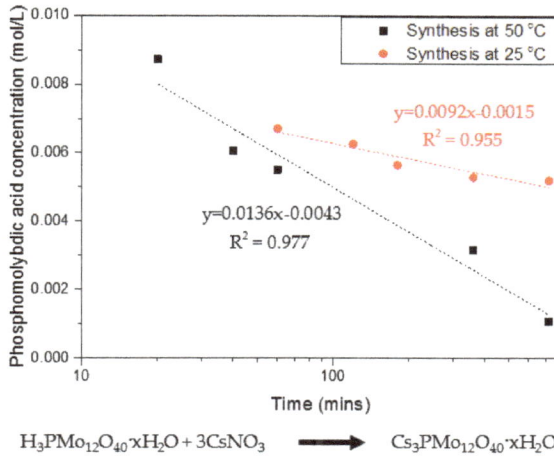

$$H_3PMo_{12}O_{40} \cdot xH_2O + 3CsNO_3 \longrightarrow Cs_3PMo_{12}O_{40} \cdot xH_2O$$

Figure 2. Rate of reaction showing first-order kinetics for the co-precipitation reaction of caesium nitrate and phosphomolybdic acid forming caesium phosphomolybdate (CPM) at 25 °C and 50 °C with corresponding reaction equation.

CPM was also synthesized at 100 °C to investigate the effect of temperature differences on its morphology. Figure 3 shows SEM images of the synthesized CPM at both 25 °C and 100 °C. The CPM synthesized at 25 °C compared well to the CPM synthesized at 50 °C (Figure 1a), suggesting that application of heat between 25 °C and 50 °C does not alter the final particle morphology significantly (although the kinetics of the reaction are slower, as shown in Figure 2.). For the CPM synthesized at 100 °C, it was found that, although spheroidal particles were also formed, there was an increase in large agglomerates and a greater range of particles varying in size and in shape, compared to the CPM synthesized at 25 °C and 50 °C. A potential reason for the differences may be that the faster reaction kinetics that occur with increase in temperature means that there is less time for the nanocrystallite clusters to form into self-similar spheroidal particles through diffusion interaction. As seen with the difference in kinetics between 25 °C and 50 °C, if there is a similar increase in the rate of precipitation at 100 °C, the CPM will precipitate almost instantaneously, leading to larger and more disordered clusters. Additionally, at 100 °C, it would be expected that formed CPM would be less stable, as it is known to be a temperature range in which its breakdown should begin to occur [30]. Hence, precipitates may partially re-solubilise, especially outer surfaces, leading to fusion of nanocrystallite clusters.

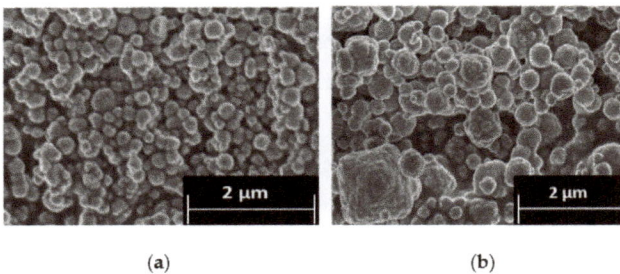

(a)   (b)

Figure 3. Scanning electron micrograph of: (**a**) caesium phosphomolybdate (CPM) synthesised at 25 °C; and (**b**) CPM synthesized at 100 °C.

The conversion of CPM to ZM-a was also qualitatively tracked, via SEMs of intermediate structures over a period of ten days, as presented in Figure 4. Through Days 0–6, it was observed that the particles remain predominantly spheroidal in their morphology and nanometre in size, representative of CPM. In addition, the solids all remained yellow in colour, visually a characteristic expected for CPM [21]. For the solids precipitating out from Day 8 to Day 10, there appeared to be a mix of yellow and white solids suggesting that ZM-a was beginning to form (Supplementary Materials, Figure S1 shows the colours of both pure CPM and ZM-a solids). The SEMs taken at Days 8 and 10 also suggest this, as the cuboidal micrometre particles that would be expected of ZM can be seen to appear, in addition to the spheroidal CPM particles coating the ZM-a. Once Day 10 was reached, the solids were washed with 1 M ammonium carbamate which dissolved the CPM particles [1], resulting in the clear cuboidal particles, as seen in Figure 1c. This sequence of SEM images demonstrates the length of time it takes for ZM to transform from CPM in a highly controlled environment, and it is evident that the conversion yield from CPM to ZM remains low (estimated to be ~30–40%). In comparison, as Figure 2 demonstrates, the CPM forms rapidly (within hours), suggesting that there may be a higher concentration of CPM in contrast to ZM within the HASTs. However, tank conditions and specific compositions of the HAL within the tanks are variable, and thus the ratio of CPM:ZM is extremely difficult to predict. This issue highlights the importance of understanding both systems individually, especially considering the amount of time that HAL is left in the tanks, which is often in the order of months.

Figure 4. Scanning electron micrographs taken at various times during the tracking of zirconium molybdate (ZM-a) synthesis from a caesium phosphomolybdate (CPM) precursor over 10 days.

2.2. Chemical Composition

Figure 5a shows the powder X-ray diffraction (PXRD) patterns for CPM, ZM-a and ZM-b. Both the ZM patterns were compared to The International Centre for Diffraction Data (ICDD) online database, where they correlated with the ICDD number 04-011-0171. ZM morphologies are reported to crystallise as a body-centred tetragonal lattice with space group *I41cd*, lattice parameters a = b = 11.45 Å, c = 12.49 Å and angles $\alpha = \beta = \gamma = 90°$. Although the morphologies of ZM-a and ZM-b are quite different, the PXRD patterns show their bulk crystal structure remains the same. The PXRD patterns for ZM-a and ZM-b also agree with those published within the literature [21,24]. Whilst the pattern for CPM was not found in the ICDD database, it is in good correlation to the patterns for CPM found within the literature [21,27]. CPM is reported to crystallise in a cuboidal lattice with space group *Pn-3m*, lattice parameters a = b = c = 11.79 Å and angles $\alpha = \beta = \gamma = 90°$.

Infrared analysis was used as a fast method for fingerprinting the synthesised compounds, following methods previously detailed in the literature [11,24]. Figure 5b, presents the Fourier Transform Infrared (FTIR) spectra for the synthesised CPM, ZM-a and ZM-b. The spectrum published by Rao et al. [11] is in good agreement to the ZM spectra shown in Figure 5b. The spectra for the ZM samples shows a band between 3000 cm^{-1} and 3300 cm^{-1}, which is representative of the O–H group, with the band at 1600 cm^{-1} representing the O–H–O bonds. The "fingerprint" region is below 1000 cm^{-1} corresponding to metal to oxygen groups. If the sample was to be anhydrous, the IR spectrum would differ with no bands within 400–700 cm^{-1} [11]. The ZM-b spectrum differs slightly from the ZM-a spectrum with more intense bands around 1000 cm^{-1}, potentially equated to the presence of more bound water molecules present in the structure. For the CPM spectrum, the higher region, from 1250 cm^{-1} upwards, is largely similar to the ZM spectra (representing the O–H and O–H–O groups with slight intensity differences). The CPM spectrum published by Ghalebi et al. [30] is in good agreement to the CPM spectrum published here. The "fingerprint" region (<1000 cm^{-1}) highlights clear differences in intensity from the ZM, due to the main metal bonding, indicating IR probe analysis may potentially be useful as an in situ technique to determine compositional differences, in CPM/ZM mixtures.

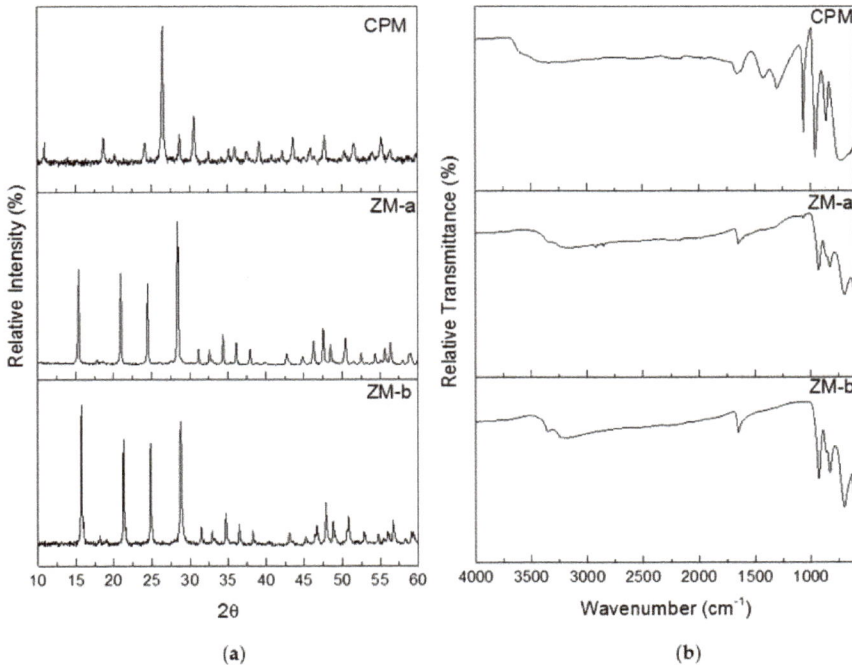

Figure 5. (a) Powder X-ray Diffraction patterns for caesium phosphomolybdate synthesised at 50 °C (CPM), and zirconium molybdate (ZM-a and ZM-b). (b) Infrared spectra of CPM synthesised at 50 °C, ZM-a and ZM-b.

Figure 6 presents the thermal gravimetric analysis (TGA) plots of CPM, ZM-a and ZM-b, over the temperature range from 30 °C to 400 °C. For CPM, the water loss begins below 100 °C and continues until ~400 °C, with a total mass loss equating to 13 moles of water, therefore resulting in a chemical formula: $Cs_3PMo_{12}O_{40} \cdot 13H_2O$. This value is in good agreement with the reported literature values, which are generally between 9 and 14 moles of water, depending on the drying method chosen [24]. In the present study, CPM was dried using an oven at 70 °C for 12 h. Therefore, it is possible some of the initial water loss could potentially have been strongly adsorbed bound water, or water with extremely low binding energy, although, given the length of drying time, it is not assumed to be from any free water. In comparison, the dehydration process for both ZM morphologies had a clearer start and end temperature. For ZM-a, the mass loss starts around 100 °C and stops just before 200 °C, equating to three moles of water and a chemical formula of $ZrMo_2O_7(OH)_2 \cdot 3H_2O$, which is the same value reported in literature [17]. The dehydration process of ZM-b is similar, but appears to begin with a slightly slower rate and lower temperature, while again the mass loss stops just before 200 °C. This value equates to a loss of four moles of water, showing a difference in bound water content between the ZM-a and ZM-b samples. This may be a result of the citric acid incorporation, although, again, differences in reaction conditions may also be a cause. The extra water present in ZM-b is also consistent with the slight variation in the FTIR spectrums (Figure 5b) with the more intense band <1000 cm^{-1} for ZM-b in comparison to ZM-a.

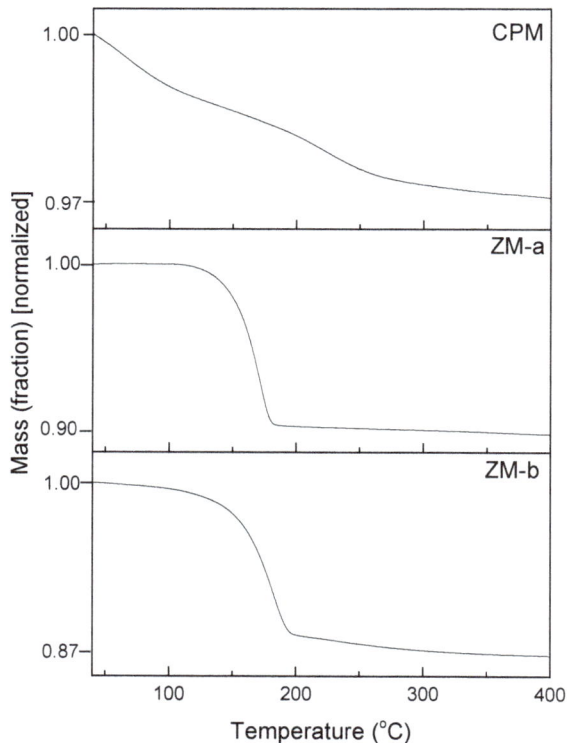

Figure 6. Thermogravimetric analysis curves of caesium phosphomolybdate synthesised at 50 °C (CPM) and zirconium molybdate (ZM-a and ZM-b). CPM shows a loss of 13 moles of water, ZM-a a loss of 3 moles of water and ZM-b a loss of 4 moles water.

2.3. Size, Stability and Settling Behaviour

The particle sizes of all simulants are shown in Figure 7, as a relative volume percentage versus size distribution. The D50 value of CPM synthesised at 50 °C is 300 nm, which corresponds well with the SEM images of CPM nanoclustersnshown in Figure 1a. In comparison, the CPM synthesised at 100 °C has a significantly higher D50 value of 249 μm, which is even larger than evident from the SEM images (Figure 3b). Although individual crystallite sizes appear (via SEM) to be fairly similar to those formed at lower reaction temperatures, it is clear they cluster to a much greater degree and become extremely agglomerated, resulting in the high D50 value. The secondary smaller peak representing the finer particles ranging between 300 nm to 20 μm, likely represents both individual particles and smaller agglomerates. The poly dispersity index (PDI) for CPM synthesised at 100 °C is 4.5 times larger than the CPM synthesised at 50 °C, which is attributed to the bimodal distribution of the CPM synthesised at 100 °C.

The TiO_2 was expected to be slightly larger than the CPM particles synthesised at 50 °C, due to agglomeration, as evidenced in the SEM images (Figure 1b) and confirmed by the D50 value of 700 nm for TiO_2 in comparison to the CPM 50 °C D50 value of 300 nm. For the ZM-a simulant, the cuboidal particles are shown to be around thirty times larger than the CPM with a D50 value of 10 μm. Unsurprisingly, when comparing SEM images (Figure 1d), ZM-b displays a larger D50 value than ZM-a with a value of 14 μm, although caution must be taken with light scattering estimations of any non-spherical particles [22]. In comparison to the literature, the CPM synthesised at 50 °C particle size is within the expected range, [21] while the D50 for the ZM-a is slightly larger than previously reported [33].

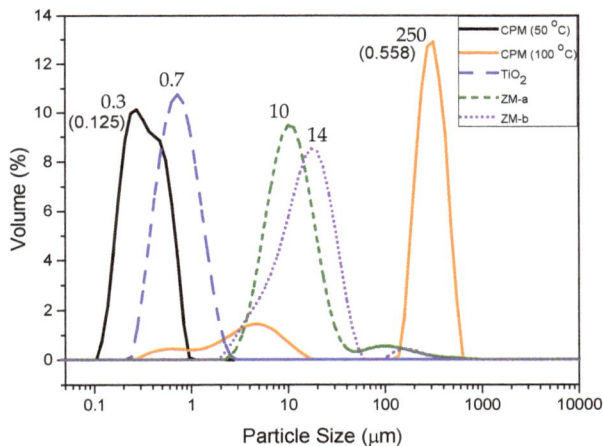

Figure 7. Particle size distributions of caesium phosphomolybdate (CPM) synthesised at 50 °C and 100 °C, titanium dioxide (TiO_2) and zirconium molybdate (ZM-a and ZM-b). Corresponding D50 values shown by each relevant peak. Polydispersity index values for CPM at 50 °C and 100 °C shown in brackets.

The zeta potentials of all simulants as a function of pH are shown in Figure 8. The Isoelectric Point (IEP), was around pH 2.5 for both ZM-a and ZM-b, which is similar to that found by Paul et al. for cuboidal and rod-like ZM particles [22]. For CPM, the IEP could not be obtained confidently due to the observed error at very low pH, likely because of the high effective counterion concentration. However, through extrapolation, the IEP appears to be in the region of pH 1–1.5, and again similar to values previously reported by Paul et al. [22]. TiO_2 had the highest IEP at ~pH 4, which compares well to previous literature on measurements of anatase and rutile mixtures [35,36]. The zeta potential data largely suggest that, in low pH conditions, such as those experienced in the HASTs, the ZM species will be positively charged, while the CPM may be close to an uncharged state. However, the high acid concentration in the processing environments will mean that there is a high effective electrolyte concentration (resulting from acid counterions), collapsing the electric double layer around the particles, despite any native charge at low pH. Thus, it is important to study dispersion stability in both acid and water environments (that latter of which may represent conditions in POCO).

Further, previous reported work by Paul et al. [22] indicated that the equilibrium pH for CPM and ZM dispersions in water may be significantly reduced over time, due to potential hydride reactions from the bound water. Therefore, the equilibrium pH after 48 h of 4 vol% dispersions was measured for all HAL simulants. The pH for CPM and ZM-b was ~1.5, while for ZM-b was slightly higher at ~2.4. Therefore, even in pure water environments, the zeta potential of the simulants may be altered, affecting their stability. It would appear from the equilibrium pH measurements that ZM-b and ZM-a will be likely positively charged (although the ZM-a may be close to its IEP) while CPM will be weakly negatively charged and approaching its IEP. It is noted that the pH of titania dispersions in water was close to neutral, although very slightly acidic due to the use of deionised Milli-Q water (~pH 5.5–6).

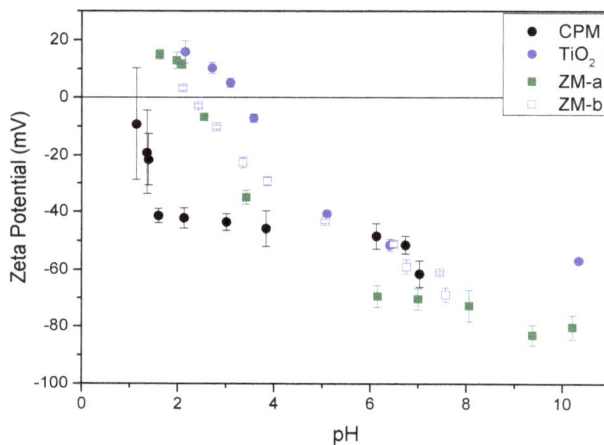

Figure 8. Zeta potential curves for caesium phosphomolybdate (CPM), titanium dioxide (TiO$_2$) and zirconium molybdate (ZM-a and ZM-b) measured at concentrations of 1000 ppm in 10^{-4} M potassium nitrate (KNO$_3$) solution.

Figure 9 shows the settling rates for all the simulants in both water and 2 M HNO$_3$ at 4 vol% (assumed to be close to relevant concentration conditions). Both ZM-a and ZM-b sediment at considerably higher rates in the 2 M HNO$_3$, indicating that the high effective electrolyte conditions lead to significant coagulation of the dispersions, likely from the collapse of the electric double layer. For the TiO$_2$ dispersions, sedimentation rates are also enhanced in acid (although to a lower degree) and, importantly, the rate in water is greater than either of the larger ZM species, suggestive also of a degree of coagulation in water conditions. While the zeta potential data indicated good stability at neutral pH, the overall values measured are an average for the mixed anatase/rutile particles, and it is known that for the anatase phase, the expected IEP is around neutral [22] (while pure rutile is ~3–4). Therefore, some degree of heterogeneity in surface charge would be expected, leading to some partial dispersion instability, which is evident from the settling data.

For CPM, the difference in settling rates is minimal for the two conditions, which would suggest similar levels of dispersion stability. As it was assumed from the equilibrium pH of CPM in water that dispersions may be close to the IEP in these conditions, the similarity of settling data also in acid indicates potential coagulation is occurring in both conditions. While the sedimentation rates for CPM are lower than for other species in acid, it is noted they have the smallest particle size. In addition, any comparison between species must be made with caution, as the 4 vol% dispersions will be within the hindered settling regime, and cannot be associated directly with expected Stokes settling velocities.

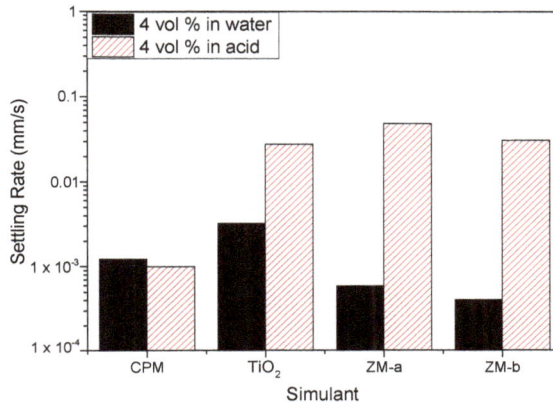

Figure 9. Settling rates for caesium phosphomolybdate (CPM), titanium dioxide (TiO_2) and zirconium molybdate (ZM-a and ZM-b) at 4 vol% concentration in both water and 2 M HNO_3 environments.

For more quantitative analysis, sedimentation of the simulants for a range of concentrations in both water and acid were analysed using the Richardson-Zaki (RZ) power-law hindered settling model, with data presented in Figure 10 [37]. Here, the natural log of the linear settling rates $ln(u)$ are given versus the log of the porosity $ln(1) - \Phi$. Exponent values associated with the fits can aid understanding on the coagulation of the simulants, in addition to the influence of particle shape. For non-agglomerated spherical dispersions, an exponent of ~4.65 would be expected [37,38]. For spheroidal CPM, its exponent values were 23.93 and 18.05, in water and acid, respectively, while for the TiO_2, its exponents were 103.17 and 42.31. These values are an order of magnitude, or more, greater than for spherical systems, inferring a high degree of aggregation, which is consistent with the 4 vol% settling data discussed in Figure 9. It is interesting that the TiO_2 exponent value in water is considerably higher than in acid, even though settling rates are lower. This behaviour may indicate that while agglomerates are smaller in water, they have a more open structure, which increases hindered settling effects.

For the ZM-a and ZM-b simulants, both show similar trends across the concentration regime measured, with their settling rates in acid being much faster than that in water, consistent with the single 4 vol% data in Figure 8. Therefore, for potential POCO environments, lower acidity wash waters may aid in stabilising ZM dispersions, reducing issues of sedimentation on transfer, whereas effects on CPM will likely be minimal. There are however some interesting differences in the power-law exponent values, especially in water. The exponent value for ZM-a in water is 7.09 while for ZM-b it is 19.48. These values are higher than expected for spherical particles, although the slow settling rates and zeta potential data suggest high stability. It is assumed that, similar to studies on stable non-spherical particles [39–42], the high exponent values occur from the enhanced drag due to their shape. Indeed, shape factor may help explain the higher value for ZM-b, as the orientation of the elongated wheatsheaf/rod-like particles may have an additional effect on the drag, which will likely be greater if they adapt a flat confirmation [43,44]. The exponent values are both similar to each other in acid, and notably less than the ZM-b in water. It may be the aggregated ZM clusters actually have a reduced drag in comparison to the elongated and stable ZM-b particles.

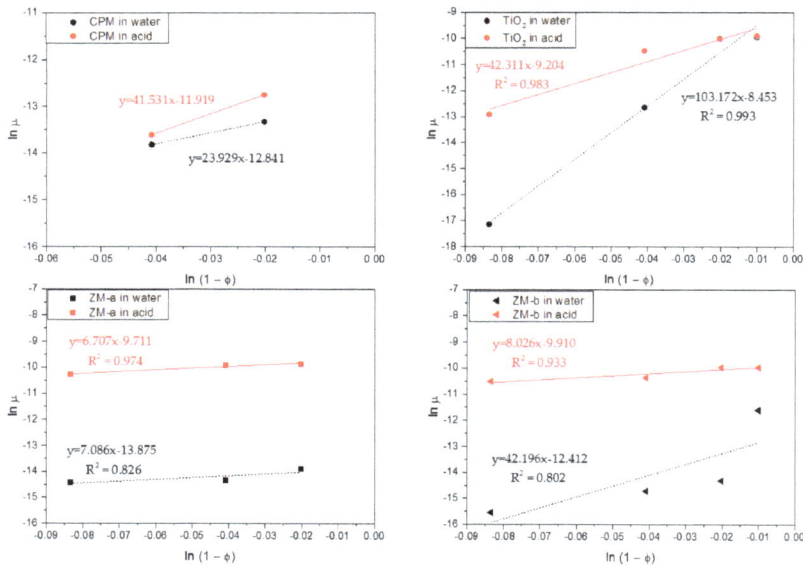

Figure 10. log–log linearised settling rates versus porosity $(1 - \Phi)$ for caesium phosphomolybdate (CPM), Titanium dioxide (TiO$_2$) and Zirconium molybdate (ZM-a and ZM-b) in both water and 2 M HNO$_3$ environments. Dashed lines indicate power-law fits.

The zero concentration intercept from the fits in Figure 10 were used to estimate the free settling velocities of the simulants in each system. Stoke's law was then utilised to determine what spherical equivalent size, the particles would be expected to be [22]. For CPM and TiO$_2$ in both water and acid, the calculated sizes were much larger than their measured D50 mediums from Figure 7 (at ≥ 1 µm), which is consistent with the hypothesis that both systems are coagulated in all conditions. In comparison, for both ZM-a and ZM-b systems, the calculated particle sizes were much smaller than their measured D50 sizes (at between 1 µm and 8 µm). Estimated size values from the ZM-a and ZM-b acid settling data were however much larger (>15–30 µm) again consistent with coagulation.

The low estimates sizes for ZM in water may be due to the fact that simple Stoke's law calculations makes several assumptions that do not apply to the ZM samples directly, such as particle sphericity and monodispersity. Considering the morphology of both ZM simulants, their representative drag coefficient will be significantly larger than that of a sphere, and Stoke's law will likely lead to underestimations of size, as is evident from the values derived from Figure 10. Additionally, the presented settling data were taken at a single threshold of 40%. A single threshold represents a certain fraction of the particles, but it does not capture complete settling data for polydisperse systems, such as the ZM. For example, when ZM-b settling data were analysed at a range of thresholds, calculated linear settling rates vary by almost an order of magnitude (comparing 10% and 80% thresholds). While the 40% threshold was chosen as a fixed value to allow comparison of all samples, it appears this likely correlates to a fraction of the dispersion that is under the medium sizes. Therefore, caution should be taken when extracting sizes from centrifugal settling data.

3. Materials and Methods

3.1. Synthesis and Materials

The synthesis routes for both CPM and ZM-a were based upon the methods published by Paul et al. [21]. Phosphomolybdic acid and caesium nitrate were mixed at a 1 to 3 molar ratio with equal volumes in 2 M HNO_3 at 25 °C, 50 °C and 100 °C over a 12 h period with constant stirring to form the yellow precipitate CPM. ZM-a was then formed from the CPM synthesised at 50 °C with the addition of zirconyl nitrate in 6 M HNO_3 at a 1 to 1 volume ratio, at a temperature of 100 °C and constant stirring for 10 days before washing with 1 M ammonium carbamate. The reactants used for these synthesis methods are given in Table 1. In addition, the National Nuclear Laboratory (NNL) provided a simulant of ZM (ZM-b) industrially synthesised by Johnson Matthey (Royston, UK). This sample had a mix of rod-like and wheatsheaf morphology, resulting from the addition of citric acid during the synthesis process. While the overall synthesis procedure is similar to that detailed by Paul et al. [21], where rod-like particles were formed, the exact reagent conditions and concentrations for the industrial sample are unknown. Titanium dioxide (TiO_2) in its anatase/rutile mixed form was purchased from Venator Materials PLC (formally Huntsmans Pigments Ltd., Wynyard, UK) with product code TS46424 to be used for comparative purposes and as a standard for some preliminary experiments.

Table 1. List of reagents used for the synthesis of both caesium phosphomolybdate (CPM) and zirconium molybdate (ZM).

Material	Formula	Purity	Supplier
Phosphomolybdic acid hydrate	$H_3PMo_{12}O_{40}$	Solid—80%	Acros Organics
Caesium nitrate	$CsNO_3$	Solid—99.9%	Aldrich
Nitric acid	HNO_3	Solution—70%	Fisher Scientific
Zirconyl nitrate	$ZrO(NO_3)_2$	Solution—35 wt. % in dilute HNO_3	Sigma-Aldrich

3.2. Ultraviolet-Visible Spectroscopy

To track the synthesis of CPM, a UV-Vis spectrophotometer Lambda XLS (PerkinElmer, Waltham, UK) was utilised to track the concentration of one of the reactants; phosphomolybdic acid, as the other reactant (caesium nitrate) was found to not be significantly UV-Vis active. Before the reaction was conducted, a calibration curve was generated for various concentrations of phosphomolybdic acid versus their absorbance taken at wavelength 458 nm. The calibration can be seen in the Supplementary Materials, Figure S2. Once the reaction had begun, regular aliquots of the solution were taken at varying time periods, which were then diluted within 2 M HNO_3 and centrifuged to remove any CPM solid formed before the remaining supernatant underwent UV-Vis spectroscopy. The absorbance value of the supernatant was then compared to the calibration curve previously taken for phosphomolybdic acid, to determine its concentration at that particular time. The corresponding concentrations were plotted against the time they were taken and the gradient taken in order to determine the rate constant.

3.3. Particle Shape, Density and Size Characterisation

Particle shape analysis was completed through the use of a SU8230 scanning electron microscope (SEM, Hitachi, Krefeld, Germany). The solid samples were prepared using a carbon based adhesive disk, in which dry ground up powder was placed before being platinum coated. This SEM was also used to take the images for the ZM-a morphology tracking; daily aliquots were taken from the reactor, and then centrifuged before the supernatant was removed and the solid dried before imaging.

A Mastersizer 2000 (Malvern, Worcester, UK) was used to determine the particle size distribution for the simulants. For each sample, a small amount of solid was allowed to form a dispersion within distilled water before being added in the Mastersizer until the required transmission value was met. It is crucial that the sample be fully dispersed in order to achieve the most accurate representative sizing measurement, avoiding agglomerates of material which would give inaccurate results. Each sample was then measured 10 times over 10 s and an average was taken. The D50 value was determined by the Mastersizer as the 50th percentile, therefore is was inclusive of any bimodal distribution.

An AccuPycTM 1330 Pycnometer (micromeritics, Norcross, GA USA) was used to take density measurements via gas pycnometry. The solid simulants with a known mass were placed into the instrument then using the pressure change of helium in a known calibrated volume the density of the simulants was determined. The following densities were determined: CPM 3.82 g/cm^3, TiO$_2$ 4.23 g/cm^3, ZM-a 3.41 g/cm^3 and ZM-b 3.41 g/cm^3.

3.4. Power X-ray Diffraction, Infrared Spectroscopy and Thermogravimetric Analysis

A D8 X-ray Diffractometer (Bruker, Coventry, UK) was used to measure the crystalline structure of the samples with an electron beam of 40 kV. The copper source (Cu Kα) has a wavelength of 1.54 Å, energy of the radiation source was 1.6 kW, and measurements were taken over the 2θ range 10–60° with a step size of 0.032° 2θ and a scan speed of 0.2 s per step. The raw data for the diffraction pattern were then extracted and normalised for comparison to The International Centre for Diffraction Data (ICDD) online database.

The functional groups of the simulants were determined through Fourier Transform Infrared (FTIR) spectroscopy, carried out using a Nicolet iS10 FTIR spectrometer (ThermoFisher, Waltham, UK) with a ZnSe ATR attachment, at a resolution of 4 cm^{-1} and 64 scans. Dry power samples were placed upon the sample holder before being clamped into place and the spectroscopy conducted. The spectra were then analysed to identify the functional groups.

Thermal dehydration of the simulants (used to determine the amount of bound water) was determined via thermogravimetric analysis (TGA), using a TGA/DSC 1100 LF (Mettler Toledo, Leicester, UK). For each test, a 0.1 g sample was inserted, and heated using a temperature profile from 30 °C to 400 °C at a heating rate of 10 °C·min^{-1} under a nitrogen atmosphere. The mass loss over this time period was then converted to the amount of water molecules lost for each simulant.

3.5. Zeta Potential Measurements and Sedimentation Experiments

The zeta potential of the simulants at various pH values were determined using a Zetasizer Nano ZS (Malvern, Worcester, UK) which directly measures the electrophoretic mobility and then uses this to calculate the zeta potential via an internal algorithm. Dispersions were prepared with concentrations of 1000 ppm in a 10^{-4} M potassium nitrate (KNO$_3$) solution. Nitric acid (HNO$_3$) and potassium hydroxide (KOH) solutions between 0.01 and 0.1 M were used to adjust the pH of the samples. For every sample, five measurements were taken and repeated on fresh samples 2 or 3 times, where an average of these results is presented. The Malvern Zetasizer Nano ZS was also used to calculate the polydispersity index—a parameter determined from a Cumulants analysis of an intensity–intensity autocorrelation function.

A LUMiSizer® (LUM GmbH, Berlin, Germany)was used to study the dispersion settling stability, where sedimentation studies for the simulants were conducted in triplicate (with an average standard deviation of 5% of the mean value) with an average of the results taken, in both deionised water and 2 M HNO$_3$. The centrifuge speed was set between 500 and 2000 RPM (depending on the sample) at 25 °C and transmission profiles taken every 10 s with the total number of profiles equalling 255. The LUMiSizer measures the solids settling rate by centrifugation using LED light sources that emit light at different wavelengths to produce-transmission profiles at set time intervals. From the produced transmission profiles, a threshold of 40% was chosen and the data converted to give suspension height vs. time, where the linear zone settling rate for each simulant at that RPM was determined (Figure S3 in Supplementary Materials shows a raw transmission profile with corresponding suspension height vs. time). A threshold of 40% was chosen as on analysis it was deemed the best representative threshold allowing for comparison of all samples. It should be noted as a limitation it represents a certain fraction of the particles at a certain size, and a single threshold cannot capture the full behaviour of a polydisperse suspension. The sedimentation rates were then back calculated to estimate the settling rate at normal gravity, assuming linear dependence on RCA, using Equations (1) and (2). Here, *RCA* is relative centrifugal acceleration, r is the radius of the instrument plate where the dispersion is housed (in cm) and RPM is the revolutions per minute of the experiment [45].

$$RCA = 1.1118 \times 10^{-5} \times r \times RPM^2 \tag{1}$$

$$\frac{Measured\ velocity}{RCA} = gravity\ velocity \tag{2}$$

4. Conclusions

Research was conducted on non-active simulants of two known precipitated fission waste products found in nuclear fuel reprocessing: caesium phosphomolybdate (CPM) and zirconium molybdate (ZM). Both CPM, with spheroidal particle shape (formed from agglomerated nanoclusters), and ZM-a, with a cuboidal morphology, were successfully synthesised and characterised using SEM, PXRD, IR and TGA. In addition, ZM-b with a rod-like/wheatsheaf morphology was also characterised, along with titanium dioxide (a commercially available alternative, morphologically similar to CPM). The reaction kinetics of CPM precipitation was investigated at various temperatures, finding it to be a first-order reaction with respect to phosphomolybdic acid, and able to form at a range of temperatures, although size and stability properties begin to change at ~100 °C. While the kinetics of CPM synthesis was very fast, ZM formation from CPM precursor substitution was slow, being observed to convert partially only after ~10 days. The ease of formation of CPM compared to ZM suggest that within the HASTs there could be a larger proportion of CPM in comparison to ZM. Additionally, as the formation of ZM is sensitive to the effects of additives, it is likely that any ZM formed will contain a range of morphologies.

The dispersion stability of the simulants in water and 2 M nitric acid was observed by comparing zeta potential and pH measurements with centrifugal sedimentation analysis. All simulants were found to have low IEP values; however, acid group leaching reduced the natural pH of water suspensions to around or below these values. Therefore, in low pH conditions such as those experienced within the HASTs, the waste products are likely to be unstable and coagulate. The CPM concentration dependence on the settling rate was found to be more pronounced due to coagulation in both water and acid environments, which were qualitatively similar to the titania. The ZM-a and ZM-b conversely appeared stable with low settling rates in water that significantly increased in acid (assumed to be caused by coagulation from the collapse of the electric double layer). Overall, results highlight the complex morphology and chemistry of these precipitated nuclear wastes, and imply their stability may be critically altered, depending on changes in acid levels as waste treatment moves to a post operational clean out phase.

Supplementary Materials: The following are available online at http://www.mdpi.com/1996-1944/11/7/1235/s1. Figure S1. Images of: (**a**) caesium phosphomolybdate (CPM) formed at 50 °C displaying a yellow coloured solid; and (**b**) zirconium molybdate (ZM-a) displaying a white coloured solid. Figure S2. UV-Vis calibration curve for phosphomolybdic acid at various concentrations. Figure S3. Raw LUMiSizer settling data showing transmission profile taken at 40% converted to suspension height vs. time, for zirconium molybdate (ZM-a) at 4 vol% in water at 500 rpm.

Author Contributions: J.S., D.H. and T.N.H. conceived and designed the experiments; J.S. performed the experiments; J.S. and T.N.H. analysed the data; T.N.H. and D.H. contributed with revisions of the article; and J.S. drafted the manuscript.

Funding: This research was jointly funded by the Engineering and Physical Sciences Research Council (EPSRC) UK, through a Direct Training Award (DTA) grant number EP/M506552/1 and Sellafield Ltd.

Acknowledgments: The authors would like to thank the University of Leeds Nuclear Fuel Cycle Centre for Doctoral Training, as part of a Direct Training Award (DTA). Thanks are given to Martyn Barnes and Geoff Randall from Sellafield Ltd., as part of ongoing support from the Sludge Centre of Expertise. Acknowledgment is also given for materials provided by the National Nuclear Laboratory and to the valuable input from Jonathan Dodds, Tracy Ward and Barbara Dunnett.

Conflicts of Interest: The authors declare no conflict of interest.

References

1. Edmondson, M.; Maxwell, L.; Ward, T.R. A methodology for POCO of a highly active facility including solids behaviour. In *Waste Management*; Waste Management Symposium: Phonenix, AZ, USA, 2012.
2. Harrison, M.T.; Brown, G.C. Chemical durability of UK vitrified high level waste in Si-saturated solutions. *Mater. Lett.* **2018**, *221*, 154–156. [CrossRef]
3. Dobson, A.J.; Phillips, C. High level waste processing in the U.K.—Hard won experience that can benefit U.S. Nuclear cleanup work. In *Waste Management*; Waste Management Symposium: Tucson, AZ, USA, 2006.
4. Evans, J.V.; Moore, W.; Shying, M.E.; Sodeau, J.M. Zirconium molybdate gel as a generator for technetium-99m. I. The concept and its evaluation. *Appl. Radiat. Isot.* **1987**, *38*, 19–23. [CrossRef]
5. Monroy-Guzmán, F.; Díaz-Archundia, L.V.; Contreras Ramírez, A. Effect of Zr:Mo ratio on 99mTc generator performance based on zirconium molybdate gels. *Appl. Radiat. Isot.* **2003**, *59*, 27–34. [CrossRef]
6. Monroy-Guzman, F.; Díaz-Archundia, L.V.; Hernández-Cortés, S. 99Mo/99mTc generators performances prepared from zirconium molybate gels. *J. Braz. Chem. Soc.* **2008**, *19*, 380–388. [CrossRef]
7. Monroy-Guzman, F.; Rivero Gutierrez, T.; Lopez Malpica, I.Z.; Hernandez Cortes, S.; Rojas Nava, P.; Vazquez Maldonado, J.C.; Vazquez, A. Production optimization of 99Mo/99mTc zirconium molybate gel generators at semi-automatic device: Disigeg. *Appl. Radiat. Isot.* **2011**, *70*, 103–111. [CrossRef] [PubMed]
8. Lind, C.; Wilkinson, A.P.; Rawn, C.J.; Payzant, E.A. Preparation of the negative thermal expansion material cubic $ZrMo_2O_8$. *J. Mater. Chem.* **2001**, *11*, 3354–3359. [CrossRef]
9. Varga, T.; Wilkinson, A.P.; Lind, C.; Bassett, W.A.; Zha, C.-S. Pressure-induced amorphization of cubic $ZrMo_2O_8$ studied in situ by x-ray absorption spectroscopy and diffraction. *Solid State Commun.* **2005**, *135*, 739–744. [CrossRef]
10. Rao, B.S.M.; Gantner, E.; Muller, H.G.; Reinhardt, J.; Steinert, D.; Ache, H.J. Solids formation from synthetic fuel reprocessing solutions charaterisation of zirconium molybdate. *Appl. Spectrosc.* **1986**, *40*, 330–336. [CrossRef]
11. Rao, B.S.M.; Gantner, E.; Reinhardt, J.; Steinert, D.; Ache, H.J. Characterization of the solids formed from simulated nuclear fuel reprocessing solutions. *J. Nucl. Mater.* **1990**, *170*, 39–49. [CrossRef]
12. Doucet, F.J.; Goddard, D.T.; Taylor, C.M.; Denniss, I.S.; Hutchison, S.M.; Bryan, N.D. The formation of hydrated zirconium molybdate in simulated spent nuclear fuel reprocessing solutions. *Phys. Chem. Chem. Phys.* **2002**, *4*, 3491–3499. [CrossRef]
13. Magnaldo, A.; Noire, M.H.; Esbelin, E.; Dancausse, J.P.; Picart, S. Zirconium molybdate hydrate precipitates in spent nuclear fuel reprocessing. In Proceedings of the ATALANTE Conference on Nuclear Chemistry for Sustainable Fuel Cycles, Nimes, France, 5–10 June 2004; pp. 1–4.
14. Magnaldo, A.; Masson, M.; Champion, R. Nucleation and crystal growth of zirconium molybdate hydrate in nitric acid. *Chem. Eng. Sci.* **2007**, *62*, 766–774. [CrossRef]
15. Usami, T.; Tsukada, T.; Inoue, T.; Moriya, N.; Hamada, T.; Serrano Purroy, D.; Malmbeck, R.; Glatz, J.P. Formation of zirconium molybdate sludge from an irradiated fuel and its dissolution into mixture of nitric acid and hydrogen peroxide. *J. Nucl. Mater.* **2010**, *402*, 130–135. [CrossRef]
16. Vereshchagina, T.A.; Fomenko, E.V.; Vasilieva, N.G.; Solovyov, L.A.; Vereshchagin, S.N.; Bazarova, Z.G.; Anshits, A.G. A novel layered zirconium molybdate as a precursor to a ceramic zirconomolybdate host for lanthanide bearing radioactive waste. *J. Mater. Chem.* **2011**, *21*, 12001–12007. [CrossRef]
17. Zhang, L.; Takeuchi, M.; Koizumi, T.; Hirasawa, I. Evaluation of precipitation behavior of zirconium molybdate hydrate. *Front. Chem. Sci. Eng.* **2013**, *7*, 65–71. [CrossRef]
18. Liu, X.; Chen, J.; Zhang, Y.; Wang, J. Precipitation of zirconium and molybdenum in simulated high-level liquid waste concentration and denitration process. *Procedia Chem.* **2012**, *7*, 575–580.
19. Arai, T.; Ito, D.; Hirasawa, I.; Miyazaki, Y.; Takeuchi, M. Encrustation prevention of zirconium molybdate hydrate by changing temperature, nitric acid, or solution concentration. *Chem. Eng. Technol.* **2018**, *41*, 1199–1204. [CrossRef]
20. Izumida, T.; Kawamura, F. Precipitates formation behavior in simulated high level liquid waste of fuel reprocessing. *J. Nucl. Sci. Technol.* **1990**, *27*, 267–274. [CrossRef]
21. Paul, N.; Hammond, R.B.; Hunter, T.N.; Edmondson, M.; Maxwell, L.; Biggs, S. Synthesis of nuclear waste simulants by reaction precipitation: Formation of caesium phosphomolybdate, zirconium molybdate and morphology modification with citratomolybdate complex. *Polyhedron* **2015**, *89*, 129–141. [CrossRef]

22. Paul, N.; Biggs, S.; Shiels, J.; Hammond, R.B.; Edmondson, M.; Maxwell, L.; Harbottle, D.; Hunter, T.N. Influence of shape and surface charge on the sedimentation of spheroidal, cubic and rectangular cuboid particles. *Powder Technol.* **2017**, *322*, 75–83. [CrossRef]

23. Paul, N.; Biggs, S.; Edmondson, M.; Hunter, T.N.; Hammond, R.B. Characterising highly active nuclear waste simulants. *Chem. Eng. Res. Des.* **2013**, *91*, 742–751. [CrossRef]

24. Clearfield, A.; Blessing, R.H. The preparation and crystal structure of a basic zirconium molybdate and its relationship to ion exchange gels. *J. Inorg. Nucl. Chem.* **1972**, *34*, 2643–2663. [CrossRef]

25. Krtil, J.; Kouřím, V. Exchange properties of ammonium salts of 12-heteropolyacids. Sorption of caesium on ammonium phosphotungstate and phosphomolybdate. *J. Inorg. Nucl. Chem.* **1960**, *12*, 367–369. [CrossRef]

26. Lento, J.; Harjula, R. Separation of cesium from nuclear waste solutions with hexacyanoferrate(ii)s and ammonium phosphomolybdate. *Solvent Extr. Ion Exch.* **1987**, *5*, 343–352. [CrossRef]

27. Bykhovskii, D.N.; Kol'tsova, T.I.; Kuz'mina, M.A. Phases of variable composition in crystallization of cesium phosphomolybdate. *Radiochemistry* **2006**, *48*, 429–433. [CrossRef]

28. Bykhovskii, D.N.; Kol'tsova, T.I.; Roshchinskaya, E.M. Cesium preconcentration by recovery from solutions in the form of phosphomolybdate. *Radiochemistry* **2009**, *51*, 159–164. [CrossRef]

29. Bykhovskii, D.N.; Kol'tsova, T.I.; Roshchinskaya, E.M. Reduction of radioactive waste volume using selective crystallization processes. *Radiochemistry* **2010**, *52*, 530–536. [CrossRef]

30. Rezaei Ghalebi, H.; Aber, S.; Karimi, A. Keggin type of cesium phosphomolybdate synthesized via solid-state reaction as an efficient catalyst for the photodegradation of a dye pollutant in aqueous phase. *J. Mol. Catal. A Chem.* **2016**, *415*, 96–103. [CrossRef]

31. Bradley, D.F.; Quayle, M.J.; Ross, E.; Ward, T.R.; Watson, N. Promoting the conversion of caesium phosphomolybdate to zirconium molybdate. In Proceedings of the ATALANTE Conference on Nuclear Chemistry for Sustainable Fuel Cycles, Nimes, France, 5–10 June 2004.

32. Jiang, J.; May, I.; Sarsfield, M.J.; Ogden, M.; Fox, D.O.; Jones, C.J.; Mayhew, P. A spectroscopic study of the dissolution of cesium phosphomolybdate and zirconium molybdate by ammonium carbamate. *J. Solut. Chem.* **2005**, *34*, 443–468. [CrossRef]

33. Dunnett, B.; Ward, T.; Roberts, R.; Cheesewright, J. Physical properties of highly active liquor containing molybdate solids. In Proceedings of the ATALANTE Conference on Nuclear Chemistry for Sustainable Fuel Cycles, Montpellier, France, 5–10 June 2016.

34. Biggs, S.; Fairweather, M.; Hunter, T.; Peakall, J.; Omokanye, Q. Engineering properties of nuclear waste slurries. In Proceedings of the ASME 2009 12th International Conference on Environmental Remediation and Radioactive Waste Management, Liverpool, UK, 11–15 October 2009.

35. Mandzy, N.; Grulke, E.; Druffel, T. Breakage of TiO_2 agglomerates in electrostatically stabilized aqueous dispersions. *Powder Technol.* **2005**, *160*, 121–126. [CrossRef]

36. Liao, D.L.; Wu, G.S.; Liao, B.Q. Zeta potential of shape-controlled TiO_2 nanoparticles with surfactants. *Colloids Surf. A Physicochem. Eng. Asp.* **2009**, *348*, 270–275. [CrossRef]

37. Richardson, J.F.; Zaki, W.N. The sedimentation of a suspension of uniform spheres under conditions of viscous flow. *Chem. Eng. Sci.* **1954**, *3*, 65–73. [CrossRef]

38. Bargieł, M.; Tory, E.M. Extension of the richardson–zaki equation to suspensions of multisized irregular particles. *Int. J. Miner. Process.* **2013**, *120*, 22–25. [CrossRef]

39. Turney, M.A.; Cheung, M.K.; Powell, R.L.; McCarthy, M.J. Hindered settling of rod-like particles measured with magnetic resonance imaging. *AIChE J.* **1995**, *41*, 251–257. [CrossRef]

40. Chong, Y.S.; Ratkowsky, D.A.; Epstein, N. Effect of particle shape on hindered settling in creeping flow. *Powder Technol.* **1979**, 55–66. [CrossRef]

41. Lau, R.; Chuah, H.K.L. Dynamic shape factor for particles of various shapes in the intermediate settling regime. *Adv. Powder Technol.* **2013**, *24*, 306–310. [CrossRef]

42. Tomkins, M.R.; Baldock, T.E.; Nielsen, P. Hindered settling of sand grains. *Sedimentology* **2005**, *52*, 1425–1432. [CrossRef]

43. Loth, E. Drag of non-spherical solid particles of regular and irregular shape. *Powder Technol.* **2008**, *182*, 342–353. [CrossRef]

44. Dogonchi, A.S.; Hatami, M.; Hosseinzadeh, K.; Domairry, G. Non-spherical particles sedimentation in an incompressible Newtonian medium by Padé approximation. *Powder Technol.* **2015**, *278*, 248–256. [CrossRef]

45. Lerche, D.; Sobisch, T. Direct and accelerated characterization of formulation stability. *J. Dispers. Sci. Technol.* **2011**, *32*, 1799–1811. [CrossRef]

Article

Compact Storage of Radioactive Cesium in Compressed Pellets of Zeolite Polymer Composite Fibers

Masaru Ooshiro [1,2], Takaomi Kobayashi [1,* and Shuji Uchida [3]

[1]   Department of Materials Science and Technology, Nagaoka University of Technology, 1603-1 Kamitomioka, Nagaoka 940-2188, Japan; ohshiro@kasai-corporation.co.jp
[2]   Kasai Co., Ltd., 578-3 Kawaguchi Akiha-ku, Niigata 956-0015, Japan
[3]   Department of Chemistry and Biochemistry, National Institute of Technology, Fukushima College, Taira-kamiarakawa Nagao 30, Iwaki 970-8034, Japan; uchidas@fukushima-nct.ac.jp
*   Correspondence: takaomi@vos.nagaokaut.ac.jp; Tel.: +81-258-47-9326

Received: 30 May 2018; Accepted: 30 July 2018; Published: 3 August 2018

Abstract: To facilitate the safe storage of radioactive Cs, a zeolite–poly(ethersulfone) composite fiber was fabricated to be a compact storage form of radioactive Cs, and an immobilization was investigated with respect to the effects of volume reduction and stability of the fiber's adsorbent matrix. Using compressed heat treatment at 100–800 °C for a zeolite polymer composite fiber (ZPCF) containing Cs, the fabrication changed its form from a fiber into a pellet, which decreased the matrix volume to be about one-sixth of its original volume. The Cs leakage behavior of the ZPCF matrix was examined in its compact pellet form for non-radioactive Cs and radioactive Cs when different fabrication conditions were carried out in the immobilization. The elution ratio of non-radioactive Cs from the matrix was minimal, at 0.05%, when the ZPCF was compressed with heat treatment at 300 °C. When using radioactive Cs for the compression at below 300 °C, the pellet form also had no elution of the pollutants from the matrix. When the compressed treatment was at 500 °C, the matrix exhibited elution of radioactive Cs to the outside, meaning that the plastic component was burning and decomposed in the pellet. A comparison of ZPCF and natural zeolite indicated that the compressed heating process for ZPCF was useful in a less-volume-immobilized form of the compact adsorbent for radioactive Cs storage.

Keywords: cesium adsorbed; radioactive cesium; safe storage; zeolite polymer composite fiber

1. Introduction

The severe accident that occurred on 11 March 2011 at the Fukushima Daiichi Nuclear Power Plant resulted from difficulties related to the catastrophic earthquake and the subsequent tsunami. The event damaged the plant, which released great amounts of radioactive ^{134}Cs and ^{137}Cs. Since then, the residual radioactivity has persisted as a hazard for local residents [1]. The huge volume of contaminants was estimated in 2013 as about $15-28 \times 10^6$ m^3 [2] and now in 2018 the radioactive Cs has been reduced in the outfields by the efforts of the decontamination process. It is known that the estimated amounts of radionuclides released into the atmosphere in 2012 in Fukushima were 6.1–62.5 PBq and 65–200 PBq for ^{137}Cs and ^{131}I, respectively. Especially, the Cs radioisotopes have been found frequently in aqueous radioactive wastes, mostly at levels exceeding the standards set for the areas. Because Cs belongs to a chemically similar group that includes sodium and potassium, ingestion of Cs radioisotopes can engender their deposition in tissues throughout the human body, thereby presenting an internal hazard to human health.

Over the years that have passed since the accident, external exposure to ^{137}Cs, which has a long half-life of 30.5 years, has come to dominate radionuclide exposure. The trapped ^{137}Cs wastes now present a health risk in Fukushima because the decontamination processes have emitted huge amounts of radioactive waste that remain in the environment. Along with increased concern related to the Cs radionuclide waste, people feel threatened in their life environments. Consequently, decontamination processes have continued to cope with the huge amounts of radioactive Cs. An effective mitigation method must be found through the consideration of attractive technologies. Among such methods, immobilization techniques have been presented for the remediation of radioactive Cs [3–5] (Awual et al., 2016; Kobayashi et al., 2016; Miah et al., 2010). For the large amounts of radioactive Cs that still exist, the development of some adsorption technology is needed in a safe form. Additionally, for the proper management and storage of radioactive waste, the volume reduction of secondary wastes has been especially important in the adsorbed wastes.

The methods for immobilizing radioactive Cs include solidification. In some studies of radioactive fly ash treatment, embedding radioactive materials in a solidified form was accomplished using a nanometallic Ca/CaO suspension for wastewater [6] (Reddy et al., 2014). Reportedly, pyrolytic carbon-coated zeolite is effective [7] (Stinton et al., 1983). Radioactive wastewater has been immobilized in a concrete matrix and in struvite ceramics [8] (Wagh et al., 2016). Moreover, immobilization has been achieved in a $HZr_2(PO_4)_3$ matrix [9] (Nakayama et al., 2003), ash-based geopolymers [10] (Cozzi et al., 2013), and in rice husk silica geopolymers [11] (Lopez et al., 2014). Nevertheless, no report of the relevant literature has described immobilization by polymeric envelopment of a radioactive species with zeolite. Flammable wastes have been incinerated to accumulate enormous amounts of fly ash containing radioactive Cs. Our groups have reported that a zeolite polymer composite fiber (ZPCF) is an effective agent for treating water contaminated with radioactive Cs [12] (Ohshiro et al., 2017) and heavy metal ions [13] (Nakamoto et al., 2017) when the diluted pollutant was concentrated by the decontamination process in the outside of the field [4] (Kobayashi et al., 2016). However, post-adsorption in the ZPCF containing radioactive Cs presents storage problems for coming decades because a greater volume of wastes has to be stored. In post-adsorption processes for such radioactive Cs, reducing the volume becomes necessary and important for later storage processes. Therefore, several methods of immobilizing radioactive Cs in a matrix have been proposed as described above. For radioactive Cs remediation, the ZPCF used has been holding it strongly in the matrix even though the radioactive Cs was concentrated from an extra-diluted solution of radionuclides. Therefore, in the present study, the required compaction technology for ZPCF is described for the fiber's volume reduction and its safe storage. The immobilization of ZPCF used for radioactive Cs was assessed in terms of the leakage of the pollutant from the matrix. Here, the tests were carried out by using an actual radioactive fly ash source. The results showed that the reduced-volume fiber has excellent capabilities for Cs immobilization.

2. Materials and Methods

2.1. Materials

Natural mordenite zeolite powder ($\leq$100 μm) was purchased from Nitto Funka Trading Co. Ltd. (Miyagi, Japan). Poly(ethersulfone) (PES) was used as received (PES, MV = 50,000; BASF Japan Ltd., Ludwigshafen, Germany). *N*-methyl-2-pyrrolidone (NMP; Nacalai Tesque Inc., Kyoto, Japan) was used for the ZPCF without purification. The radioactive Cs adsorbed into the ZPCF was prepared as shown in Figure 1. Here, the ZPCF contained 59 wt% zeolite and 41 wt% was PES, which formed porous fibers. An aqueous solution containing radioactive Cs was prepared using hydrothermal extraction of fly ash for 2 h at 200 °C and 1.5 MPa. Here, the radioactive Cs fly ash was sampled in Namie, Fukushima (Figure 1). After the supernatant dispersed with fly ash was filtrated, the radioactive aqueous solution was used for experiments to fill the ZPCF. Then, the Cs-adsorbed fibers were prepared. The total radioactivity in the weighted fibers was measured in Becquerel per kilogram (Bq/kg) units. As Figure 1 shows, after Cs was bound to the ZPCF, a procedure for immobilization by compressed heat treatments of different temperatures was tested to create pellets. Then, Cs release was evaluated in addition to the matrix properties of the heat-treated matrix. The Cs adsorption process was conducted as follows by using radioactive fly ash (4 kg) having about 30,000 Bq/kg with Cs. The fly ash was dispersed in water (16 L) and filtered. Then, ZPCF (4 kg) was immersed in the Cs solution for 12 h. After this binding process, the ZPCF was used to measure the radioactive Cs concentration. For example, the radioactive Cs that remained in the fibers was 13,100–33,000 Bq/kg after the immersion process.

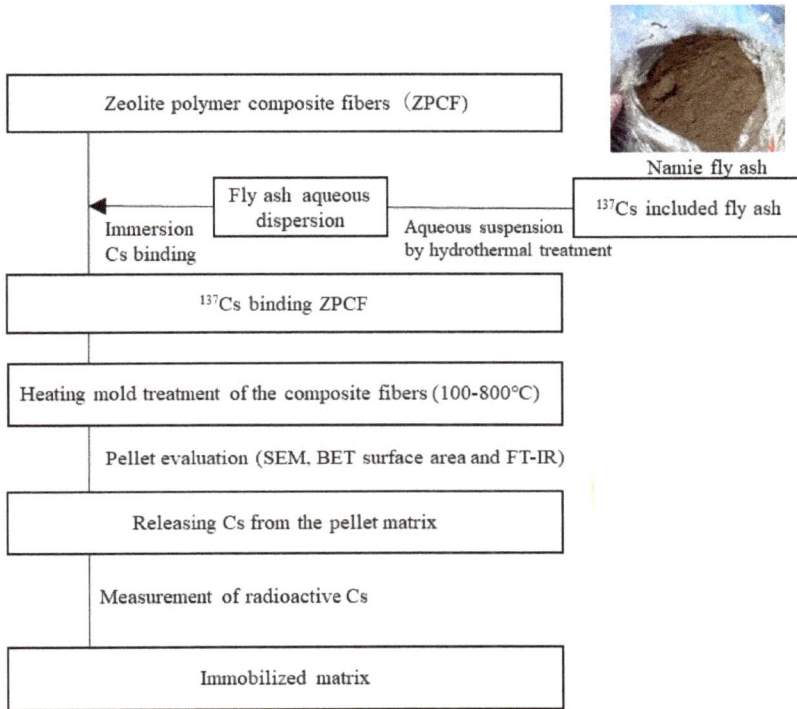

Figure 1. Flowchart of experiment procedure for Cs immobilization by heat treatment of zeolite polymer composite fibers and their Cs release processes. BET, Brunauer–Emmett–Teller.

2.2. Characterization of Zeolite Polymer Composite Fibers

To fabricate the ZPCF pellets enveloping the radioactive Cs, processes of immobilization and volume reduction were included in the heat molding process. The heat molding process is depicted in Figure 2. After the fibers (10 g) were pressed inside of a cylindrical stainless steel tube (50 mm height, 40 mm diameter, and 2.5 mm thickness), they were heated by a surrounding ribbon heater for 2 h at different temperatures of 100, 200, 300, 400, 600, and 800 °C. The temperature was measured using a thermocouple thermometer (FINE THERMO DG2N 100; HAKKO Ltd., Nagano, Japan). Then, the cover was pressed using a hydraulic press machine (P-16B Air Valve; Riken Seiki, Ojiya, Niigata, Japan) at 200 kg/cm^2 for 6 h. After heat molding processing under pressure, the Brunauer–Emmett–Teller (BET) surface area and weight loss of adsorbents were measured. The pellet form was prepared using heat mold processing. For experimental procedures to assess radioactive Cs leakage from the fiber pellet matrix, a 10 g pellet matrix was immersed in 100 mL water at 20 °C. The fibers or the pellets, after accurate measurement of their weight, were washed 10 times with water and were immersed in 100 mL of water for 6 h and 24 h with stirring at 200 rpm. Then, the matrix and the solution were used to measure the resident Cs with a germanium semiconductor detector. The remaining and eluted Cs were estimated and then the value of the Cs eluted rate (%) was calculated using the following equation:

$$\text{Cs eluted rate (\%)} = (C \times V/C_0 \times W) \times 100$$

where C is the Cs concentrations of eluted water, V denotes the eluted water volume, C_0 represents the Cs concentrations of the fiber or pellet, and W is the fiber or pellet weight. Then, the released radioactive Cs was evaluated by measuring the Bq/kg amounts of the pellet and the washed water. The thermal analyses were carried out by Differential Scanning Calorimetry (DSC) (Thermo plus EVO2 DSC8231; Rigaku, Japan), Thermomechanical Analysis (TMA) (TMA-60; SHIMADZU, Kyoto, Japan), and Thermogravimeter-Differential Thermal Analysis (TG-DTA) (DTG-60; SHIMADZU, Kyoto, Japan). Here, the upper limitation of the elevated temperature in DSC was at 400 °C. The temperature of the TG-DTA was 480 °C in an aluminum pan. Unfortunately, setting DSC to 480 °C is impossible because at over 400 °C it damages the equipment with the aluminum pan.

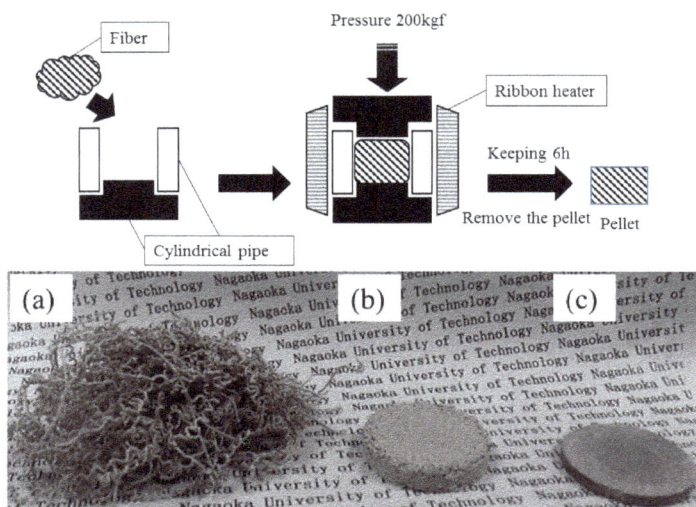

Figure 2. Schematic illustration of heating mold processes for preparation of the pellet matrix. The picture shows, respectively, fibers (**a**) before and (**b**,**c**) after heating mold processes at 100 °C and 300 °C [14].

In addition, scanning electron microscopy (SEM) images were taken (JSM-5310LVB; JEOL, Tokyo, Japan) of the ZPCF and the compacted pellet. Additionally, the sample was sputtered at 0.1 Torr for 40 s with Au for SEM measurements. A Fourier transform infrared spectrometer (FT-IR, IR Prestige-21 FTIR 8400s; Shimadzu Corp., Kyoto, Japan) was used with the KBr method. The N_2 adsorption was analyzed using the Brunauer–Emmett–Teller (BET) surface area (Tristar II 3020; Micrometric Inc., Sarasota, FL, USA).

3. Results and Discussion

3.1. Immobilization of Radioactive Cesium by Heating Mold Processing of Composite Fibers

As shown in Figure 2, for the schematic illustration of heating mold processes, the radioactive Cs was bound by the ZPCF in the dispersed fly ash solution. Then, the compacted pellet was shaped from the fibrous sample by heating it at different temperatures under pressure. The DSC curves of the PES fiber are shown in Figure 3 for the ZPCF and PES without zeolite. Endothermic peaks were found at 220 °C and 227 °C for the PES and the ZPCF in each DSC curve. This indicated that heating at over 230 °C led to the PES being melted. In the ZPCF, the observed temperature was found to be a little bit higher relative to that of the PES. It was noted that in the higher 220 °C and 227 °C range, the heat molding was reasonable to process and enabled the pellet form, while a pellet prepared at lower than 220 °C was only compressed by pressure. Additionally, it was apparent that the pellet form could be made to different densities at the 100 °C and 300 °C process temperatures. As shown in Figure 4, this was also supported by TMA measurement data. As a result of TMA measurement, the PES was found to be softened and melted from around 234–250 °C. Therefore, the TMA measurements were automatically stopped at over 250 °C. Here, the thermal DSC and TMA analyses could not obtain results at over 400 °C.

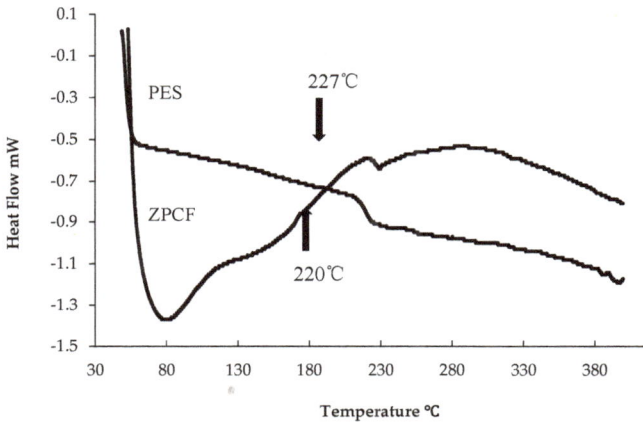

Figure 3. Results of Heat analysis of zeolite polymer composite fiber (ZPCF) and PES.

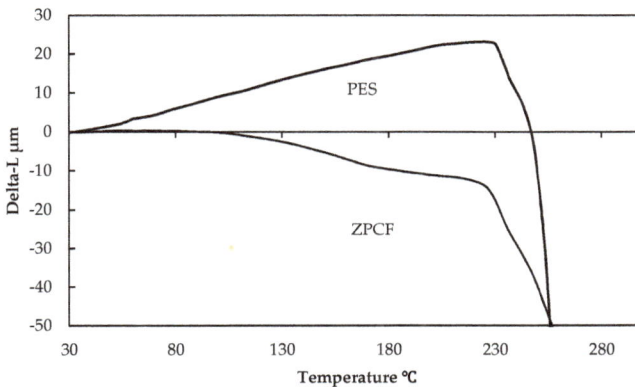

Figure 4. TMA measurement of PES and ZPCF at different temperatures.

Figure 5 presents the relation between the volume (cm^3) and density (g/cm^3) changes of the pellets. The ZPCF volume was decreased from 40 cm^3 to 12.5 cm^3 under the compressed heat molding process at 100 °C. Then, at over 300–800 °C, the volume changed by 6.3 cm^3. The comparison indicated that the molding at 300 °C showed significant volume reduction to about one-sixth in the fabricated pellet. In addition, the pellet density increased from 0.25 to 1.52 g/cm^3 after compression processing was implemented in the range of room temperature to 300 °C. This exactly was due to PES melting in the dense pellet formation as supported by Figures 3 and 4. When the temperature was changed from 300 °C to 500 °C and 800 °C, the pellet density was decreased from 1.52 g/cm^3 to 1.13 g/cm^3 and then 1.1 g/cm^3, respectively. Absolutely, the decrease was attributable to the polymer decomposition of the organic plastic PES. This was also supported in Figure 4 by the gradual decrease of the TMA results at over 300 °C in both curves.

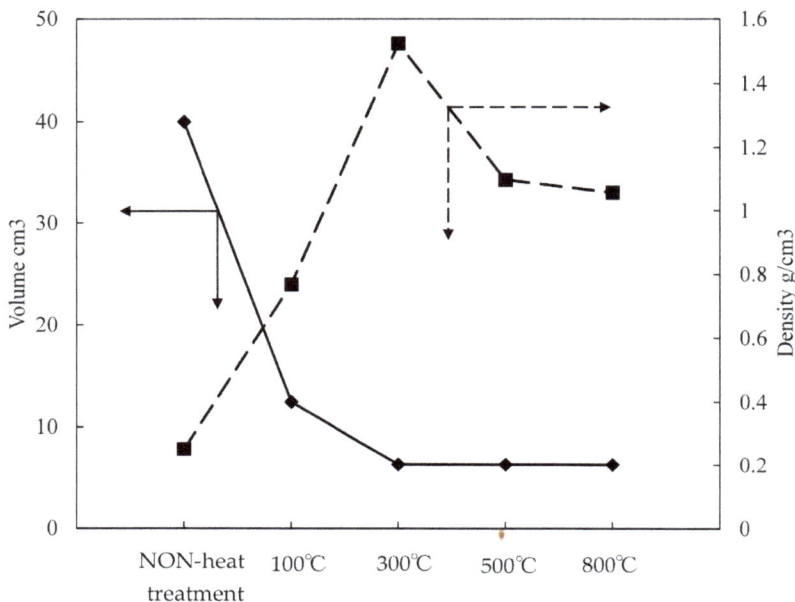

Figure 5. Relation between heating mold press temperature and the pellet volume and density.

Figure 6 portrays SEM images of the surface and cross sections of ZPCF treated at different temperatures for the molding processing. Low-temperature molding at 100 °C revealed that the zeolite powders were embedded in the PES medium as they were in the case of the non-heating mold. However, at temperatures higher than 300 °C, the images showed that the PES amounts decreased concomitantly with increasing temperature. At temperatures higher than 500 °C, the fibers were in a brittle state. Moreover, the polymer layer disappeared, reflecting the organic PES matrix's decomposition. A cross-section view (Figure 6 left) of a pellet revealed the dense structure of the compressed fiber at 500 °C (d) and 800 °C (e). The FT-IR spectral patterns for the 800 °C sample (Figure 8) presented no peaks at 1580, 1485, and 895 cm^{-1} corresponding to PES in the ZPCF and peaks at 3460, 1658, and 1060 cm^{-1} corresponding to zeolite. Table 1 presents the respective assignments for their FT-IR peaks. Actually, band broadening was apparent in spectra obtained at 500 °C and 800 °C. Moreover, PES peaks were nonexistent. It was inferred from these results that the SiO$_2$ component remained in the zeolite. The appearance of the 1658 cm^{-1} peak in the heated pellets might be assigned to water adsorbed from the atmosphere after heating. It is noteworthy that the appearance of the peaks at 1580 and 1485 cm^{-1} is temperature-dependent. At 800 °C, both PES peaks disappeared, meaning that the PES in the pellet was burned out of the material. Therefore, the spectra retained broad peaks at 1060 cm^{-1} and at 802 cm^{-1} for the Si–O–Si and Al–O bands of zeolite, respectively. As seen in Figure 7 for the TG-DTA measurement of ZPCF in the range of 50–500 °C, an endothermic peak and depletion of TG considered to be dehydrated from zeolite were observed up to 250 °C. After 450 °C, the exothermic peak and a decline of TG can be confirmed in the 350–500 °C range. This was presumed to be caused by the combustion of PES.

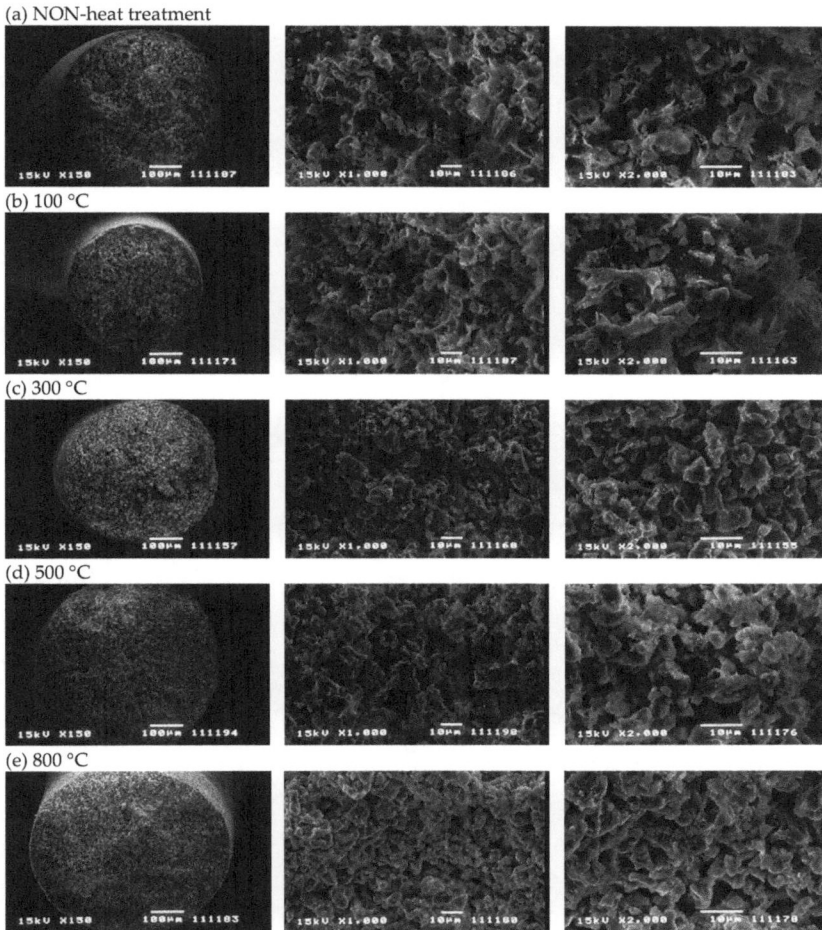

(a) NON-heat treatment

(b) 100 °C

(c) 300 °C

(d) 500 °C

(e) 800 °C

Figure 6. External view of pellets heated at different temperatures and corresponding SEM images [14]. (**a**) NON-heat treatment; (**b**) 100 °C; (**c**) 300 °C; (**d**) 500 °C and (**e**) 800 °C heat treatment.

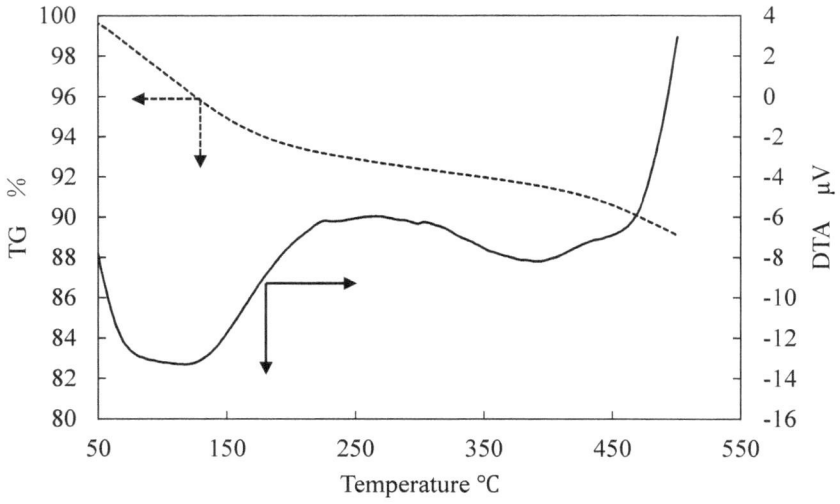

Figure 7. TG-DTA measurement of ZPCF at different temperatures.

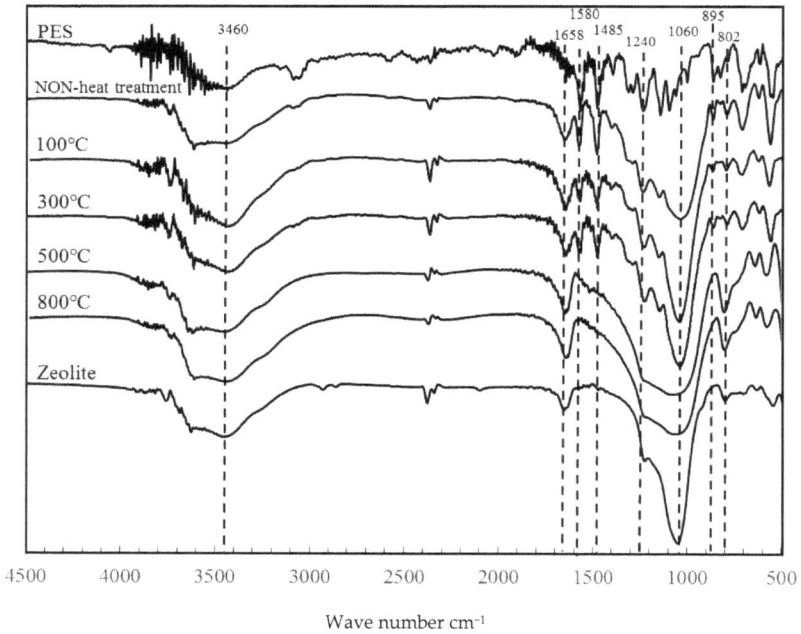

Figure 8. FT-IR spectra of ZPCF treated for 6 h at different temperatures.

Table 1. Infrared adsorption peaks of zeolite and PES.

Wavenumber (cm^{-1})	Component in ZPCF	Assignment
3460	Zeolite	OH st.
1658	zelite	Si–O–Si st.
1580	PES	C=O st.
1485	PES	C–O st.
1240	PES	SO st.
1060	zeolite	Si–O st.
895	PES	Polymer group C–C st.
802	zeolite	Al–O st.

st.: stands for stretching.

To evaluate the porous and dense properties of the fibers and pellets, nitrogen (N_2) adsorption and desorption were measured at different pressures. Figure 9 portrays the N_2 adsorption isotherm of the PES and ZPCF in the absence (a) and presence (b) of heat molding processing. According to the isotherms, their samples presented that the adsorption behavior followed that of the type 2 isotherm, reflecting the presence of a macroporous structure [5] (Kobayashi et al., 2016). Compared to the PES and zeolite shown in Figure 9a, it was apparent that the PES fiber had a lower capacity for N_2 adsorption amounts relative to the zeolite powder. As Figure 9b shows, the pellets molded at 500 °C and 800 °C showed similar isotherm curves to that of zeolite, as was true also for the non-heat treatment. In the different pellet samples at 500 °C and 800 °C, the amounts of adsorption of N_2 were changed, although the shapes of the isotherms were almost similar. However, it was seen that the values of the pellet at 800 °C were lower than those at 500 °C. This strongly suggested that the numbers of the mesopores in the pellet at 800 °C were much lower than the numbers of mesopores in the pellet at 500 °C. This meant that the 800 °C molding destroyed the porous structure of the pellet. However, it is noteworthy that the curve of the pellet heated at 300 °C had lower amounts of N_2 adsorption. Consequently, the melted PES penetrated and covered the zeolite pores causing the zeolite volume to decrease. As seen in Figures 4 and 7, this was strongly suggested by the TMA and DTA results. Also, the FT-IR results of different temperatures showed decomposition of the PES components at temperatures over 500 °C. It is noteworthy that the N_2 adsorption amounts were somewhat higher at the temperatures of 500 °C and 800 °C, which indicates that the PES component in the pellets decomposed gradually because the temperature was higher than around 300 °C and that it increased concomitantly with increasing temperature to 800 °C. Therefore, the PES component decomposition occurred gradually with increasing temperature. The results obtained at 800 °C showed that the zeolite surface was exposed without envelopment by the PES layer.

Table 2 presents the BET surface area and weight reduction of the ZPCF pellets heated at different temperatures. The BET surface area of the ZPCF was 32 m^2/g before heat molding processing. The value was remarkably lower at 300 °C because the melted PES enveloped the porous structure of the zeolite. The pellet structure became denser. At 500 °C, however, the surface area value increased to 32 m^2/g, suggesting that little zeolite remained after heating. Furthermore, since the heating treatment decomposed the PES component and thereby exposed the zeolite surface, the experimental results of the ZPCF weight reduction at 500 °C and 800 °C were, respectively, 30.8% and 33.3% after heat molding processing. This was due to the polymer being well-decomposed at over 500 °C. However, the value of weight reduction at 300 °C was 2.17 wt %, meaning that the PES had decomposed only a little. Figure 4 depicts the TMA results, which show that the PES of the thermoplastic polymer was melted at temperatures higher than 300 °C. Then, the melting PES seemed to cover the zeolite powder. Therefore, from the weight reduction results, one can reasonably infer that the melted PES surrounding the zeolite powders produced an enveloping layer that decreased the zeolite surface area. Table 2 contains the adsorbent concentration the radioactive Cs of the pellet matrix. At 500 °C and 800 °C, the values of the radioactive Cs were 21,200 and 22,400 Bq/kg. The results of radioactive Cs concentration were higher at 500 °C and 800 °C than those at 100–300 °C. This was due to weight reduction by the PES decomposition.

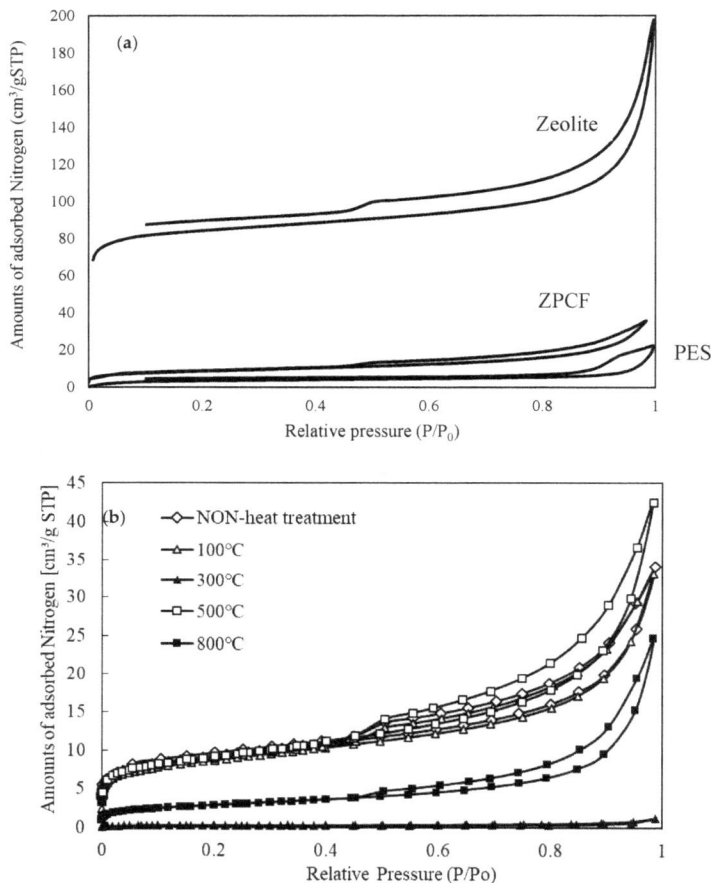

Figure 9. N_2 adsorption isotherms of (**a**) PES and ZPCF and (**b**) pellets obtained with heat mold treatment at different temperatures.

Table 2. Relation between the BET surface area, weight reduction, and radioactive Cs concentration of zeolite polymer composite fibers [14].

Heat Treatment Temperature	BET Surface Area m²/g	Weight Reduction %	Adsorbent Concentration Bq/kg	
			Before	After
NON-heat treatment	32.5	0	14,000	14,200
100 °C	30.6	0.67	14,000	13,100
300 °C	0.9	2.17	14,000	15,300
500 °C	32.8	30.8	14,000	21,200
800 °C	10.1	33.3	14,000	22,400

3.2. Immobilization of Radioactive Cesium in ZPCF Pellets

Table 3 presents values of the radioactive Cs amounts observed with and without heat molding processing. The ZPCF adsorbed up to 14,000 Bq/kg of Cs. The Cs concentration amounts were increased as the temperature increased after heat molding processing. It was seen that the increment was observed between 300 °C and 500 °C for radioactive Cs concentration. The increased concentration derived from the fact that the PES weight was lower at higher temperatures. Also, the weight reduction between 300 °C and 500 °C was remarkable. Table 3 contains values for the eluted concentration of radioactive Cs from the pellets when they were washed in distilled water. The values of the eluted Cs in water phase were measured. It was noted that the values of the radioactive Cs concentration from the pellet were less than 1 Bq/L at 100 °C and 300 °C. In contrast, the values were 60 and 62 Bq/L for the washed pellet molded at 500 °C. This result strongly indicated that adsorbed radioactive Cs was released from the matrix to the water.

The ZPCF's form can be changed to a pellet by heat molding processing. Therefore, it was interesting to observe the radioactive Cs immobilization's efficiency in the matrix. To measure changes in the radioactive Cs concentration, Cs release tests were conducted in water for each sample. Figure 10 shows the elution rate (%) of non-radioactive Cs. After the pellet was washed with diluted water for 6 h, the elution rate (%) was evaluated. Furthermore, Table 3 shows the rate of radioactive Cs elution from pellet matrixes molded at different temperatures. These results indicated that elution of the Cs from the fibers and pellets depends strongly upon the heat molding process temperature. The rate of elution of the Cs was extremely low, about 0.05%, at 300 °C relative to 0.95% at 100 °C and 2.49% at 500 °C. Therefore, it was apparent that the melted PES influenced the Cs release. These results exhibited the same tendency as that for results of the surface area as presented in Table 2. Therefore, for the 300 °C treatment, envelopment within the polymer layer was effective for Cs immobilization in the pellet matrix. At 800 °C, because heat processing greatly decomposed the PES layer surrounding the zeolite powders, a compacted pellet might be effective to fix the Cs component with the zeolite powder.

Table 3. Elution rate of radioactive Cs from heat treatment of ZPCF [14].

Sample	Adsorbents Concentration of ZPCF Bq/kg (Weight Reduction Rate %)			Eluted Solution Concentration of Water Bq/L		Elution Rate %
	0 h	6 h	24 h	6 h	24 h	
NON-heat treatment	14,200	14,000 (0.1%)	14,000 (0.1%)	<1	<1	0.07
100 °C	13,100	13,300 (0.0%)	13,300 (0.1%)	<1	<1	0.08
300 °C	15,300	15,500 (0.0%)	15,500 (0.1%)	<1	<1	0.07
500 °C	21,200	29,800 (29.1%)	29,800 (29.2%)	60	62	4.10
800 °C	22,400	33,000 (39.8%)	33,000 (30.0%)	25	26	1.73
zeolite	24,500	22,000 (0.1%)	22,000 (0.0%)	274	280	11.42

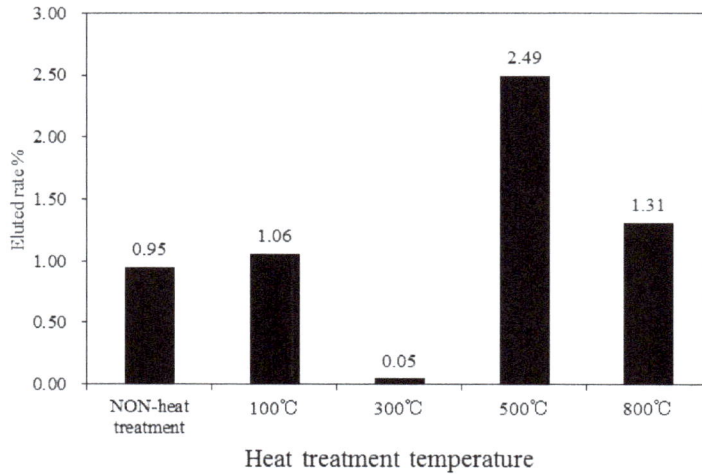

Figure 10. Elution rates of non-radioactive Cs from heat molding treatment of ZPCF [14].

Table 3 presents results of radioactive Cs release tests conducted for 6 h and 24 h using different Becquerel per kilogram amounts of ZPCF. The table shows radioactive amounts as evaluated at 0 h and at different times before and after water washing. The elution rates (%) are values obtained at 24 h. The elution rates were similar to those found for non-radioactive Cs. For example, the ZPCF had 14,200 Bq/kg before washing for the non-heat treatment sample. The sample was 14,000 Bq/kg at 6 h and 24 h after washing in water. For the pellet sample prepared at 300 °C, the value of 15,300 Bq/kg after washing represented almost no change from the original 15,500 Bq/kg. However, after mold treatment at 500 °C, the value of 21,200 Bq/kg was increased to 29,800 Bq/kg at 6 h and 24 h. Table 3 shows that the observed weight reduction was about 30% after the elution tests for 500 °C and 800 °C, demonstrating that the component remaining after burning the PES from the pellet was a soluble substance, such as calcium salt. The radioactive Cs concentration was increased in the pellet. Furthermore, the case at 800 °C showed a similar phenomenon. Figure 11 summarizes the relationship between the rate of Cs release and the surface area m²/g. Apparently, the lower elution was due to the lesser surface area. When the molding temperature was increased from 300 to 500 °C, the surface area was also increased. The increase was due to decomposition of the PES and then exposure to the porous zeolite site. This meant that the high porosity of the PES plastic shape had kept the elution of Cs and decomposition much lower from the heat molding system on the experimental time scale. At 300 °C, the reason for the lower density was the melting of the PES that covered the porous zeolite site, meaning that a covering effect decreases Cs elution. In the case of the 100 °C molding treatment, the PES was not melted, keeping a higher surface area. Thus, the leaking of the radioactive Cs, which was equivalently the same level as that at 300 °C, becomes of a lower concentration of the immobilization, at about 20,000 Bq/kg. However, that of the non-radioactive Cs concentration was at 86.1 g/kg, meaning that it was extremely higher than that of radioactive Cs. Therefore, the leaking Cs occurred at 1.06%.

In conclusion, heat molding processing of ZPCF containing Cs efficiently immobilized radioactive Cs in pellets formed at temperatures lower than 300 °C. However, at temperatures higher than 300 °C, the immobilization effect was less pronounced. Furthermore, the results showed that the heat molding processing reduced the pellet volume. The results presented herein suggest that the ZPCF was a better matrix material for fixing radioactive Cs.

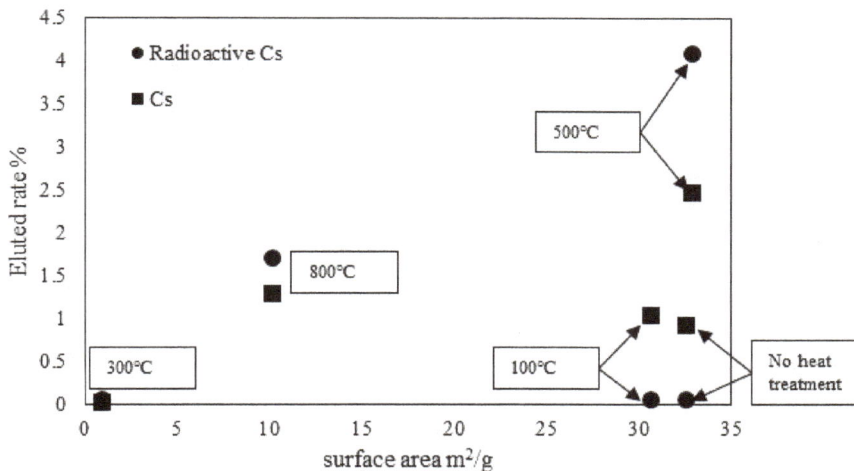

Figure 11. Relationship between elution rates of radioactive Cs and non-radioactive Cs and surface area from different heat-molding temperatures of ZPCF.

4. Conclusions

This study examined the immobilization of radioactive Cs by heated mold treatment of ZPCF for safe storage. The properties of pellets formed at different temperatures using heat molding processing were compared. An effective decrease of the eluted Cs from the pellet was observed at 300 °C, suggesting that the melted PES at 300 °C enveloped the zeolite, thereby fixing the radioactive Cs. This inference was supported by several observations of the ZPCF pellet's surface area, which decreased because of envelopment by the melted polymer surrounding the zeolite powders. Actually, elution of the radioactive Cs was extremely low: less than 0.07–0.08%. Moreover, a volume reduction to 1/6 was achieved by heat molding of the ZPCF.

Author Contributions: M.O., T.K. and S.U. conceived and designed the experiments; S.U. performed the experiments for detection of radioactive cesium; M.O. and T.K. analyzed the data; M.O. and T.K. contributed with revisions of the article; and T.K. drafted the manuscript.

Funding: This research was partially funded by Demonstration Tests on Decontamination and Volume reduction Technology of the Ministry of Environment, Japan on 2015, as titled with Verification of Removal of Radioactive Cesium from Incineration Ash by Hydrothermal Extraction, Volume Reduction, and Stabilization of Radioactive Substances.

Acknowledgments: The authors would like to thank Shinichi Kasai and Shinichi Kuboki for useful discussions of the project. We also thank Hiroaki Iimura and Tsuneo Nakanishi for cooperating their expertise on the device design. This research was partially supported by the decontamination and volume reduction of decontamination waste Ministry of the Environment government of Japan, "Verification of Removal of Radioactive Cesium from Incineration Ash by Hydrothermal Extraction, Volume Reduction, and Stabilization of Radioactive Substances" (No. 1).

Conflicts of Interest: The authors declare no conflicts of interest.

References

1. Morino, Y.; Ohara, T.; Watanabe, M.; Hayashi, S.; Nishizawa, M. Episode analysis of deposition of radiocesium from Fukushima Daiichi Nuclear Power Plant. *Environ. Sci. Technol.* **2013**, *47*, 2314–2322. [CrossRef] [PubMed]
2. Parajuli, D.; Tanaka, H.; Hakuta, Y.; Minami, K.; Fukuda, S.; Umeda, K.; Kamimura, R.; Hayashi, Y.; Ouchi, M.; Kawamoto, T. Dealing with the aftermath of Fukushima Daiichi Nuclear Accident: Decontamination of radioactive cesium enriched ash. *Environ. Sci. Technol.* **2013**, *47*, 3800–3806. [CrossRef] [PubMed]

3. Awual, M.D.R.; Miyazaki, Y.; Taguchi, T.; Shiwaku, H.; Yaita, T. Encapsulation of cesium from contaminated water with highly selective facial organic–inorganic mesoporous hybrid adsorbent. *Chem. Eng. J.* **2016**, *291*, 128–137. [CrossRef]

4. Kobayashi, T.; Ohshiro, M.; Nakamoto, K.; Uchida, S. Decontamination of Extra-Diluted Radioactive Cesium in Fukushima Water Using Zeolite-Polymer Composite Fibers. *Ind. Eng. Chem. Res.* **2016**, *55*, 6996–7002. [CrossRef]

5. Miah, M.Y.; Volchek, K.; Kuang, W.; Tezel, F.H. Kinetic and equilibrium studies of cesium adsorption on ceiling tiles from aqueous solutions. *J. Hazard. Mater.* **2010**, *183*, 712–717. [CrossRef] [PubMed]

6. Reddy, S.; Mitoma, Y.; Okuda, T.; Sakita, S.; Simon, C. Preferential removal and immobilization of stable and radioactive cesium in contaminated fly ash with nanometallic Ca/CaO methanol suspension. *J. Hazard. Mater.* **2014**, *279*, 52–59. [CrossRef]

7. Stinton, D.P.; Lackey, W.J.; Angelin, P. Immobilization of radioactive cesium in pyrolytic-carbon-coated Zeolite. *J. Am. Ceram. Soc.* **1983**, *66*, 389–392. [CrossRef]

8. Wagh, A.S.; Sayenko, S.Y.; Shkuroptenko, V.A.; Trasov, R.V.; Dykiy, M.P.; Svitlychniy, Y.O.; Virych, V.D.; Ulybkina, E.A. Experimental study of cesium immobilization in struvite structures. *J. Hazard. Mater.* **2016**, *302*, 241–249. [CrossRef] [PubMed]

9. Nakayama, S.; Itoh, K. Immobilization Technique of cesium to HZr$_2$(PO$_4$) using an autoclave. *J. Nucl. Sci. Technol.* **2003**, *40*, 631–633. [CrossRef]

10. Cozzi, A.D.; Bannochie, C.J.; Burket, P.R.; Crawford, C.L.; Jantzen, C.M. Immobilization of Radioactive Waste in Fly Ash Based Geopolymers. In Proceedings of the 2011 World of Coal Ash (WOCA) Conference, Denver, CO, USA, 9–12 May 2011; Available online: http://www.flyash.info/ (accessed on 24 March 2017).

11. Lopez, F.J.; Sugita, S.; Kobayashi, T. Cesium-adsorbent Geopolymer Foams Based on Silica from Rice Husk and Metakaolin. *Chem. Lett.* **2014**, *43*, 128–130. [CrossRef]

12. Ooshiro, M.; Kobayashi, T.; Uchida, S. Fibrous zeolite-polymer composites for decontamination of radioactive waste water extracted from radio-Cs fly ash. *Int. J. Eng. Technol. Res.* **2017**, *7*, 2454–4698.

13. Nakamoto, K.; Ohshiro, M.; Kobayashi, T. Mordenite zeolite-polyethersulfone composite fibers developed for decontamination of heavy metal ions. *J. Environ. Chem. Eng.* **2017**, *5*, 513–525. [CrossRef]

14. The project report of Decontamination and volume reduction of decontamination waste Ministry of the Environment government of Japan. In *Verification of Removal of Radioactive Cesium from Incineration Ash by Hydrothermal Extraction, Volume Reduction, and Stabilization of Radioactive Substances (No.1)*; Nagaoka University of Technology: Nagaoka, Japan, 2015.

materials

MDPI

Article

Treatment Technology of Hazardous Water Contaminated with Radioisotopes with Paper Sludge Ash-Based Geopolymer—Stabilization of Immobilization of Strontium and Cesium by Mixing Seawater

Zhuguo Li [1], Mariko Nagashima [2] and Ko Ikeda [1,*]

[1] Department of Architectural Design and Engineering, Graduate School of Science and Technology for Innovation, Yamaguchi University, Ube 755-8611, Japan; li@yamaguchi-u.ac.jp
[2] Department of Earth Sciences, Graduate School of Science and Technology for Innovation, Yamaguchi University, Yamaguchi 753-8512, Japan; nagashim@yamaguchi-u.ac.jp
* Correspondence: k-ikeda@yamaguchi-u.ac.jp; Tel./Fax: +81-836-22-8296

Received: 28 July 2018; Accepted: 19 August 2018; Published: 24 August 2018

Abstract: Long-term immobilization ratios of strontium (Sr^{2+}) and cesium (Cs^+) in paper sludge ash-based geopolymer (PS-GP) were investigated in one year. PS-GP paste specimens were prepared in the conditions of 20 °C and 100% R.H., using two kinds of paper sludge ash (PS-ash). Two kinds of alkaline solution were used in the PS-GP as activator. One was prepared by diluting aqueous Na-disilicate (water glass) with seawater. Another was a mixture of this solution and caustic soda of 10 M concentration. When seawater was mixed into the alkaline solution, unstable fixations of Sr^{2+} and Cs^+ were greatly improved, resulting stable and high immobilization ratios at any age up to one year, no matter what kind of PS-ash and alkaline solution were used. Element maps obtained by EPMA exhibited nearly even distribution of Cs^+. However Sr^{2+} was biased, making domains so firmly related to Ca^{2+} presence. The mechanism that seawater stabilizes immobilization of Sr^{2+} and Cs^+ was discussed in this study, but still needs to further investigation. Chemical composition analyses of PS-GP were also conducted by SEM-EDS. Two categories of GP matrix were clearly observed, so called N-A-S-H and C-A-S-H gels, respectively. By plotting in ternary diagrams of SiO_2-$(CaO + Na_2O)$-Al_2O_3 and Al_2O_3-CaO-Na_2O, compositional trends were discussed in view of 'plagioclase gels' newly found in this study. As a result, it is suggested that the N-A-S-H and C-A-S-H gels should be strictly called Na-rich N-C-A-S-H and Ca-rich N-C-A-S-H gels, respectively.

Keywords: geopolymer; paper sludge ash; radionuclide; hazardous water; immobilization; seawater; strontium; cesium; chlorine

1. Introduction

Conventionally, by using minerals as immobilization media, various attempts have been made so far to remove hazardous elements from water contaminated with heavy metals. Representative minerals are zeolite, apatite, etc. Among others, tobermorite is peculiar [1]. Cementitious materials are generally used to treat contaminated sludge with less water. However, Portland cement paste or concrete has incomplete immobilization capabilities, especially for lead and sometimes zinc [2,3]. On the other hand, geopolymer as other kind of immobilization media has attracted attention in recent years [3–11]. The immobilization principle of geopolymer is the polycondensation of silicic acid monomers, in which the foreign ions are incorporated into the siloxane bonds of tetrahedra to promote polymerization of monomer [SiO_4]-complex associated with other supplemental coordination sites to maintain charge neutrality. Therefore, it is possible to fix toxic metals into geopolymer. However, geopolymer is versatile, it is unsuitable for immobilization of arsenic [9]. The present situation of immobilization of toxic elements using geopolymer are reviewed recently in the literature [12].

Radionuclide-contaminated waste is classified into high dose and low dose pollutant waters. In the former, it is vitrified by melting to place into containers called canister and are permanently stored in deep underground, which is called geological disposal. In the latter, it is solidified by pitch or cement, then put into drums, and finally discarded underground, which is called shallow burial. The urgent problem is an unexpected leakage of radionuclides, happening in the nuclear reactors of Fukushima Daiichi Nuclear Power Plant due to meltdown caused by the large earthquake and subsequent tsunami. The leaked radionuclides have been spreading out in contaminated water due to infiltration of groundwater to cause serious environmental problem. Even though the radionuclides' doses are not so extremely high, it is a tedious work to treat a huge amount of the contaminated water. Current treatment is in a two-stage plant equipped with SARRY and ALPS. In the first stage cesium is adsorbed with zeolite and Ti-silicate. In the second stage, the rest 62 radionuclides are eliminated in addition to cesium uncaptured and leaked from the first stage by a flowline installed iron coprecipitation, carbonate coprecipitation, titanium oxide adsorbent columns, etc. However, since radioactive tritium cannot be removed, the treated water has been still stored in several hundred tanks. The storage tanks continue to increase day by day. Furthermore, in the long term, there is a concern about leakage from the water tanks due to the metal tanks corroding. Therefore, it is urgently necessary to treat the contaminated water rapidly in bulk quantities. Incidentally, though tritium is radioactive, it is said that tritium does not cause serious health problems, since ingested tritium would be discharged together with urine promptly. Therefore, it is theoretically no problem to run out the treated water into open sea. However, fishermen around the Fukushima Daiichi Nuclear Power Plant are fiercely opposed out of fear of a rumor that the water would poison fish, so the treated water has been kept in the storage tanks.

For this reason, the authors developed the effective method as described in the previous reports to solve this problem [4,5]. That is, papermaking sludge incineration ash (PS-ash) is mixed with the alkaline solution to make geopolymer. The alkaline solution is prepared by adding the contaminated water. Since PS-ash is porous, it can absorb a very large amount of liquid, when using it as active filler or precursor of GP. It has been confirmed that the PS-ash-based geopolymer (PS-GP) may treat hazardous water contaminated with radionuclides. One ton of PS-ash can treat more than one ton of contaminated water, and the immobilization ratios of Sr^{2+} and Cs^+ are generally very high, showing more than 90%. However, some serious issues have been encountered after the previous works [4,5] as well as after continuous measurement up to one year later. That is, the immobilization ratios of Sr^{2+} and Cs^+ in the PS-GP added with non-radioactive strontium nitrate and cesium nitrate as surrogates are unstable and fluctuating.

Now the damaged nuclear reactors are cooled by fresh water. However, in the beginning of the Fukushima Daiichi Nuclear power plant disaster, seawater was actually pumped up from the port in front of the premises to cool down the damaged nuclear reactors as an emergency measure. Hence, in this context, we used seawater to prepare alkali silicate solution to produce PS-GP, and found that the immobilization ratios of Sr^{2+} and Cs^+ became stable even in long-term. In this study, we aim to clarify the long-term immobilization ratios of Sr^{2+} and Cs^+ in PS-GP and the immobilization mechanism in view of polycondensation products of PS-GP, when the alkali silicate solution is prepared by adding sea water.

2. Materials and Methods

2.1. Preparation of Alkali Silicate Solutions

Commercially available water glass, called JIS (Japanese Industrial Standards) No. 1 sodium disilicate aqueous solution (nominal composition, $Na_2O \cdot 2SiO_2 \cdot aq$), was diluted with deionized water to prepare a sodium silicate stock solution with specific gravity (S.G.) of 1.54. Then, seawater was added to the stock solution to obtain the first alkali silicate solution with S.G. 1.27, which is called GP-liquor #1SW. Furthermore, the GP-liquor #1SW was mixed with caustic soda aqueous solution of 10 mole concentration by a volume ratio of 3:1 to obtain another solution with S.G. 1.30, which is called GP-liquor #0SW. Details are shown in Table 1.

The used seawater was retrieved from the outside of the Yakeno Coast breakwater of the Seto Inland Sea, Sanyo-Onoda City, Yamaguchi Prefecture, Japan, at the time of high tide. Prior to using, the seawater was percolated by filter paper of no. 131. The general S.G. of Pacific seawater is in the range of 1.02–1.03, but the Seto Inland Sea is slightly heavy, as measured 1.04 of S.G.

Table 1. Specifications of GP-liquor used.

Aqueous Solution	Specific Gravity	Salt Concentration (%)	
	(S.G.)	Bulk	Chlorine
Seawater	1.04	3.445 [1]	1.898 [1]
JIS No. 1 stock solution	1.54	-	-
GP-liquor #0SW [2]	1.30	1.357	0.748
GP-liquor #1SW [2]	1.27	1.389	0.765

[1] Standard concentration of Pacific Ocean, [2] Seawater mixing.

2.2. Preparation of Specimens and Strength Test

Chemical compositions of two kinds of PS-ash used as active fillers of PS-GP are shown in Table 2 together with their physical properties, which were determined by conventional techniques, including X-ray fluorescent analysis (XRF, MagixPro, Royal Philips, Amsterdam, Holland), and Blaine specific surface area measurement (Marubishi Kagaku, Tokyo, Japan). The PS-ash, called OTo3, is characterized by high content of Al_2O_3 component, whereas N45 has high CaO and MgO components. Details are kindly referred to the previous works in addition to constituent minerals of the PS-ashes [4,5].

Table 2. Chemical compositions determined by XRF and physical constants of air-dried PS-ash [4,5].

PS-ash	SiO_2	TiO_2	Al_2O_3	Fe_2O_3	MnO	CaO	MgO	Na_2O	K_2O	P_2O_5	SO_3	Cl	Others [1]
OTo3	30.94	1.55	37.37	1.88	0.03	18.57	3.58	0.33	0.81	1.56	2.71	0.41	0.28
N45	21.87	0.77	13.75	2.58	0.37	34.95	10.44	0.80	0.78	3.56	9.25	0.42	0.45

PS-ash	Key	Apparent Density, g/cm³	Specific Surface Area, cm²/g, Blaine	Total of XRF Analysis
OTo3	1	2.50	6460	100.02 %
N45	3	2.26	5680	99.99 %

[1] Including ZnO, CuO, BaO, SrO, NiO, PbO, ZrO_2, CeO_2, Cr_2O_3, Bi_2O_3, etc. SrO is 0.023 and 0.066% for OTo3 and N45, respectively. Cs_2O is not detected at all. Loss on ignition (LOI) heat-treated at 1000 °C for 2 h is 6.00 (1.66)% and 22.10 (12.49)% for OTo3 and N45, respectively, where in parentheses is indicated H_2O (-) dried at 105 °C for 2 h. LOI and H_2O (-) are excluded from the total of XRF analysis.

PS-GP bodies were respectively prepared from the two kinds of PS-ash. As shown in Table 3, the liquor/filler ratio (L/F) varied with the type of PS-ash. 100 grams of the PS-ash sample was weighed, and the GP-liquor was mixed to the limit amount, at which bleeding did not occur. Then, the mixture was hand-mixed in a 500 mL plastic beaker and cast into a metallic mold consisting of three prismatic cells. Each cell has 20 × 20 × 80 mm dimension. Grease was preliminarily smeared to the interior of the mold for easily demolding. In order to reduce the dry shrinkage of PS-GP before demolding, the GP specimens were placed in the sealed plastic chamber that was 20 °C and had shallow water in its bottom so that it had nearly 100% R.H. After being cured in the humid air for 24 h, the GP specimens were demolded, the curing was continued in the same conditions until 28 days age. The bulk densities of the PS-GP bodies were measured right after the 28 days curing on the basis of mold volume, and the three-point bending test was conducted. The span of specimen in the bending test was 50 mm and the loading speed was 0.2 mm/min. Either flexural strength and bulk density was an average of three specimens. Extra PS-GP bodies were continued to cure in the ambient air of 20 ± 3 °C. Then, at the ages of 6 (4 + 2), 12, 24, and 52 weeks ($\approx$1 year), bulk densities were measured again to determine the strontium and cesium contents of the PS-GP samples.

Table 3. Results on bulk density and flexural strength of PS-GP.

Hardened Body	Seawater Mixed	Active Filler	L/F [1]	Flexural Strength (MPa)	Bulk Density (g/cm^3)				
					Age (Week)				
Series	GP liquor	-	-	4 weeks	4	6 (4 + 2) [2]	12	24 [5]	52 [5] (1 year)
(a) PS-GP using seawater-mixed GP-liquors									
0-1-SCSW [3]	#0SW	OTo3	1.94	0.68	1.09	0.71	0.67	0.70	0.65
0-3-SCSW	"	N45	1.50	0.81	1.48	1.10	0.95	1.07	1.02
1-1-SCSW [4]	#1SW	OTo3	1.50	1.69	1.56	1.04	1.09	1.06	1.02
1-3-SCSW	"	N45	1.50	0.92	1.53	1.02	0.98	1.01	0.96
(b) PS-GP using non-seawater-mixed GP-liquors									
0-1-SC [3]	#0	OTo3	1.50	1.22	1.29	0.84	0.82	-	-
0-3-SC	"	N45	1.20	0.99	1.49	1.08	0.97	-	-
1-1-SC [4]	#1	OTo3	1.50	1.07	1.58	1.05	1.02	-	-
1-3-SC	"	N45	1.20	1.19	1.54	1.05	0.98	-	-

[1] Liquor/Filler ratio by mass; [2] Refer to the text; [3] Marked foaming; [4] Slight foaming; [5] In case of PS-GP using non-seawater-mixed GP-liquors, no marked change of bulk density was observed after 12 weeks so that the same data at 12 weeks were applied for 24 and 52 weeks.

2.3. Dissolution Test

The water discharged from the Fukushima Daiichi Nuclear Power Plant is contaminated mainly by ^{90}Sr and ^{137}Cs in addition to ^{134}Cs. Specifically, the former two radionuclides have very long half-lives of nearly 30 years. In this study, nonradioactive nitrate reagents ($Sr(NO_3)_2$, $CsNO_3$) were added as surrogates at a ratio of 1% by mass, respectively, to the PS-ash. As the dosages of nitrate reagents were very small, they were excluded from the calculation of liquor/filler ratio (L/F) for convenience. The radiological dosage of contaminated water of Fukushima Daiichi nuclear power plant is estimated to be 10^8 to 10^9 Bq/L. In contrast, the surrogate nitrates added to the PS-GP correspond to a level of 10^{12} Bq/L, thus the addition of nitrates was sufficient. The same addition of nitrates was also done to the PS-GP without mixing seawater in previous works, in which the dissolution test of Sr^{2+} and Cs^+ was conducted only up to 24 weeks [4,5]. The previous test results are shown again in this paper, together with the data newly obtained in this study at 52 weeks.

The dissolution test method is as follows: the PS-GP body was firstly pulverized to the size under 4 mm. Then, the 12.5 g sample was filled into a 250 mL wide-mouth plastic bottle and the acid water was also poured into a bottle. The water mass was 10 times that of the GP sample. Hydrochloric acid or glacial acetic acid is generally used as leaching fluid, well-known as JLT-13 (pH 6.3) and TCLP (pH 2.88) respectively, but for the convenience of our laboratory facility, a standard buffer solution of phthalate salt of pH 4.01 was used in this study [4]. The bottle was cap-sealed and rotated for 6 h under a rotating speed of 60 rpm. The mixture in the bottle was then filtered with a qualitative filter paper no. 131, and the filtered leachate was further diluted up to 10 times by volume with pure water to use as the sample of the dissolution test. Finally, the dissolution concentrations of Sr^{2+} and Cs^+ were measured by induction coupled plasma atomic emission spectrometry (ICP-AES, Optima 8300, PerkinElmer, Waltham, MA, USA).

The dissolution ratio was calculated as following procedure: Firstly, comparing the bulk densities at each age with the four weeks to determine the PS-ash amount used to produce the 12.5 g GP sample, the decrease of density of the test sample is simply due to evaporation of constituent water. Then, the amounts of the surrogates included into the 12.5 g GP sample were determined. Secondly, dissolution ratios were obtained on the basis of the ICP results and the charged fluid amount that is 125 grams. Finally, the immobilization ratio was obtained simply as

$$\% \text{ immobilization ratio} = 100 - \% \text{ dissolution ratio} \tag{1}$$

More details can be kindly referred to in the previous works [4,5].

2.4. XRD Analysis

In order to clarify the phases of PS-GP body, XRD analysis was carried out for the specimens with one-year age, employing RIGAKU RINT-2250 (Rigaku, Tokyo, Japan). Measuring conditions were $40 \text{ kV} - 200 \text{ mA}$ monochromatic CuKα radiation, $1° - 1° - 0.3$ mm slit system, 0.02 degree step scan, and $4°/\text{min}$ scanning speed.

2.5. EPMA Analysis

Element mapping analysis was carried out for the one-year old specimens, employing an electron probe micro analyzer (EPMA, JEOL JXA-8230, Jeol, Tokyo, Japan). Measuring conditions were 15 kV acceleration voltage and 20 nA irradiation current. Peak intensity positions of the elements analyzed in this study were determined by using the standard samples in advance. The PS-GP sample was gathered by cutting the PS-GP body with a 0.5 mm thick diamond blade, thoroughly dried after washed with ethanol, and then subjected to impregnation by embedding with epoxy resin. After preparing a thin slice, which was stuck to a slide glass as conventionally done in petrology discipline, then mirror finish was applied with diamond paste and finally subjected to vapor deposition with carbon.

3. Results and Discussion

In the present study, generally accepted terminologies of 'N-A-S-H' and 'C-A-S-H' are still used to describe the resultant matrix gels for convenience [13–15], though strictly speaking, they should be called 'Na-rich N-C-A-S-H' and 'Ca-rich N-C-A-S-H' gels, respectively, as concluded later. Moreover, the term of 'GP-minerals' is used to express crystalline phases presented finally in hardened GP bodies except amorphous N-A-S-H and C-A-S-H gels, where C, N, A, and H denote CaO, Na_2O, Al_2O_3, SiO_2, and H_2O, respectively.

3.1. Strength and Density of PS-GP Body

As shown in Table 3, the PS-GP using the seawater had a larger liquor/filler ratio (L/F), compared to the series without mixing the seawater. That is to say, the addition of seawater makes the PS-GP able to uptake much contaminated water. Just because the PS-GP using the seawater-containing GP-liquor #0SW had a higher L/F, its flexural strength was smaller than the counterparts of non-seawater mixing PS-GP. Series 1-1-SCSW, using the GP-liquor #1SW and the PS-ash OTo3, exceptionally shows a higher flexural strength. The reason is unknown at this moment.

In the case of adding the seawater, the bulk density of PS-GP was not significantly different from the non-seawater mixing PS-GP at all the age, despite the L/F of the former was larger. However, the bulk density of the series 0-1-SCSW, using the GP-liquor #0SW and the PS-ash OTo3, exceptionally had a lower bulk density. Remarkable foaming and more porous features were observed in 0-1-SCSW specimens, resulting in lower bulk density. Thus, water evaporation easily occurred through continuous pores, which were confirmed by the floating test of specimen in water vessel. In the beginning, the specimen floated, but soon after it sank to the bottom of water vessel. Slight foaming, not to affect the density markedly, was also observed in 1-1-SCSW specimens. It is considered that the foaming is resulted from metallic aluminum included in the PS-ash, which generates hydrogen gas in alkaline GP-liquors. Refuse derived fuel (RDF) is used in some incineration plants. Thereby, aluminum foil appliances, sometimes found in refuse collected from households, may be not completely oxidized during the incineration process and may evaporate to precipitate into the PS-ash as fine particles.

Next, some increase of the bulk density with the elapsed time, found in some seawater-mixed specimens at 24 weeks, should be mentioned. The increase may be firstly caused by damp in rainy season around the 24 weeks. Not only porous characteristics but also hygroscopic nature of chlorine may promote moisture absorption. Secondly, the abnormally high increase of 0-3-SCSW may be caused by the presence of fissures filled with chlorine or chloride which may play an increased role in moisture absorption. The same reason was thought for 1-1-SCSW at 12 weeks when it was not in rainy season. Presumably, the fissures formed between 12 and 24 weeks for 0-3-SWSC and between 6 and 12 weeks for 1-1-SWSC, judging from the sudden increase of the bulk density. The fissures can be kindly referred to Section 3.4.

3.2. Dissolution Test of Strontium and Cesium

Two calculating examples of dissolution and immobilization ratios at 52 weeks are shown in Table 4 for the seawater-mixed PS-GP and the non-seawater-mixed PS-GP specimens, respectively. All the results of dissolution and immobilization ratios of the seawater-mixed PS-GP are summarized in Table 5, in comparison with those of non-seawater-mixed PS-GP. The SrO component was also detected in PS-ashes, but it was not taken into account since the SrO amount was less than a few-hundredth percent orders (see Table 2).

For the seawater-mixed PS-GP, the immobilization ratios of Sr^{2+} and Cs^+ were stable within one year, regardless of the types of GP-liquor as well as the types of PS-ash used, and the over-scale (O. S.) was not met, which occurred in some of the non-seawater-mixed PS-GP specimens. However, at 52 weeks, a slight decrease in the immobilization ratios of Sr^{2+} and Cs^+ were found in the PS-GP, which used the GP-liquor #0SW. As an overall trend, the PS-GPs, which used the GP-liquor #1SW, gave good results of the immobilization ratios of Sr^{2+} and Cs^+. In addition, the GP-liquor #1SW has an advantage of relatively low cost over the GP-liquor #0SW that contains caustic soda.

On the other hand, the immobilization ratios of Sr^{2+} and Cs^+ of non-seawater-mixed PS-GP were unstable, fluctuating with the elapsed time. This phenomenon was found especially from series 1-3-SC using PS-ash N45. However, series 1-1-SC using PS-ash OTo3 was an exception and the immobilization ratios were relatively stable within the experimental age. The sort of PS-ash probably is another influencing factor of the immobilization ratios.

Table 4. Exemplified data for calculating immobilization ratios of Sr^{2+} and Cs^+ in PS-GP.

Age	% Filler	12.5 g Sample	Surrogates (mg)			ICP, 421 nm	ICP, 459 nm
52 Weeks		(g)	As Nitrate	Sr^{2+}	Cs^+	Sr^{2+} (ppb)	Cs^+ (ppb)
			(a) Seawater-mixed PS-GP				
Liquor #0SW							
0-1-SCSW	57.04	7.13	71.3	29.5	48.6	610	23,410
0-3-SCSW	58.04	7.26	72.6	30.1	49.5	1770	19,900
Liquor #1SW							
1-1-SCSW	61.18	7.65	76.5	31.7	52.2	3340	0
1-3-SCSW	63.75	7.97	79.7	33.0	54.4	7840	0
	125 g Leaching Solution			Dissolution Ratio		Immobilization Ratio	
	Sr^{2+} (µg)	Cs^+ (µg)		Sr^{2+} (%)	Cs^+ (%)	Sr^{2+} (%)	Cs^+ (%)
Liquor #0SW							
0-1-SCSW	76.25	2926.25		0.26	6.02	99.74	93.98
0-3-SCSW	221.25	2487.50		0.74	5.02	99.26	94.98
Liquor #1SW							
1-1-SCSW	417.50	0		1.32	0	98.68	100
1-3-SCSW	980.00	0		2.97	0	97.03	100

Table 4. Cont.

Age	% Filler	12.5 g Sample	Surrogates (mg)			ICP, 421 nm	ICP, 459 nm
52 Weeks		(g)	As Nitrate	Sr^{2+}	Cs$^+$	Sr^{2+} (ppb)	Cs$^+$ (ppb)
(b) Non-seawater-mixed PS-GP							
Liquor #0							
0-1-SC	62.93	7.87	78.7	32.6	53.7	510	0
0-3-SC	61.44	7.68	76.8	31.8	52.4	380	0
Liquor #1							
1-1-SC	70.41	8.80	88.0	36.4	60.0	4420	26,680
1-3-SC	71.43	8.93	89.3	37.0	60.9	O.S.[1]	22,060

	125 g Leaching Solution		Dissolution Ratio		Immobilization Ratio	
	Sr^{2+} (µg)	Cs$^+$ (µg)	Sr^{2+} (%)		Sr^{2+} (µg)	Cs$^+$ (µg)
Liquor #0						
0-1-SC	63.75	0	0.20	0	99.80	100
0-3-SC	47.50	0	0.15	0	99.85	100
Liquor #1						
1-1-SC	552.50	3335.00	1.52	5.56	98.48	94.44
1-3-SC	O.S.[1]	2757.50	O.S.[1]	4.53	O.S.[1]	95.47

[1] Over-scale (O.S.) took place due to too much concentrations of testing leachate to measure and no more measurements were conducted by further dilutions. Table 5 is the same.

Table 5. Dissolution test results of PS-GP.

Hardened Body	Seawater Mixed	Active Filler (Key)	Immobilization Ratio (%) for Each Age (Week)							
Series	**GP-Liquor**		6 (4 + 2)		12		24		52 (1 Year)	
			Sr^{2+}	Cs$^+$	Sr^{2+}	Cs$^+$	Sr^{2+}	Cs$^+$	Sr^{2+}	Cs$^+$
(a) PS-GP using seawater mixed GP-liquors										
0-1-SCSW	# 0SW	OTo3 (1)	99.72	96.99	99.59	100	99.70	100	99.74	93.98
0-3-SCSW	"	N45 (3)	99.61	93.06	99.72	100	99.41	100	99.26	94.98
1-1-SCSW	# 1SW	OTo3 (1)	99.30	93.08	97.61	100	98.84	100	98.68	100
1-3-SCSW	"	N45 (3)	97.70	95.22	95.34	100	97.30	100	97.03	100
(b) PS-GP using non-seawater-mixed GP-liquors										
0-1-SC	# 0	OTo3 (1)	99.79	97.24	99.80	O.S.	99.89	97.48	99.80	100
0-3-SC	"	N45 (3)	99.83	71.54	99.83	O.S.	99.88	98.30	99.85	100
1-1-SC	# 1	OTo3 (1)	98.23	96.58	98.27	90.95	98.59	98.77	98.48	94.44
1-3-SC	"	N45 (3)	O.S.	71.06	91.73	46.16	O.S.	95.65	O.S.	95.47

However, the most important factor may be the issue that the stability of C-A-S-H and N-A-S-H gels with progress of material age. As elucidated in literatures [16,17], these two kinds of gels are far separated in a ternary diagram SiO$_2$-CaO-Al$_2$O$_3$. With progress of material age, these gels come closer and line up alongside SiO$_2$-CaO line, as also found by Yamaguchi et al. [18]. Since the chemical compositions of these gels are instable, the fixation of strontium and cesium may become unstable accordingly. It is strongly estimated that the seawater, most probably chlorine, suppresses the instability of these gels, thus more stable gels may form at earlier age. More detailed discussion will come up in Section 3.4 about this issue.

3.3. XRD Results

XRD results are represented in Figure 1 for seawater-mixed and non-seawater-mixed PS-GP at the age of 52 weeks. Two categories of PS-GP showed the similar patterns in XRD diagram. Firstly, faujasite formation is peculiar to 0-1-SCSW and 0-1-SC samples as well as presumably in 1-3-SCSW and 1-3-SC samples. Although faujasite peak is unclear in the XRD chart of 1-3-SC at 52 weeks, it was clearly detected at 4 weeks [5]. Therefore, 1-3-SC was instable with elapsed time. Secondly, magnesian calcite was found in both 0-1-SCSW and 0-1-SC samples. Thirdly, pirssonite formed in 0-3-SCSW as well as presumably in 0-3-SC.

Remaining quartz, calcite, and talc were still observed in the PS-GP samples, which are constituent minerals of the raw PS-ashes used in this study, designated as "minerals of primary origin" in our past paper [5]. A few remained ettringite was observed in 0-1-SCSW sample, which is also one of the constituent minerals of the raw PS-ash, designated as "minerals of secondary origin" [5].

In our past studies [4,5], we considered that talc and ettringite completely disappeared from the PS-GPs at four weeks, and carbonate ettringite was produced by the reaction of ettringite and calcite. However, according to the XRD and SEM-EDS results at 52 weeks, as described later in Section 3.5, it was found that part of talc did not react completely and still remained in the PS-GPs. Thus, the formation of carbonate ettringite is implausible. This misunderstanding was caused by overlapping the main peak positions ($\approx 9°$, 2θ) between talc and carbonate ettringite. Moreover, forsterite seems to gradually react and will be exhausted with the elapsed time, and its trace was only found in 1-1-SC sample at 52 weeks.

Crystalline GP-minerals, including thenardite, Na_2SO_4, and burkeite, $Na_6(CO_3)(SO_4)_2$ previously called "minerals in PS-ash based geopolymers" [5], were fundamentally identified in different PS-GP samples, as seen in Figure 1. There was faujasite, $(Na_2, Ca, Mg)_{3.5}(Al_7Si_{17}O_{48})\cdot 32H_2O$, only in the 0-1 series samples, and its trace may be found in 1–3 series samples. If compared with the XRD charts of 0-1 series samples, seawater addition seems to promote faujasite formation, as seen in the chart of 0-1-SCSW. On the other hand, pirssonite, $Na_2Ca(CO_3)_2\cdot 2H_2O$, presented only in the 0-3 series samples, especially, it was clearly found in the seawater-mixed PS-GP sample of 0-3-SCSW. As amorphous minerals, N-A-S-H and C-A-S-H gels were observed as a hump in the range approximately 20–40°, 2θ, especially, they were clear in the 0-1 series samples, no matter whether the seawater was mixed or not.

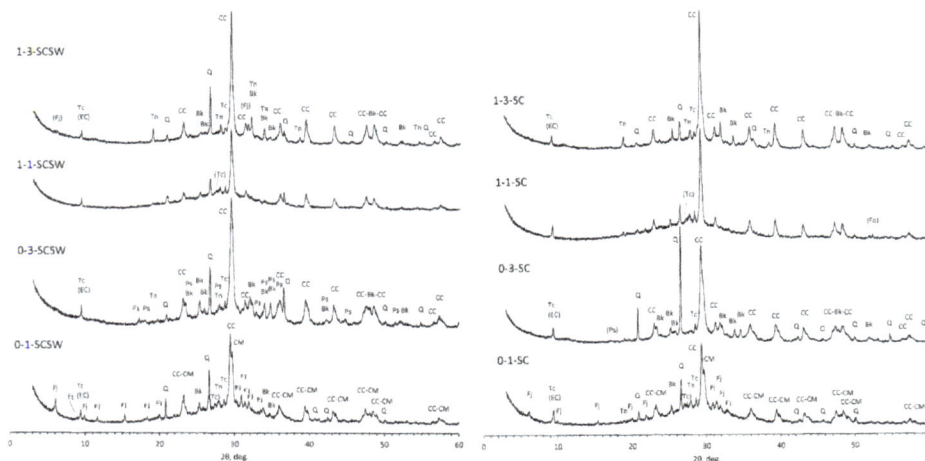

Figure 1. XRD diagrams of seawater-mixed and non-seawater-mixed PS-GP at 52 weeks. CC: calcite; CM: magnesian calcite; Q: quartz; Tc: talc; Fo: forsterite; Fj: faujasite; Bk: burkeite; Tn: thenerdite; Ps: pirssonite; Et: ettringite; EC: carbonate ettringite; ( ): uncertain.

3.4. EPMA Analysis of One Year Old Seawater-Mixed PS-GP

3.4.1. Back Scattered Images and Al-Si Distributions

Back scattered electron images of EPMA are shown in Figure 2, together with Al-distribution maps, which well describe GP matrix textures. Large pores resulted from foaming were able to observe with the naked eye in 0-1-SCSW specimen, as studied previously without using seawater [4,5], and small round-shaped voids were also noticed in the back scattered image, which were presumably resulted from air trapping during GP mixing. 1-3-SCSW sample had no visible big pores generated by foaming, but there were small round-shaped voids caused by air trapping too. On the other hand, other samples, 0-3-SCSW and 1-1-SCSW, had crescent lake-like fissures, which are thought to be caused by delayed foaming after setting. However, the destruction of hardened GP bodies, caused by the delayed foaming, did not take place.

From the back scattered electron images, it is very easy to recognize matrix formation. Dark wide areas, looking like sea, are so-called N-A-S-H gels, whereas bright areas, looking like islands, are C-A-S-H gels. It is noted that N-A-S-H and C-A-S-H gels can be very easily distinguished in 0-3-SCSW and 1-1-SCSW specimens, because in these two specimens remained PS-ash particles are small in number. Conversely, for 0-1-SCSW and 1-3-SCSW specimens, the discrimination between the two categories of matrix gel is not so easy, because there are a relatively large quantity of remained PS-ash particles in these specimens, exhibiting sharp and elongated shapes. More details will be explained again in Section 3.5.

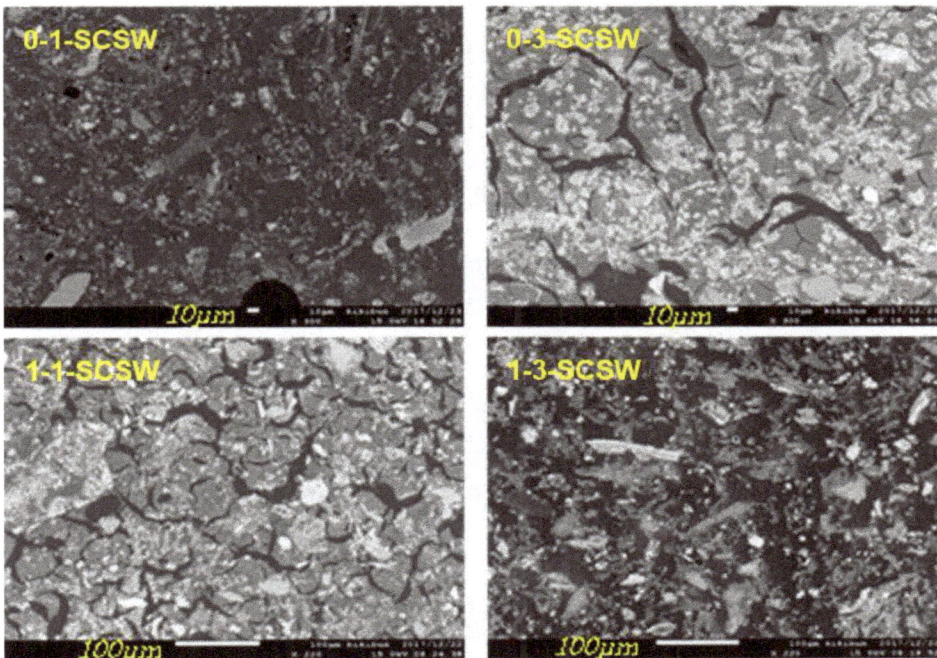

(**a**)

Figure 2. *Cont.*

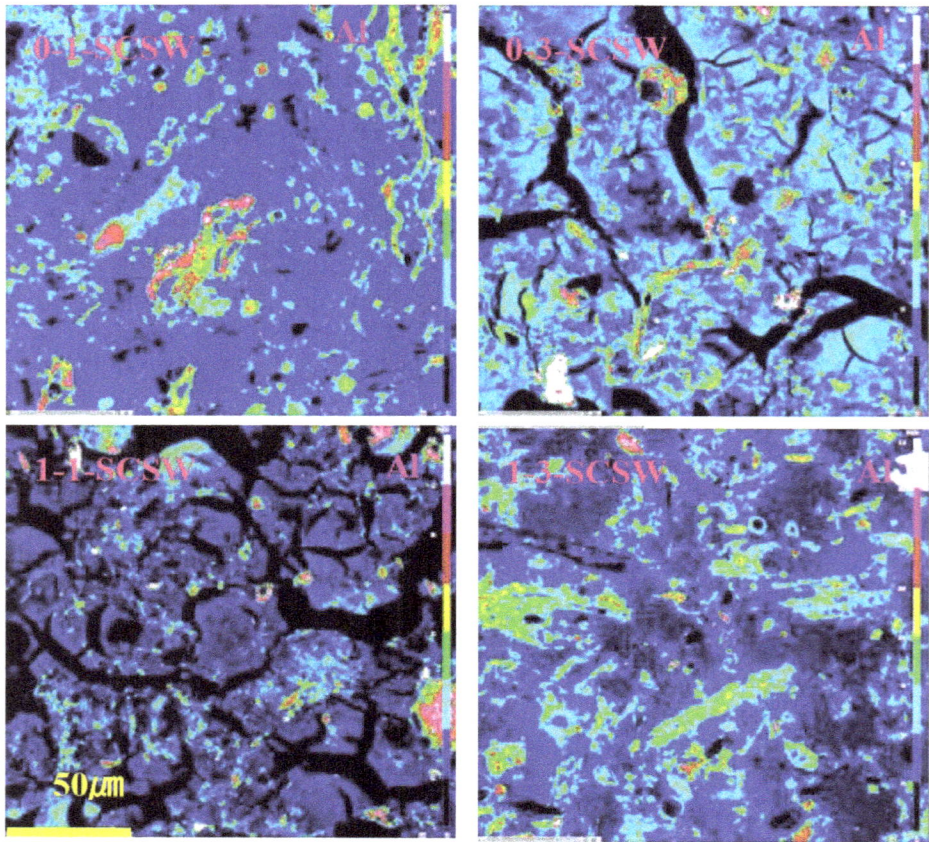

Figure 2. Back scattered electron images (**a**), and Al-distribution maps; (**b**) of PS-GP taken by EPMA. All the scale bars are 50 mm in length for (**b**).

The concentration of aluminum included in the matrix gels is generally very low, as indicated by blue and light blue color codes, corresponding to N-A-S-H and C-A-S-H gels, respectively. There was a reversal concentration of Al between so-called C-A-S-H gels (island pattern) and so-called N-A-S-H gels (sea pattern), indicated by Al-color codes. This result was in contradict with the results of SEM-EDS point analysis, as mentioned in Section 3.5. It is probably due to the influence of highly Al-bearing relicts of host mineral observed in C-A-S-H gels. The low incorporation tendency of Al_2O_3 component is consistent with other literature data, that is alkali-free C-A-S-H gels are generated with Al_2O_3 in the range of only 6–13 mol %, as plotted in SiO_2-CaO-Al_2O_3 ternary diagram [19]. For the geopolymers prepared from urban refuse incineration ash slags (U-slags) cured at 80 °C, it was found that the Al_2O_3 content in GP matrix gels was 10–13 mol % too, when CaO, Al_2O_3, and SiO_2 components were looked over as main compositions of GP matrix composed of N-C-A-S-H, and were plotted in a SiO_2-CaO-Al_2O_3 ternary diagram in dry base [18]. Another study [17] shows in the hybrid cement geopolymers prepared from fly ash and Portland cement mixture cured at 21 °C for one year, the Al_2O_3 content is also very low, falling in the range of 0–18 mol %, mainly 1–10 mol %. More details will be mentioned in Section 3.5.

As mentioned above, PS-ash particles decomposed and converted to the GP-minerals other than N-A-S-H and C-A-S-H gels. Among the GP-minerals, quartz remained intact, as seen in the Si-maps (Figure 3), in which the quartz particles are indicated by red and white color codes.

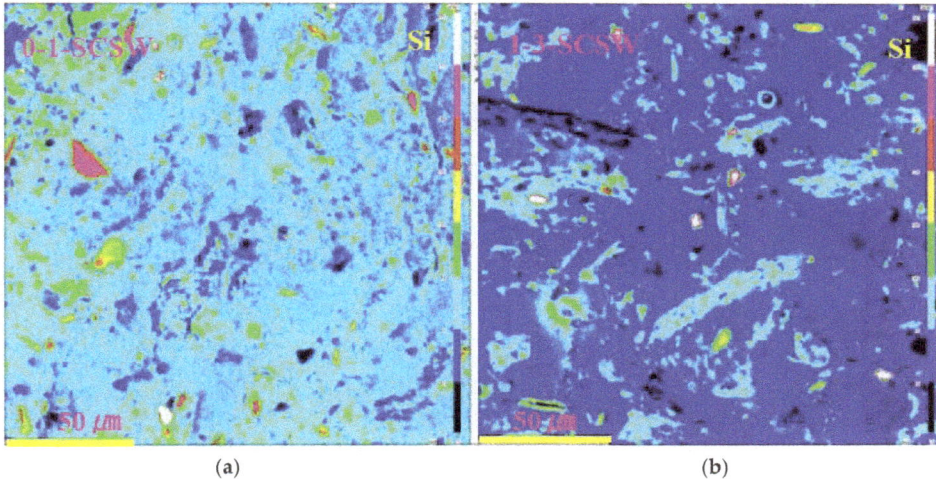

| (a) | (b) |

Figure 3. Selected Si-distribution maps of PS-GP taken by EPMA. (**a**) Series 0-1-SCSW; (**b**) Series 1-3-SCSW.

3.4.2. Chlorine Distribution

As shown in Figure 4, chlorine is distributed overall, and its distribution strongly correlates with that of sodium. In other words, chlorine has a preference to incorporate into N-A-S-H rather than C-A-S-H gels. This was also confirmed by point analyses using SEM-EDS, as shown in Tables 6 and 7. These two categories of geopolymer gel constituting the matrices of GP body can also be seen in the Na-maps as well as the Al-maps, as clearly shown by light and shade patterns of color codes. The chlorine preference of N-A-S-H may be due to its zeolite-like structure having sodalite cages as sub-cells which can accommodate much chlorine. According to a study reporting on LCFA (low calcium fly ash)-based GP, a zeolite-like lattice image taken by a high-resolution electron microscope (HREM) shows that there are partial precipitates of zeolite A-like crystals with sodalite cages in the matrix N-A-S-H gel [20]. From this result, it is estimated that N-A-S-H gel may have sodalite cages in its gel structure.

In present study, chlorine was also found in the fissures generated due to the delayed foaming. It seems that the presence of chlorine in the fissures was a result of seeping out from the GP matrix. However, counterpart sodium was scarcely found in the maps. SEM-EDS analysis at certain point shows a pronounced presence of chlorine up to 70–80 mol % in addition to 7–8 mol % SiO_2. More details are now under investigation and the results will be reported in the near future.

3.4.3. Strontium and Cesium Distribution

As shown in Figure 4, strontium is concentrated at several domains, while cesium is entirely scattered though there are some domains. From Figures 4 and 5, it is concluded that the presence of strontium strongly correlates with calcium, while cesium is greatly associated with sodium. Now, we cannot yet conclude certainly the compatibility of elements or compounds, because the dosages of strontium and cesium are smaller. Speculations on the basis of these maps and the literatures will be explained below.

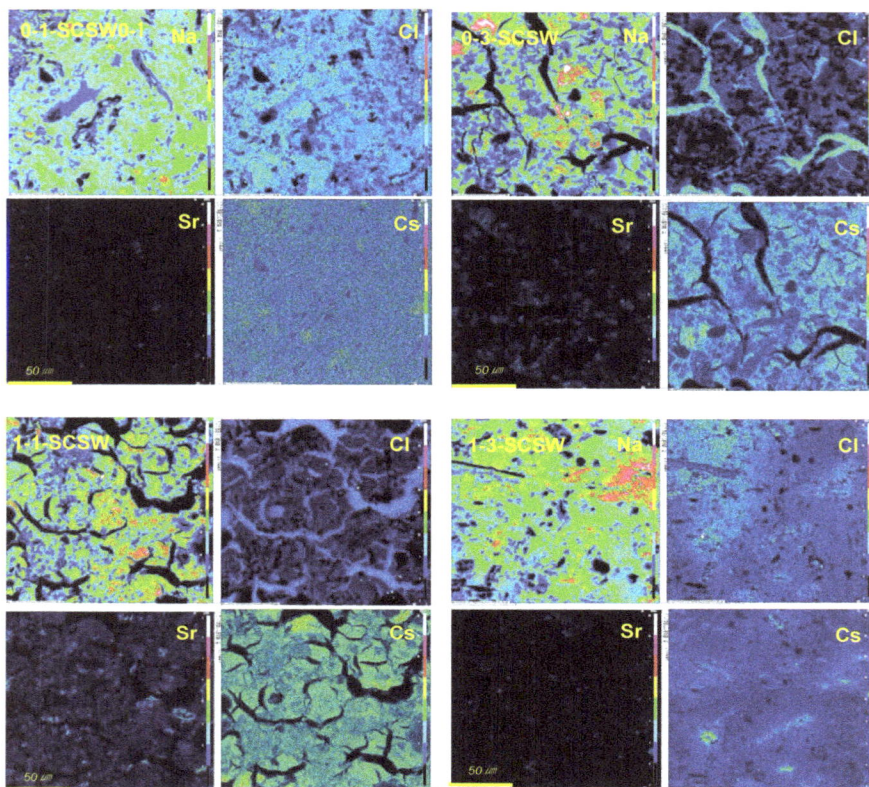

Figure 4. Element distribution comparison of Na with Sr and Cs as well as Cl, taken by EPMA. All the scalar bars are 50 μm in length.

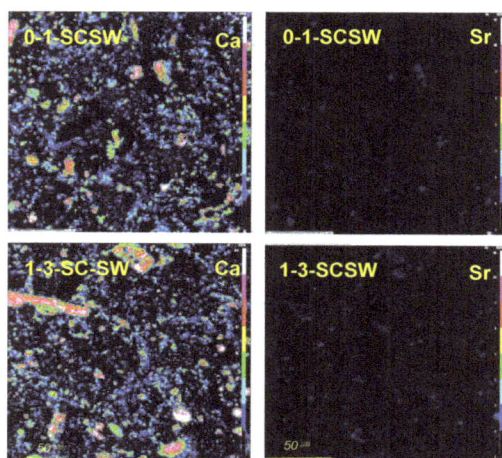

Figure 5. Exemplar of Ca and Sr distribution maps taken by EPMA. Picture field is the same as Figure 4, and all the scalar bars are 50 μm in length.

Regarding compatible partnership of strontium, the first candidate is calcite. According to the literatures [21,22], Sr^{2+} can be incorporated into calcite. Thereby, simultaneous incorporation of Mg^{2+} would promote the Sr^{2+} incorporation to compensate the lattice distortion due to the gap of ionic radii of Ca^{2+} as expressed $(0.131 + 0.072)/2 = 0.10$ nm. The ionic radii are Sr^{2+} (0.131 nm—IX), Mg^{2+} (0.072 nm—VI) and Ca^{2+} (0.10 nm—VI), respectively [23,24], in which the Roman numerals indicate coordination numbers. Therefore, calcite has a high potential to accommodate strontium and other divalent cations.

The second candidate is calcite-aragonite overgrowth. According to literature [25], strontium plays an important role in the growth of calcite-aragonite alternate layer triggered by seasonal change of hot spring water such as temperature, pH and so on. This fact was found in a hot spring in Japan. Thereby, Sr^{2+} acts as nucleation agent of metastable formation of aragonite overgrowth on calcite. Present PS-ash fillers have sufficient Mg-potentials, as found in Table 1.

Based on the above research data, we consider the reason of the stable immobilization of strontium as follows. The calcite is soluble in acidic solution. However, it might be covered by geopolymer matrix gels so that its dissolution is hindered. Accordingly, the strontium combined in the calcite does not easily dissolute even in acidic environment. We identified the calcite by XRD in the PS-GP specimens, but we have not yet detected the calcite itself as well as the calcite-aragonite overgrowth from the SEM-EDS maps of all the PS-GP specimens. Maybe this is because the calcite was scarce and in slanted distribution in the scanned fields of specimens. Otherwise, the calcite was extremely small in size beyond the resolution power of SEM apparatus.

The third candidate is the C-A-S-H gel that is rich in calcium. Specifically, simultaneous incorporation of magnesium may be important in the case that calcite compensates lattice distortions. Magnesium source is talc and forsterite from the PS-ashes. Actually, relatively high content of MgO has been detected in some C-A-S-H gels, as shown in Tables 6 and 7.

The fourth candidate is so-called 'plagioclase gels', presumably comprising faujasite, $(Na_2CaMg)_{3.5}(Al_7Si_{17}O_{48})\cdot 32H_2O$, of zeolite family, of which presence is limited solely to 0–1 and 1–3 series samples until one year age, as seen in Figure 1. Therefore, incorporation of strontium into Ca-rich plagioclase gels themselves and/or faujasite-Ca is plausible for the PS-GP specimens besides the candidates mentioned above. The details of plagioclase gels can be referred to Section 3.5.

Discrete formation of strontianite, $SrCO_3$, was reported in a literature about metakaolin-based geopolymer studied by HREM [26]. However, this is not directly applicable to present study, because present PS-ash fillers have multi-phases rather than single phase filler. Metakaolin only produces N-A-S-H gels, which have little compatibility to strontium incorporation, as mentioned above. Accordingly, strontium is obliged to appear as strontianite in metakaolin, combined with carbon dioxide.

Next, we continue to discuss incorporation of cesium.

About compatible partnership of cesium, the first candidate is N-A-S-H gel, as clearly seen in 0-3-SCSW and 1-1-SCSW specimens without faujasite. Incidentally, incorporation of cesium into matrix gels was elucidated in metakaolin-based and LCFA-based geopolymers, respectively [7,10]. These geopolymers, using single sort of active filler, have one kind of matrix gel.

Besides, faujasite-Na would be the third candidate of cesium accommodation. Cs-faujasite was encountered in XRD identification file, ICDD 01-079-1887, $Cs_{39.36}Na_{40.80}Al_{96}Si_{96}O_{384}\cdot H_2O_{164.48}$, so that faujasite is versatile to gather monovalent and divalent cations into its structure. Another versatile mineral to accommodate cesium and strontium is herschelite, which has been renamed chabazite-Na, $(Na_2, K_2, Ca, Sr, Mg)_2[Al_2Si_4O_{12}]_2\cdot 12H_2O$. Chabazite-Na formation has been reported in studies [27–29]. However, it was not detected at all from our specimens prepared in this study as well as our previous studies [4,5].

Table 6. Results of SEM-EDS point-analysis for seawater-mixed PS-GP using NaOH-containing GP-liquor #0SW at 52 weeks.

	Phase		Specimen 0-1-SCSW										
EDS	Screen	Point	SiO_2	TiO_2	Al_2O_3	Fe_2O_3	CaO	MgO	Na_2O	K_2O	P_2O_5	SO_3	Cl
	C-A-S-H [1]	S7	25.76	0.61	10.62	0.34	39.81	3.87	16.29	0.17	1.11	0.38	1.06
	C-A-S-H	S9	38.06	0.34	10.00	0.26	36.61	2.61	9.27	0.85	0.91	0.25	0.83
	N-A-S-H [2]	S11	55.47	0.17	13.95	0.29	11.62	1.52	11.47	0.62	0.35	0.47	4.09
	N-A-S-H	S12	56.21	0.22	14.72	0.29	9.32	1.34	12.00	0.76	0.30	0.58	4.27
	N-A-S-H	S13	54.48	2.16	12.83	0.39	9.53	2.12	12.53	0.65	0.40	0.65	4.27
	N-A-S-H	S15	52.51	0.28	13.31	0.38	15.53	2.32	10.93	0.58	0.36	0.45	3.34
	N-A-S-H	S16	51.56	0.16	12.18	0.40	15.14	1.71	12.28	0.58	0.61	0.62	4.74
	N-A-S-H	S17	52.67	0.38	13.30	0.47	10.51	1.33	14.81	0.63	0.61	0.54	4.75
	C-A-S-H	S18	31.91	2.40	12.80	0.61	27.41	4.39	14.83	0.31	2.89	0.83	1.61
PL [3]	C-A-S-H	S19	50.02	0.20	25.21	0.28	3.04	4.10	14.25	0.38	0.59	0.23	1.72
PL	C-A-S-H	S20	44.42	0.49	27.03	0.26	7.76	3.48	13.22	0.33	1.08	0.31	1.62
PL	C-A-S-H	S21	47.23	0.17	30.09	0.18	2.02	1.08	16.96	0.28	0.47	0.27	1.25
PL	C-A-S-H	S22	43.04	1.46	24.79	0.87	16.72	4.26	5.59	0.09	0.66	1.82	0.71

	Phase		Specimen 0-3-SCSW										
EDS	Screen	Point	SiO_2	TiO_2	Al_2O_3	Fe_2O_3	CaO	MgO	Na_2O	K_2O	P_2O_5	SO_3	Cl
	N-A-S-H	S7	60.13	0.14	7.54	0.25	4.23	2.19	21.77	0.90	0.15	0.27	2.42
	N-A-S-H	S8	61.42	0.11	8.16	0.21	4.18	2.66	19.03	0.67	0.14	0.38	3.03
	N-A-S-H	S9	64.49	0.11	7.95	0.15	4.45	1.99	16.12	0.96	0.17	0.46	3.15
	N-A-S-H	S10	61.75	0.13	7.66	0.34	5.19	2.85	18.34	0.87	0.26	0.22	2.40
	C-A-S-H	N9	24.07	0.41	4.01	0.12	55.35	5.38	8.85	0.18	0.75	0.43	0.45
	C-A-S-H	N10	21.06	0.00	3.04	0.25	61.58	3.71	8.96	0.12	0.52	0.40	0.36
	C-A-S-H	N13	18.72	0.21	3.49	1.08	55.43	11.93	3.55	0.18	3.80	0.88	0.74
	C-A-S-H	N14	19.59	0.10	2.65	0.09	64.29	3.79	8.26	0.14	0.34	0.35	0.39
	C-A-S-H	N15	35.46	0.17	5.43	0.64	37.81	13.77	4.60	0.21	0.65	0.33	0.93
	N-A-S-H	N17	64.04	0.05	7.83	0.36	5.75	2.46	14.76	0.67	0.22	0.31	3.55
	N-A-S-H	N19	66.32	0.11	8.40	0.24	5.33	3.89	11.44	0.58	0.23	0.39	3.07
	N-A-S-H	N20	70.21	0.10	8.29	0.28	4.56	3.61	8.63	0.72	0.29	0.27	3.03

[1,2] Conventionally accepted nomenclatures are used in this table. Actually, they are Na-rich N-C-A-S-H for N-A-S-H and Ca-rich N-C-A-S-H for C-A-S-H, respectively. [3] Plagioclase gels.

Table 7. Results of SEM-EDS point-analysis for seawater-mixed PS-GP using non-NaOH-containing GP-liquor #1SW at 52 weeks.

	Phase		Specimen 1-1-SCSW										
EDS	Screen	Point	SiO_2	TiO_2	Al_2O_3	Fe_2O_3	CaO	MgO	Na_2O	K_2O	P_2O_5	SO_3	Cl
	N-A-S-H	S1	65.33	0.03	6.26	0.12	7.35	2.04	12.62	0.89	0.18	0.35	4.82
	N-A-S-H	S2	50.17	0.38	4.92	0.41	18.55	4.99	15.36	0.54	0.45	0.51	3.72
	N-A-S-H	S3	59.35	0.00	5.88	0.46	10.34	2.95	13.26	0.62	0.26	0.49	6.39
	C-A-S-H	S4	22.28	0.32	5.73	0.81	55.29	10.11	2.66	0.16	1.18	0.65	0.82
	C-A-S-H	S5	35.14	0.06	4.81	0.40	38.31	10.05	8.57	0.21	0.92	0.67	0.86
	C-A-S-H	S7	29.58	0.15	5.13	0.49	44.67	13.19	3.80	0.18	0.69	0.77	1.34
	C-A-S-H	S9	26.15	0.22	5.27	0.12	50.04	9.85	6.17	0.12	0.64	0.77	0.64

	Phase		Specimen 1-3-SCSW										
EDS	Screen	Point	SiO_2	TiO_2	Al_2O_3	Fe_2O_3	CaO	MgO	Na_2O	K_2O	P_2O_5	SO_3	Cl
	N-A-S-H [2]	S1	56.01	0.61	15.08	0.42	11.34	2.40	9.98	0.56	0.51	1.27	1.81
	N-A-S-H	S2	61.91	0.09	13.53	0.33	5.27	2.36	12.12	0.30	0.66	1.65	1.78
	N-A-S-H	S3	60.49	0.16	13.56	0.22	5.63	2.49	12.99	0.37	0.66	1.36	2.09
	N-A-S-H	S8	61.60	0.11	14.04	0.23	4.33	3.02	12.67	0.27	0.52	1.54	1.65
	N-A-S-H	S9	61.09	0.04	13.52	0.23	5.20	2.56	13.10	0.41	0.41	1.18	2.25

Table 7. *Cont.*

EDS	Phase Screen	Specimen 1-3-SCSW Point	SiO_2	TiO_2	Al_2O_3	Fe_2O_3	CaO	MgO	Na_2O	K_2O	P_2O_5	SO_3	Cl
PL [3]	C-A-S-H [1]	S4	58.26	0.22	17.41	0.63	6.64	3.18	8.98	1.87	0.44	1.48	0.89
PL	C-A-S-H	S7	45.89	0.46	22.98	0.23	17.04	2.83	8.03	0.24	0.77	1.27	0.25
PL	C-A-S-H	S11	60.37	0.83	17.04	0.27	9.61	3.08	6.58	0.47	0.42	1.06	0.26
PL	C-A-S-H	S12	48.41	0.37	20.42	3.15	14.30	3.42	7.86	0.39	0.64	0.66	0.36
PL	C-A-S-H	S13	56.32	2.85	18.79	0.27	4.60	3.45	11.30	0.58	0.43	0.78	0.62
PL	C-A-S-H	N1	38.39	1.09	23.80	0.47	22.08	5.15	6.37	0.15	1.07	1.12	0.33
PL	C-A-S-H	N2	46.66	0.29	30.87	0.26	2.99	1.86	14.05	0.37	0.49	1.54	0.63
	C-A-S-H	N3	30.11	1.26	16.98	0.42	42.38	4.61	0.93	0.02	1.34	1.14	0.80
	C-A-S-H	N4	28.57	1.47	14.48	0.27	50.18	3.94	0.66	0	0.15	0.18	0.08
	N-A-S-H	N7	52.79	0.08	12.15	0.25	11.84	2.76	15.37	0.25	0.63	1.66	2.23
	N-A-S-H	N9	64.04	0.19	13.39	0.14	3.88	1.43	12.93	0.40	0.22	1.13	2.25
	N-A-S-H	N10	57.02	0.11	12.53	0.26	7.46	2.14	15.76	0.36	0.38	1.67	2.31
	N-A-S-H	N11	59.17	0.18	13.16	0.30	5.91	2.15	13.96	0.32	0.59	1.97	2.30

[1,2,3] are the same as Table 6.

3.5. SEM-EDS Analysis

3.5.1. N-A-S-H and C-A-S-H Issue

All the analytical data of components of N-A-S-H and C-A-S-H gels are summarized in Tables 6 and 7. Figures 6 and 7 show a ternary diagram of SiO_2-(CaO + Na_2O)-Al_2O_3 and Al_2O_3-CaO-Na_2O, respectively, by molar ratio. As exemplified in Figure 2 that was the results of EPMA, N-A-S-H and C-A-S-H gels can be seen in sea and island patterns. Exemplified SEM images under high magnification power are shown in Figure 8, in which the sea and island patters appear more vividly, and sponge-like texture of C-A-S-H gel, a proof of topotactic precipitation, is also clearly observed.

It is thought that N-A-S-H gels precipitated in the GP-liquors by way of so-called "through solution process", while C-A-S-H gels were formed by way of so-called 'topotactic process' precipitating on the surfaces of active filler particles. It is likely that most of the constituent minerals of the PS-ashes except quartz turned into N-A-S-H and C-A-S-H gels in the 0-3-SCSW and 1-1-SCSW. To the contrary, in the 0-1-SCSW and 1-3-SCSW, a considerable number of PS-ash particles remained even in the one-year-old specimens, maintaining original elongated shapes in back scattered electron images of EPMA. On the other hand, in the SEM images, the sea and island texture patterns can be well clarified. In addition, it should be stressed that the grains exhibited bright in color that is white to light gray so as to look like C-A-S-H gels at a glance on the display screen. However, some of them actually possessed Al-rich N-A-S-H gel-like compositions. This kind of gel is temporarily called 'plagioclase gels' here, noted as PL in Table 7, and surrounded with a dotted oval circles in Figures 6 and 7. The plagioclase gels appear generally in elongated shape, which may be pseudomorphs after the PS-ash minerals connect most probably with faujasite formation, specifically anorthite included in the PS-ashes [4,5]. However, they do not always appear in elongated shape, but sometimes in round shape.

Moreover, according to the Al_2O_3-CaO-Na_2O diagram of Figure 7, and the XRD results of Figure 1 where faujasite peaks are clear for 0-1-SCSW but are slight for 1-3-SCSW, some of the plagioclase gels maybe converted to crystalline faujasite in form of solid solution consisting of faujasite-Ca and faujasite-Na end-members. There may be some participation of faujasite-Mg, but MgO-component is omitted from the ternary diagrams.

From the SiO_2-(CaO + Na_2O)-Al_2O_3 diagram shown in Figure 6, alignments alongside CN-Ab join can be seen for the plots of series 0-3-SCSW (red marks) and series 1-1-SCSW (blue marks), where CN denotes CaO + Na_2O. However, the alignments are somewhat different between 0-1-SCSW (black marks) and 1-3-SCSW (purple marks). Specifically, compositional ratios of their C-A-S-H draw out largely from the CN-Ab join, probably due to the formation of plagioclase gels. The plagioclase gels were only found in 0-1-SCSW and 1-3-SCSW specimens, both of which had faujasite. Hence, part of the plagioclase gels would be crystalline faujasite.

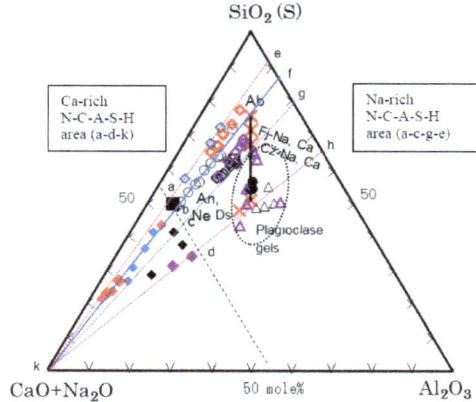

Figure 6. Ternary diagram of so-called N-A-S-H and C-A-S-H matrix gels in terms of SiO_2-(CaO + Na_2O)-Al_2O_3. ◇-◆-△: 0-1-SCSW, in black. ◇-◆: 0-3-SCSW, in red. ◇-◆: 1-1-SCSW, in blue. ◇-◆-△: 1-3-SCSW, in purple. ○: Yamaguchi et al., 2013. □-■: Yip et al., 2005. ●: Iwahiro et al., 2002. Unfilled marks except triangles are for Na-rich N-C-A-S-H. Filled Marks are for Ca-rich N-C-A-S-H. Unfilled triangles are for plagioclase gels. Cross marks in red indicate positions of Ab: albite; An: anorthite; Ne: nepheline; Ds: davidsmithite; Fj: faujasite; and Cz: chabazite, respectively. Solid blue line: main track of trend line (CN-Ab). Dotted blue line: sub-track of trend line. Red lines: frontiers of plots. Black broken line: estimated boundary between Na-rich N-C-A-S-H (N-A-S-H) and Ca-rich N-C-A-S-H (C-A-S-H).

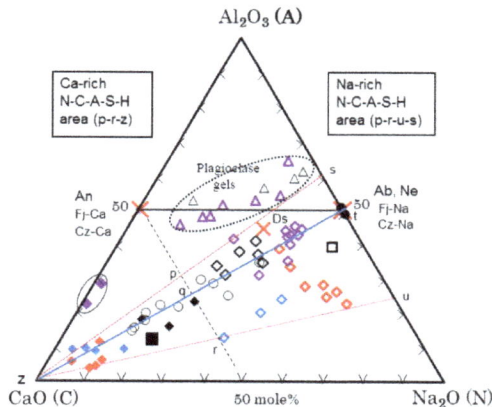

Figure 7. Ternary diagram of so-called N-A-S-H and C-A-S-H matrix gels in terms of Al_2O_3-CaO-Na_2O. The solid oval circle indicates nearly pure C-A-S-H gels. Other notations are the same as Figure 6.

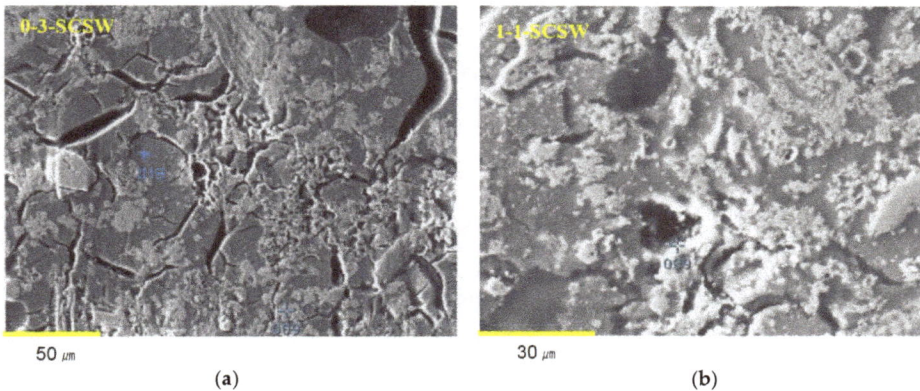

Figure 8. Fine SEM images for showing N-A-S-H and C-A-S-H gels in dark and bright contrast. The numbers of analyzed points correspond with Tables 6 and 7. (**a**) Series 0-3-SCSW; (**b**) Series 1-1-SCSW.

The Al_2O_3-CaO-Na_2O diagram (refer to Figure 7) is explained here. This diagram was prepared to separate the CN into two components that are CaO and Na_2O, respectively. The plots are concentrated alongside C-Ab join, where C denotes apical CaO. However, large deviations from the C-Ab join were recognized for N-A-S-H gels of 1-1-SCSW (blue marks) and 0-3-SCSW (red marks), of which N-A-S-H gels were plotted toward Na-rich region. The plots of 0-3-SWSC may make a trend line alongside N-Ds join, where N denotes Na_2O and Ds is davidsmithite, $(Ca, \square)_2Na_6Al_8Si_8O_{32}$, where the box is vacancy. On the other hand, it is noted that C-A-S-H gels of 1-3-SCSW (purple marks) contain very small amounts of N-components so that they can be regarded as nearly genuine C-A-S-H gels.

Lastly, 'plagioclase gels' are discussed. Their characteristics were explained in Section 3.4.3 as well as in the middle of this subsection. The plots of these gels locate in the vicinity of Ab-An join. Otherwise, that is also equivalent to (Fj-Na)-(Fj-Ca) join and (Cz-Na)-(Cz-Ca) join in the Al_2O_3-CaO-Na_2O diagram. The trend is somewhat upward and downward against the exact Ab-An join. Presumably, some of these gels partially converted to crystalline faujasite as mentioned, especially in case of 0-1-SCSW.

3.5.2. Literature Data for Supplement

When using metakaolin (MK) and Na-silicate GP-liquors to prepare GP, it is quite naturally considered that N-A-S-H gels form as binding matrices in geopolymer. Low calcium fly ash (LCFA) is the same. On the other hand, geopolymer prepared from ground granulated blast-furnace slag (GGBS or GGBFS), generally together with LCFA corresponding to ASTM class F, may generate C-A-S-H gels other than N-A-S-H gels. However, the difference between these two extreme gels has been not yet well understood whether solid solutions such as plagioclase exist in the two extreme gels or not. However, there have been so many studies on N-A-S-H and C-A-S-H gels as seen in literatures (e.g., [6,7,18–20,26–33]).

On the other hand, according to Yamaguchi et al. [18], urban waste incineration ash slag (U-slags) has a wide range of compositions, depending on incineration plants. Their compositions are plotted alongside the first hydraulic line in the SiO_2-CaO-Al_2O_3 ternary diagram. They prepared monolithic geopolymer materials by using the U-slag powder activated with Na-silicate GP-liquor corresponding to the GP-liquor #0 used in present study and curing at 80 °C. They also made point-analyses of matrix compositions by SEM-EDS, using polished specimens. Their results provided useful information on the N-A-S-H and C-A-S-H issue. Thus, we rearranged their data in this study and the results obtained are also shown in Figures 6 and 7.

The principal components of SiO_2, CaO, Al_2O_3, and Na_2O were considered here. These oxide components account for 85~95% of the total compositions. The 10 points based on 10 analytical data from 5 different specimens are plotted in Figures 6 and 7, together with the data of this study. The data in the literature [18] strongly indicate that the N-A-S-H gels comprise the CaO-component in which the plots run from the apical CN point toward Ab point. In other words, the N-A-S-H gel of Ab composition in dry base changes to be N-C-A-S-H gel as a result of CaO incorporation. Furthermore, this terminal Ab point was actually confirmed, when possessing N/A molar ratio 1.0 in the attached Al_2O_3-CaO-Na_2O diagram, in which extending trend line from apical CN point to Ab point is clearly noted for reflecting the substitution of Na with Ca. However, independent presence of C-A-S-H gels was not encountered at all in their SEM images (unpublished). Since only U-slag was used as single-phase active filler in the preparation of monolithic geopolymer materials, it is quite natural to consider that only one kind of matrix gel was yielded. Otherwise, it is probable that C-A-S-H components were incorporated into N-A-S-H gels due to high temperature curing at 80 °C.

In addition, genuine N-A-S-H gels synthesized from fluidal reagents are not composed of albite composition but composed of nepheline-rich compositions, as plotted along nepheline (Ne)-albite (Ab) join [30]. The data in the literature were picked up for the gels synthesized only under the alkaline conditions over pH 12. When plotted along the kalsilite–orthoclase join, the same tendency was able to find for the K-analog of K-A-S-H [31]. Accordingly, there is a discrepancy between genuine synthetic gels and in-situ matrix gels presumably due to the difference in kinetics of solution and suspension of source materials. Incidentally, albite was detected as main phase in GGBS-LCFA based geopolymer pastes heated at 1150 °C [32] so that the matrices have a high albite potential in their compositions in this case.

Other literature data obtained from MK-GGBFS based geopolymer matrices cured at 40 °C [33] are recalculated into oxide components and also plotted in Figures 6 and 7, in which original oxygen data were neglected, since energy dispersive spectroscopy (EDS) is incompetent for quantitative analysis of oxygen [34]. Copresence of the two discrete gels, designated as phase A and phase B in the literature, would be described as follows: Phase A is Na-rich composition, $N_{20.04}$-$C_{3.01}$-$A_{14.81}$-$S_{62.14}$-(H) as dry base that may be MK-origin, and Phase B is Ca-rich, $N_{11.33}$-$C_{33.66}$-$A_{6.16}$-$S_{48.85}$-(H) as dry base that may be GGBFS-origin. Accordingly, they are plotted separately from each other.

Therefore, conventionally named N-A-S-H and C-A-S-H gels should be called Na-rich N-C-A-S-H and Ca-rich N-C-A-S-H gels. The formation areas of N-A-S-H gels encompass a-c-g-e and p-r-u-s trapezoids, and those of C-A-S-H gels are a-d-k and p-r-z triangles in the SiO_2-(CaO + Na_2O)-Al_2O_3 and the Al_2O_3-CaO-Na_2O ternary diagrams in Figures 6 and 7, respectively. It is noted that, when viewing from the apical CN point, no N-A-S-H gels were found beyond Ab-An join. The boundary between the two extreme gel phases is estimated to be $S_{55}CN_{45}$-$CN_{45}A_{55}$ line and $A_{50}C_{50}$-$C_{50}N_{50}$ line, when GP is cured at low temperature up to 40 °C. However, at high temperature probably high above 40 °C, both gels would be incorporated to yield a single-phase gel.

Although similar results as in Figure 6 were also obtained for N-A-S-H and C-A-S-H gels of one year old hybrid cement body, as shown in the literature [17], the big difference from the present study is whether to consider the Na_2O component or not. In their study, geopolymers of the hybrid cement aged four weeks showed separate distribution between N-A-S-H and C-A-S-H gels and eventually they lined up after one year and the distribution gap between them disappeared, which is very similar to the results described in [18]. This is an issue of non-equilibrium and equilibrium or unstable and stable argument of the two extreme gels with the elapsed time. Our results on PS-GP showed a small gap between N-A-S-H and C-A-S-H gels for one year old samples and were different from the results of the one year old hybrid cement body which showed no gap. Further investigation is required to solve this problem.

4. Conclusions

In this study, a novel technology was developed for treating hazardous water contaminated with radionuclides by geopolymer technique. The geopolymers (GP) were prepared by two kinds of PS-ash as active fillers and two kinds of alkaline activator called GP-liquor adding seawater. Non-radioactive

Sr-nitride and Cs-nitride were added as surrogates as much as 1% by mass, respectively, of the PS-ashes. All the geopolymers were cured in the ambient air. The immobilization ratios of strontium and cesium were measured at several ages up to one year, and immobilization mechanism as well as PS-ash based geopolymer's compositions and textures were investigated in detail. The main results obtained are summarized as follows:

When using the GP liquors without seawater, immobilization ratios of strontium and cesium were unstable, fluctuating with material age. However, when the seawater was mixed into the GP-liquors, this deficiency was extremely improved. Specifically, the inexpensive GP-liquor #1, which did not contain caustic soda, effectively stabilized the immobilization ratios, compared to the GP-liquor #0 with NaOH.

When mixing seawater into the GP-liquors, the liquor/filler ratio (L/F) of PS-ash based geopolymer (PS-GP) could be set larger, compared to the PS-GP without seawater-mixing. Higher L/F, however, yielded a lower strength of PS-GP. Since the PS-ashes contain free metallic aluminum, hydrogen was generated during the solidification of PS-GP. Delayed foaming after setting was observed in some PS-GP.

Strontium distribution is closely related to calcium, whereas cesium is widespread. Strontium may be accommodated into a calcite, calcite-aragonite complex, Ca-rich N-C-A-S-H gels, conventionally so-called C-A-S-H, and plagioclase gels—faujasite-Ca complex. Magnesium, originated from talc and forsterite, may promote the incorporation of strontium with these minerals. On the other hand, cesium may be incorporated into Na-rich N-C-A-S-H gels, conventionally so-called N-A-S-H, and plagioclase gels—faujasite-Na complex.

Based on the SEM-EDS analysis, generally so-called N-A-S-H and C-A-S-H gels should be called Na-rich N-C-A-S-H and Ca-rich N-C-A-S-H gels, respectively. When plotted, the analytical data in the ternary diagrams of SiO_2-$(CaO + Na_2O)$-Al_2O_3 and Al_2O_3-CaO-Na_2O, most of the two gels concentrate alongside the CN-Ab join, which can be regarded as trunk trend line. However, some deviations from this line are noted for geopolymer specimens possessing "plagioclase gels", which comprise crystalline faujasite to yield plagioclase gels-faujasite complex. The faujasite is formed at ambient temperature, and its most probable source mineral is anorthite.

5. Patents

The authors have already applied for a patent in Japan for the technology described in this paper for stabilizing of immobilization of strontium and cesium in PS-GP.

Author Contributions: Z.L., M.N., and K.I. designed the experiments. M.N. collected EPMA data. Z.L. and K.I. collected other experimental data. Z.L. wrote the paper with K.I.

Funding: This study was financially supported by Yamaguchi University, a national university under the jurisdiction of Japanese government.

Acknowledgments: Many thanks are given to doctoral student, Sha Li, for her help in the SEM-EDS analysis.

Conflicts of Interest: The authors declare no conflict of interest.

References

1. Komarneni, S.; Roy, D.M.; Roy, R. Al-substitute tobermorite: Shows cation exchange. *Cem. Concr. Res.* **1982**, *12*, 773–780. [CrossRef]
2. Pariatamby, A.; Subramaniam, C.; Mizutani, S.; Takatsuki, H. Solidification and stabilization of fly ash from mixed hazardous waste incinerator using ordinary Portland cement. *Environ. Sci.* **2006**, *13*, 289–296. [PubMed]
3. Minaricova, M.; Škvara, F. Fixiation of heavy metals in geopolymeric materials based on brown coal fly ash. *Ceramics-Silikaty* **2006**, *50*, 200–207.
4. Ikeda, K.; Li, Z. Development of paper sludge ash-based geopolymer and application to the solidification of nuclear waste water. In Proceedings of the 14th ICCC, Beijing, China, 13–16 October 2015.

5. Li, Z.; Ohnuki, T.; Ikeda, K. Development of paper sludge ash-based geopolymer and application to treatment of hazardous water contaminated with radioisotopes. *Materials* **2016**, *9*, 633. [CrossRef] [PubMed]

6. Cheng, T.W.; Lee, M.L.; Ko, M.S.; Ueng, T.H.; Yang, S.F. The heavy metal adsorption characteristics on metakaolin-based geopolymer. *Appl. Clay Sci.* **2012**, *56*, 90–96. [CrossRef]

7. Provis, J.L.; Walls, P.A.; van Deventer, J.S.J. Geopolymerization kinetics. 3. Effect of Cs and Sr salts. *Chem. Eng. Sci.* **2008**, *63*, 4480–4489. [CrossRef]

8. Zhang, J.; Provis, J.L.; Feng, D.; van Deventer, J.S.J. Geopolymers for immobilization of Cr^{6+}, Cd^{2+}, and Pb^{2+}. *J. Hazard. Mater.* **2008**, *157*, 587–598. [CrossRef] [PubMed]

9. Fernández-Jiménez, A.; Palomo, A. Fixing arsenic in alkali-activated cementitious matrices. *J. Am. Ceram. Soc.* **2005**, *88*, 1122–1126. [CrossRef]

10. Fernández-Jiménez, A.; Macphee, D.E.; Lachowski, E.E.; Palomo, A. Immobilization of cesium in alkaline activated fly ash matrix. *J. Nucl. Mater.* **2005**, *346*, 185–193. [CrossRef]

11. Fernández-Jiménez, A.; Lachowski, E.E.; Palomo, A.; Macphee, D.E. Microtructural characterization of alkali-activated PFA matrices for waste immobilization. *Cem. Concr. Compos.* **2004**, *26*, 1001–1006. [CrossRef]

12. Vu, T.H.; Gowripalan, N. Mechanism of heavy metal immobilization using geopolymerization techniques. *J. Adv. Concr. Technol.* **2018**, *16*, 124–135. [CrossRef]

13. Palomo, A.; Krivenko, P.; Garcia-Lodeiro, I.; Kavalerova, E.; Maltseva, A.; Fernández-Jiménez, A. A review on alkaline activation: New analytical perspectives. *Mater. Constr.* **2014**, *64*, 022. [CrossRef]

14. Provis, J.L. Geopolymers and other alkali activated materials, why, how, and what? *Mater. Struct.* **2014**, *47*, 11–25. [CrossRef]

15. Provis, J.L.; Bernal, S.A. Geopolymers and related alkali-activated materials. *Annu. Rev. Mater. Res.* **2014**, *44*, 299–327. [CrossRef]

16. García-Lodeiro, I.; Palomo, A.; Fernández-Jiménez, A.; Macphee, D.E. Compatibility studies between N-A-S-H and C-A-S-H gels. Study in the ternary diagram Na_2O-CaO-Al_2O_3-SiO_2-H_2O. *Cem. Concr. Res.* **2011**, *41*, 923–931. [CrossRef]

17. García-Lodeiro, I.; Fernández-Jiménez, A.; Palomo, A. Variation in hybrid cements over time. Alkaline activation of fly ash-portland cement blends. *Cem. Concr. Res.* **2013**, *52*, 112–122. [CrossRef]

18. Yamaguchi, N.; Nagaishi, M.; Kisu, K.; Nakamura, Y.; Ikeda, K. Preparation of monolithic geopolymer materials from urban waste incineration slags. *J. Ceram. Soc. Jpn.* **2013**, *121*, 847–854. [CrossRef]

19. Kwan, S.; LaRosa-Thompson, J.; Grutzeck, M.W. Structures and phase relations of aluminum-substituted calcium silicate hydrate. *J. Am. Ceram. Soc.* **1996**, *79*, 967–971. [CrossRef]

20. Uehara, M. Committee Report JCI-TC155. In *Reaction Mechanism and Definition of Geopolymer*; Japan Concrete Institute (JCI): Tokyo, Japan, 2017; pp. 2–21, ISBN 978-4-86384-090-4-.

21. Pingitore, J.; Nicholas, E.; Lytle, F.W.; Davies, B.M.; Eastman, M.P.; Eller, P.G.; Larson, E.M. Mode of incorporation of Sr^{2+} in calcite. *Geochim. Cosmochim. Acta* **1992**, *56*, 1531–1538. [CrossRef]

22. Littlewood, J.L.; Shaw, S.; Peacock, C.L.; Bots, P.; Trivedi, D.; Burke, T. Mechanism of enhanced strontium uptake into calcite via an amorphous calcium carbonate crystallization pathway. *Cryst. Growth Des.* **2017**, *17*, 1214–1223. [CrossRef]

23. Shannon, R.D.; Prewitt, C.T. Effective ionic radii in oxides and fluoride. *Acta Cryst.* **1969**, *25*, 925–946. [CrossRef]

24. Shannon, R.D. Revised effective ionic radii and systematic studies of interatomic distances in halides and chalcogenides. *Acta Cryst.* **1976**, *32*, 751–767. [CrossRef]

25. Sunagawa, I.; Takahashi, Y.; Imai, H. Strontium and aragonite-calcite precipitation. *J. Mineral. Petrol. Sci.* **2007**, *102*, 174–181. [CrossRef]

26. Blackford, M.G.; Hanna, J.V.; Pike, K.J.; Vance, E.R.; Perera, D.S. Transmission electron microscopy and nuclear magnetic resonance studies of geopolymers for radioactive waste immobilization. *J. Am. Ceram. Soc.* **2007**, *90*, 1193–1199. [CrossRef]

27. Palomo, A.; Alonso, S.; Fernández-Jiménez, A.; Sobrados, I.; Sanz, J. Alkaline activation of fly ashes: NMR study of the reaction products. *J. Am. Ceram. Soc.* **2004**, *87*, 1141–1145. [CrossRef]

28. Fernández-Jiménez, A.; Palomo, A. Alkali-activated fly ashes: Properties and characteristics. In Proceedings of the 11th ICCC, Durban, South Africa, 11–16 May 2003.

29. Yamamoto, T.; Kikuchi, M.; Otuka, H. *Physicochemical Properties of Reaction Phase of Geopolymer Prepared from FA-GGBS-SF Activated with NaOH*; Japan Concrete Institute (JCI): Tokyo, Japan, 2017; pp. 11–17, ISBN 978-4-86384-090-4-C3050.
30. Iwahiro, T.; Komatsu, R.; Ikeda, K. Chemical compositions of gels prepared from sodium metasilicate and aluminum nitrate solutions. In Proceedings of the Geopolymers, Melbourne, Australia, 28–29 October 2002.
31. Vallepu, R.; Fernández-Jiménez, A.; Terai, T.; Mikuni, A.; Palomo, A.; Mackenzie, K.J.D.; Ikeda, K. Effect of synthesis pH on the preparation and properties of K-Al-bearing silicate gels from solution. *J. Ceram. Soc. Jpn.* **2006**, *114*, 624–629. [CrossRef]
32. Ichimiya, K.; Akinaga, F.; Harada, K.; Ikeda, K. Influence of alkaline water ratio of fly ash based geopolymer on flow value and shape and strength change by high temperature heating. In Proceedings of the Japan Concrete Institute, Kobe, Japan, 4–6 July 2018. (In Japanese)
33. Yip, C.K.; Lukey, G.C.; van Deventer, J.S.J. The coexistence of geopolymeric gel and calcium silicae hydrate at the early stage of alkaline activation. *Cem. Concr. Res.* **2005**, *35*, 1688–1697. [CrossRef]
34. West, A.R. *Solid State Chemistry and Its Application*, 2nd ed.; John Wiley: Chichester, UK, 2014; pp. 271–272. ISBN 978-4-06-154390-4.

materials **MDPI**

Article

Preliminary Assessment of Criticality Safety Constraints for Swiss Spent Nuclear Fuel Loading in Disposal Canisters

Alexander Vasiliev [1,*]**, Jose Herrero** [1]**, Marco Pecchia** [1]**, Dimitri Rochman** [1]**, Hakim Ferroukhi** [1] **and Stefano Caruso** [2]

[1]   Paul Scherrer Institute (PSI), 5232 Villigen PSI, Switzerland; jjh@enusa.es (J.H.); marco.pecchia@psi.ch (M.P.); dimitri-alexandre.rochman@psi.ch (D.R.); hakim.ferroukhi@psi.ch (H.F.)
[2]   National Cooperative for the Disposal of Radioactive Waste (NAGRA), 5430 Wettingen, Switzerland; stefano.caruso@nagra.ch
*   Correspondence: alexander.vasiliev@psi.ch

Received: 18 December 2018; Accepted: 28 January 2019; Published: 5 February 2019

Abstract: This paper presents preliminary criticality safety assessments performed by the Paul Scherrer Institute (PSI) in cooperation with the Swiss National Cooperative for the Disposal of Radioactive Waste (Nagra) for spent nuclear fuel disposal canisters loaded with Swiss Pressurized Water Reactor (PWR) UO_2 spent fuel assemblies. The burnup credit application is examined with respect to both existing concepts: taking into account actinides only and taking into account actinides plus fission products. The criticality safety calculations are integrated with uncertainty quantifications that are as detailed as possible, accounting for the uncertainties in the nuclear data used, fuel assembly and disposal canister design parameters and operating conditions, as well as the radiation-induced changes in the fuel assembly geometry. Furthermore, the most penalising axial and radial burnup profiles and the most reactive fuel loading configuration for the canisters were taken into account accordingly. The results of the study are presented with the help of loading curves showing what minimum average fuel assembly burnup is required for the given initial fuel enrichment of fresh fuel assemblies to ensure that the effective neutron multiplication factor, k_{eff}, of the canister would comply with the imposed criticality safety criterion.

Keywords: spent nuclear fuel; geological repository; criticality safety; burnup credit; loading curves

1. Introduction

The Swiss National Cooperative for the Disposal of Radioactive Waste (Nagra) plans to submit a general licence application for a deep geological repository for the disposal of spent fuel and high-level waste (HLW repository) and for low- and intermediate-level waste (L/ILW repository) by 2024. One of the requirements for the design of the HLW repository is the safety of the installations (encapsulation facility and repository) from the point of view of a possible criticality excursion over a 1,000,000-year lifetime. Were it to occur, criticality would affect the properties of the engineered barrier system, namely the canister, and the backfill material and the near-field of the host rock.

For the above reasons, the criticality safety issue for the disposal of spent fuel canisters was preliminarily investigated by Nagra already in 2002 in the context of the safety assessment of a repository for spent fuel and high-level waste in the Opalinus Clay [1]. The results of that study were not considered sufficiently developed for the detailed design of canisters and for a systematic and comprehensive application of burnup credit (BUC) to all Swiss spent nuclear fuel (SNF) assemblies. However, the project came to the important conclusion that a combination of burnup credit and canister design modifications could ensure subcriticality in all cases. Furthermore, the application of BUC was identified as necessary only for the case of PWR SNF but not for the case of Boiling Water Reactor (BWR) [2]. In fact, such findings were basically in line with the studies performed in the USA in relation to the Yucca Mountain deep geologic repository project, as can be seen in Reference [3]. Therefore, a detailed criticality safety analysis for BWR fuel was not considered in the follow-up PSI/Nagra collaboration; however, internal Nagra activities were carried out to demonstrate compliance with the subcriticality criteria without the application of a burnup-credit approach, namely by considering ideal fresh (un-irradiated) fuel, and without any credit from burnable neutron poisons (e.g., gadolinium). Also, the case of a degraded canister/fuel configuration was outside the scope of the project phase described in this paper. This is a certain limitation of the preliminary results presented here since the effects on the criticality evaluations of such medium to very long-term processes as materials corrosion, alteration and dissolution of the fuel matrix [4], as well as canister deformation or even potential geochemical separation of Plutonium [5] have not been yet considered. Such studies are thus planned for investigations in subsequent phases of the PSI/Nagra collaboration.

At the present stage, a dedicated calculation methodology for criticality safety evaluations (CSE) related to interim dry storage and long-term waste disposal is under development at PSI. The CSE+BUC assessments require two coupled calculations: first, fuel depletion and decay calculations to obtain the isotopic compositions of the fuel after the discharge and cooling period and then, criticality calculations using these compositions, for different initial enrichments and final burnups in order to create loading curves. All the relevant details of the calculation methodology applied and of the studies performed are reported below. The numerous validation studies that supported the described methodology development and qualification are not presented here in detail as they have been the subject of many previous publications. In particular, the validation studies for the fuel depletion calculations with the stand-alone CASMO code [6,7] and with the CASMO/SIMULATE codes using PSI proprietary post-irradiation examination (PIE) data [8], for the reactor core-follow simulations with the CASMO/SIMULATE codes using the in-core reactor measurements [9] and for the criticality calculations with the Monte Carlo N-Particle® MCNP® or MCNPX Software (The registered trademarks are owned by Triad National Security, LLC, manager and operator of Los Alamos National Laboratory (see https://mcnp.lanl.gov/; assessed on 29 January 2019).) using criticality benchmark experiments [10,11] should be mentioned as examples.

The paper is structured as follows. Section 2 presents the approach used for defining the bounding SNF case, including the description of the employed tools and models, as well as the criticality safety criterion selected for the bounding case assessments. Section 3 presents and discusses the results obtained for the bounding SNF case definition. Section 4 describes the methodology applied for the loading curve derivation. The integration of the most penalising burnup profiles and random uncertainty components in the CSE+BUC are demonstrated in Section 5. The loading curves for the disposal canisters are finally derived in Section 6. These loading curves are preliminary, since there is still room for improvement, in particular, in relation to the treatment of uncertainties, as discussed later. Furthermore, the effect of degraded canister/fuel configurations on criticality has to be considered in future work for the development of the final loading curves. However, at the same time, these preliminary loading curves can be used to guide future advanced studies and to serve for verification of the updated results. Sections 7 and 8 provide a discussion of the results and the conclusions on the work performed, respectively.

2. Definition of the Bounding SNF Case for Loading in Disposal Canisters

For the specific case in hand in this paper, namely the application of burnup credit to the long-term disposal of PWR spent nuclear fuel, this work was focused on the simulation of the Nagra conceptual canister model [12] loaded with the fuel assemblies (FAs) corresponding to the Gösgen nuclear power plant (KKG) fuel designs. To be on a conservative side, optimum moderation conditions must be assumed in criticality safety assessments. The determination of the most reactive moderation was done in the previous preliminary assessments [1] for canisters loaded with PWR/BWR UO_2 or UO_2/MOX

(5% Pu-fiss) fuels when k_{eff} was calculated as a function of the water density. It was found in those studies that the maximum water density corresponds to the highest k_{eff} values. Thus, in the given work, the loaded canister is also assumed to be flooded with water entering through a postulated breach, as is the case for the Swedish design and related criticality assessment [13]. Additional verification of the optimal moderation conditions will be performed at future stages of the work when the canister design will have been fixed.

The determination of final loading curves for SNF to be loaded in disposal canisters was preceded by a preliminary study, namely the evaluation of the bounding fuel type and/or conservative conditions, which are discussed in the given chapter. The CSEs were realised for PWR FAs using realistic irradiation conditions for different enrichments, burnup levels, FA designs and fuel compositions. Note that the KKG nuclear power plant (NPP) has used the highest possible enrichment of fuel as compared with two other Swiss PWR reactors Beznau 1 and 2 (KKB1 and KKB2).

2.1. Calculation Tools

For the depletion phase, the two-dimensional (2-D) fuel assembly depletion code CASMO5 [14] was employed with feedback from the three-dimensional (3-D) reactor code SIMULATE-3 [15] using the in-house BOHR tool [8,16], as illustrated later in Section 2.3.

Between the depletion calculations and the criticality calculations, a decay phase is introduced for the study of the long-term evolution of the fuel composition. The decay module of the code SERPENT2 [17] has been selected as the most preferable based on the analysis of the available options [18].

For the criticality calculation case, the MCNP6 code [19] has been employed as the most validated among the Monte Carlo codes available at PSI.

2.2. Development of Models

The disposal canister is basically a carbon steel cylinder, is almost 5 metres long and is designed to fit 4 PWR FAs in 4 separate inserted and welded carbon steel boxes (see Figure 1). The preliminary dimensions selected for the present analysis were R_{in} = 41 cm, R_{out} = 55 cm, box centre-centre (C-C) separation = 17.9 cm, A ≈ 21.5 cm (side of the FA top head) and B ≈ 23.5 cm (inner side of the FA box). The analysis performed refers entirely to this disposal canister concept.

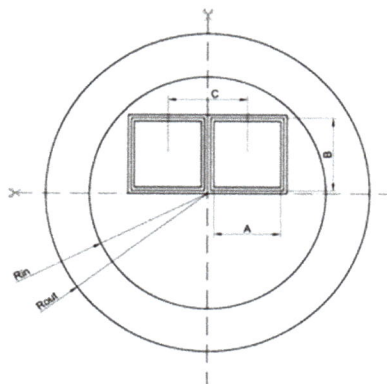

Figure 1. A sketch of the carbon steel disposal canister.

No uncertainties related to the underlying nuclear data, depletion calculations and corresponding fuel compositions are considered at this stage, but they will be included later to develop the fuel loading curves. Three types of fuel assemblies were considered for the analysis:
1. UO$_2$ 4.94 w/o U-235

The UO$_2$ assembly is formed by a 15 × 15 array of fuel pins (with 20 guide tubes) which contain fuel homogeneously enriched at 4.94 weight percent (w/o) of U-235 and operated up to 5 cycles, reaching discharge burnups of 17.61, 33.82, 50.47, 61.92 and 72.75 GWd/tHM.

2.   Mixed Oxide Fuel (MOX) 4.80 w/o Pu$_{fiss}$

The MOX fuel assembly contains a distribution of three slightly different enrichments inside the assembly, with the central rod empty, i.e., flooded with water. The content of ^{239}Pu and ^{241}Pu fissile isotopes in the plutonium fuel fraction is approximately 64%. The chosen assembly was operated to burnups of 18.10, 34.78, 44.96 and 51.72 GWd/tHM.

3.   Enriched Reprocessed Uranium (ERU), 4.599 w/o U-235 Equivalent

The ERU fuel assembly is similar in structure to the UO$_2$ fuel assembly and was operated to burnups of 17.27, 34.58, 50.10, 56.04 and 61.72 GWd/tHM.

The following different loading configurations of the canister were considered for the bounding case analysis:

- 4 similar UO$_2$ FAs
- mixed burnup UO$_2$ fuel (3 FAs with the same burnup + 1 FA with lower burnup)
- 4 similar ERU FAs
- 4 similar MOX FAs
- 1 MOX FA and 3 similar UO$_2$ FAs
- 3 similar UO$_2$ FAs and an empty position

Configurations with more than 1 MOX FA loaded in a canister are not considered feasible, as their contributions to the total heat load is too high according to the repository constraints based on the Nagra safety assessment. In fact, a heat load of 1.5 KW is considered the maximum for the disposal canister to ensure the functionality of the engineered barrier system (canister-bentonite-Opalinus Clay). The mixed loading of one MOX with 3 UO$_2$ is, however, considered possible for the Nagra conceptual design.

The calculational model is bounded by a 35 cm layer of bentonite clay, saturated with water. Vacuum boundary conditions are employed at the outer surface of the bentonite clay. However, the impact of the bentonite on the system k_{eff} is negligible for a flooded canister. Other conservative assumptions applied to the model are low material temperature and the presence of water without diluted minerals (these assumptions lead to better neutron moderation and less neutron absorption, both leading to an increasing k_{eff} value).

The study was performed with representative assemblies, selected arbitrarily from each assembly batch considered among the assemblies reaching the highest burnups at the End of Life. The irradiation history was reconstructed using real plant operating data, as was described above, with the help of the BOHR tool. Therefore, the best estimate axial burnup profiles were utilised for the bounding fuel case analysis. In this sense, the bounding assessments performed, described in the next section, are based on the best estimate modelling of the fuel operating history.

Figure 2 shows the MCNP model describing the canister (not in full detail as compared with Figure 6 given later; the model's outer boundaries are cut as they are not significant for presentation) and the fuel assemblies with the discrete axial spent fuel specifications (see further details in Section 4.2.1).

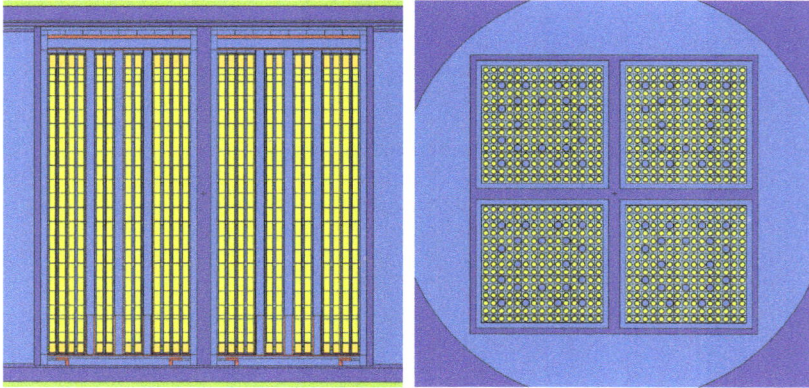

Figure 2. The axial (**left**) and radial (**right**) views of the canister MCNP model loaded with fresh UO$_2$ fuel.

2.3. The Calculation Route

The computational scheme developed and implemented at PSI [16] is based on the suite of reference CASMO5/SIMULATE-3 core models, continuously developed and validated for all Swiss reactors and all operated cycles within the PSI Core Management System (CMSYS) platform [20]. These core models are based on the real reactor operating histories. The core nodal-wise thermal-hydraulic conditions are calculated with the SIMULATE-3 code using the plant-provided boundary conditions on the inlet coolant temperature and core exit coolant pressure during the reactor cycle.

A principal limitation of the SIMULATE-3 code is that it is not suited for providing detailed spent fuel composition (SFC), which is required for the criticality calculations, because such data are not needed for the two-group nodal diffusion approximation employed in SIMULATE-3 for full core neutron transport calculations. The detailed fuel isotopic composition is actually required at the stage of preparation of the neutron cross section libraries with the CASMO5 code, which involves a higher order method of characteristics for the neutron transport solution at the spatial level of the FA axial slice. To overcome the limitations of the presently used CASMO5/SIMULATE-3 system of codes with respect to the BUC application calculation needs for which this system of codes was not designed, a specific calculation scheme was developed called BOHR (which stands for "Bundle Operating History Reconstruction"). BOHR can provide CASMO5 with the realistic operating history conditions, which can be retrieved from SIMULATE-3 full core calculations, to rerun CASMO5 depletion calculations and obtain the pin-by-pin detailed burnt fuel isotopic composition for an axial slice of the fuel assembly. Thus, CASMO5 depletion calculations are performed up to the "discharge burnup" for each particular axial FA slice, with the actual operating history which is known from the SIMULATE-3 core-follow calculations (when conservative axial burnup profiles are applied, the burnup does not correspond to realistic discharged burnup but is defined based on the conservative assumptions, as explained later in Section 4.2.1). Thus, the power history in the CASMO calculations following BOHR procedures (third step in Figure 3) is exactly the same as was used in the original SIMULATE calculations, i.e., detailed (e.g., daily-wise) power history provided by the plant operator.

In the next step, the burnt compositions are translated into the input file of SERPENT2 to compute the change in the isotopic concentrations during different decay periods. In other words, the fuel depletion is simulated in the given approach with the CASMO code, while discharged spent fuel decay is done with SERPENT (recall Section 2.1). The coupling between the basic MCNP canister model and the CASMO5/SERPENT results is performed with the help of the COMPLINK tool [21], which imports detailed SNF compositions for every assembly to define a complete canister loading.

The MCNP model of the canister loaded with burnt fuel assemblies is then used to compute the k_{eff} of the system at different time steps during the decay, aiming to assess if the system remains subcritical in the postulated case of a flooded canister. In order to ensure criticality safety, the calculated k_{eff} must be assessed with the Nuclear Criticality Safety (NCS) Criterion. Figure 3 provides an illustration of the scheme and related processes described in the following subsections. Here, the symbols T_f, T_c, r_c, C_B and P mean respectively the fuel temperature, the coolant temperature, the coolant density, the soluble boron concentration in the coolant and the coolant pressure in the computation node (i,j,k) (the axial slice of a fuel assembly in the SIMULATE-3 full core 3-D model) in the x,y,z system of coordinates at time t.

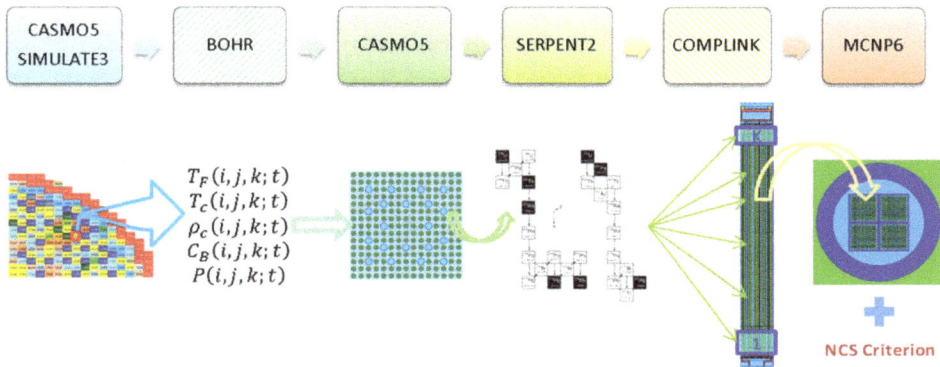

Figure 3. The computational scheme for the burnup credit in a disposal application.

2.3.1. Retrieval of the Nodal History

The sequence of calculations begins with the generation of the data library using CASMO5 2D lattice calculations. The nuclear data library employed in this step was based on the ENDF/B-VII.0 library [22]. The spacers are smeared along the full axial length of the FA, which complies with the SIMULATE-3 model requirements. Every FA type is computed, and the initial isotopic composition is taken from the final state of the previous cycle.

The reactor cycle operation is computed with SIMULATE-3. From these results, the values for the state parameters are retrieved for every FA and axial elevation at different cycle instants. The explicit spacer model for the neutronics solution is activated in SIMULATE-3.

The BOHR tool is employed for the extraction of the required values from SIMULATE-3, which provides the values for the nodal power, the fuel temperature, the coolant temperature and density and the boron concentration for every axial and radial location in the nodal calculation. The presence of inserted control rods during the operation is also taken into account.

2.3.2. Lattice Calculations for Discharge Composition Estimation

In order to extract and make use of the composition for every fuel pin at the discharge burnups and for every axial FA slice, new 2-D lattice calculations are performed with CASMO5. During every time interval, the actual irradiation history (fuel and coolant temperatures and densities, positions of control rods, etc.) is employed based on the core cycle depletion/operation calculations with SIMULATE-3. The nuclear data library used at this stage was upgraded to ENDF/B-VII.1 (The new release of the nuclear data library for CASMO5 became available at the time of the study and was therefore employed for the new calculations; however, the original CASMO5/SIMULATE-3 models described in the previous section were based on the older ENDF/B-VII.0 version. Nevertheless, this inconsistency should not be of any practical significance for the present analysis.) [23].

The spacer mass is, again, smeared along the whole axial length, introducing an approximation to realistic conditions. The additional approximations are that the irradiation-induced changes of the FA structures and the materials are currently not taken into account. More importantly, it must also be underlined that the CASMO5 reconstructed depletion calculations are performed using single assembly reflected models, i.e., without accounting for a realistic leakage term representative of the 3-D environment under which the assembly was irradiated. The impact of this approximation and the ways to improve it still need to be analysed. It should be noticed, however, that the effects of the intra-fuel assembly pin-by-pin horizontal burnup distributions are taken into account later at the stage of the full-scale MCNP criticality calculations for the loaded canister, as presented in Sections 4.2.2 and 5.2. It is foreseen that the application of the newer Studsvik codes SIMULATE5 and SNF may allow for the improvement of the presently developed BUCSS-R methodology (see later Section 7.1).

2.3.3. Decay Calculations after Discharge

From the detailed burnup results, the compositions obtained are decayed over a one-million-year period using the Transmutation Trajectory Analysis algorithm programmed in the burnup module of the code SERPENT2 [17]. The decay data from ENDF/B-VII.1 were employed. The concentrations were computed at the following times: 0, 1, 2, 5, 10, 20, 40, 60, 80, 100, 120, 150, 200, 300, 500, 1000, 2000, 5000, 8000, 10,000, 15,000, 20,000, 25,000, 30,000, 40,000, 45,000, 50,000, 100,000, 500,000 and 1,000,000 years. The fuel pin compositions in the FA decay individually at every axial elevation.

2.3.4. Criticality Calculations for the Disposal Canister

Each particular MCNP model, which includes the SFC for every discharge burnup at the end of each fuel operation cycle and for each decay period, is generated starting from a base input file with the canister model loaded with FAs of equal uniform composition corresponding to fresh fuel (illustrated in Figure 2). The tool COMPLINK is used for this purpose. The SFC is provided at the pin-by-pin and axial node-wise level, as further explained in Section 4.2.1.

Only the isotopes usually accounted for in burnup credit calculations are included in the fuel composition. Two sets of isotopes are considered following the specifications given in Reference [24], customarily termed the actinides only (AC) and the actinides plus fission products (AC+FP) groups. The AC group includes U-233, U-234, U-235, U-236, U-238, Pu-238, Pu-239, Pu-240, Pu-241, Pu-242, Am-241, Cm-242, Cm-243, Cm-244, Cm-245 and Cm-246, and the AC+FP group includes in addition to the isotopes in the AC group Np-237, Am-242^m, Am-243, Mo-95, Tc-99, Ru-101, Rh-103, Ag-109, Cs-133, Nd-143, Nd-145, Sm-147, Sm-149, Sm-150, Sm-151, Sm-152, Eu-151, Eu-153 and Gd-155. Note that compared with the list of isotopes considered in Reference [24], here, the curium isotopes are also taken into account.

The nuclear data from ENDF/B-VII.1 coming with the MCNP6 software distribution have been used in the calculations. The canister and fuel geometry are considered to be constant over time, which is a common approach in this type of preliminary criticality safety assessments [13]. The material temperature is 293.16 K everywhere, and corresponding densities and dimensions are employed.

The MCNP criticality calculations for most of the analysed cases were performed with 80 inactive cycles and 60 active cycles. The number of inactive cycles was chosen following the MCNP recommendation based on the Shannon entropy estimation of the fission distribution [25] for the AC+FP case with mixed MOX and UO$_2$ fuels, as this is the most challenging in terms of initial fission source convergence. All the cycles were run with 200,000 neutron histories each. The resulting k_{eff} statistical uncertainty (k_{eff} standard deviation reported by the MCNP code) was approximately ± 25 pcm.

2.4. Criticality Criterion Selected for the Bounding Case Assessments

The criterion for nuclear criticality safety assessment, conventionally termed the Upper Subcritical Limit (USL) hereafter, which accounts for the burnup credit of the spent nuclear fuel, is discussed in detail later in Section 4.1. Here, it must only be mentioned that the criticality safety criterion may vary depending on the methodology definition and on the national regulatory requirements. However, one particular component is basically common to all countries and methodologies, namely the so-called "administrative margin" (also referred to as "an arbitrary margin to ensure the subcriticality" [26]), which will be denoted here as Δk_{eff}^{AM}. Normally this margin equals 0.05 in terms of the absolute k_{eff} value [13,27], and it is typically dominant compared to all other margins and imposed uncertainty components. The last comment is especially true for the fresh fuel case, but even in the case of spent fuel, the depletion-related uncertainties and margins, e.g., assessed in Reference [13], are noticeably smaller as compared with 0.05. Thus, the USL values obtained with different methodologies by different organisations and countries for the same type of applications may vary but are normally not too far from the k_{eff} = 0.95 value. Therefore, for the sake of a better representability and an easier comparison with similar studies, it makes sense to consider the k_{eff} = 0.95 value as the simplified criticality criterion selected for the consequent bounding case assessments, while for the final loading curve derivations, the PSI methodology-specific USL value will be employed.

3. Results of the Bounding Fuel Case Assessments

3.1. Fresh Fuel

To start, a set of calculations has been performed with fresh fuel configurations. The configurations considered are the following:

- The canister in dry conditions (actually filled with helium gas)
- The canister flooded with water at 293.16 K
- The canister flooded with the FAs displaced diagonally towards the centre of every fuel box (within the design tolerances)
- The canister flooded with the FAs displaced diagonally towards the outer part of the fuel box

When several assemblies of the same type are considered (UO_2, ERU and MOX), they are assumed to be fully identical (originating from the same batch, i.e., with the same nuclear and mechanical design). The calculated k_{eff} values for each of these cases are presented in Table 1.

Table 1. The k_{eff} values for the canister configurations with fresh fuel.

Assumed Conditions	4 UO_2	4 ERU	4 MOX	1 MOX + 3 UO_2	1 Empty + 3 UO_2
Helium filled	0.21146	0.20772	0.26259	0.20743	0.17861
Flooded/centred	1.09513	1.08022	0.96180	1.07035	1.02971
Flooded/inwards	1.12903	1.11227	0.98601	1.10079	1.04864
Flooded/outwards	1.04355	1.02920	0.91990	1.02301	0.99541

The canister in dry conditions is clearly subcritical for any combination of fresh fuel. However, in all other cases (except the case of MOX/flooded outwards), the calculated k_{eff} values are above 0.95 and the k_{eff} increases notably if a less favourable position of the assemblies inside the canister is considered (inwards). The importance of the distance between assemblies for the k_{eff} is clearly reflected in the table. This also means that the material of the wall boxes has a small impact on decoupling the neutron fluxes of each assembly, so the loading position of the assemblies into the canister seems to be important for the k_{eff} values in flooded conditions.

3.2. Discharged Fuel

The burnup credit approach can reduce the system k_{eff} (In fact, the major change in the system k_{eff} which occurs due to taking into account the burned fuel composition is associated with the change in the infinite multiplication factor of the fuel lattice (k_∞), while the changes in the neutron leakage should be the effects of the second order. Nevertheless, for the sake of rigor, all presented evaluations were done for realistic 3-D canister models, and therefore, only the results for the system k_{eff} are discussed in this work.) as a consequence of the presence of neutronic poisons (fission products and some non-fissile actinides), as well as due to the depletion of multiplicative materials (major actinides) (Although some amount of fissile major and minor actinides is produced in the reactor operation, the net k_{eff} (primary $k\infty$) effect with fuel burnup is negative.). The impact on k_{eff} has been evaluated for both the AC and AC+FP cases.

For these calculations, the same configurations (as the fresh fuel case) are used; however, in this case, the canister is loaded with an SFC with different burnups and at different cooling times, covering the one-million-year period.

For the burnt fuel configurations, the analyses were conducted for each of the assembly-averaged burnup levels reached after every reactor cycle. The spent fuel compositions keep changing after irradiation due to the decay processes, and therefore, the fuel neutronic properties ("reactivity") are changing as well. The calculated curves of the k_{eff} evolution are plotted in Figure 4 for a flooded canister at different instants during the decay of the isotopes for both AC and AC+FP cases.

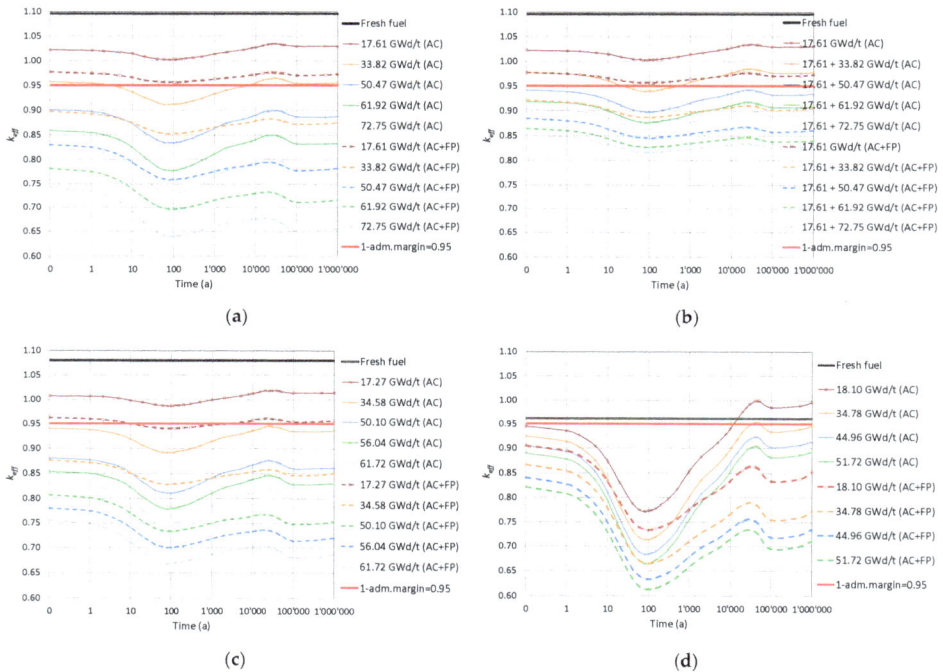

Figure 4. *Cont.*

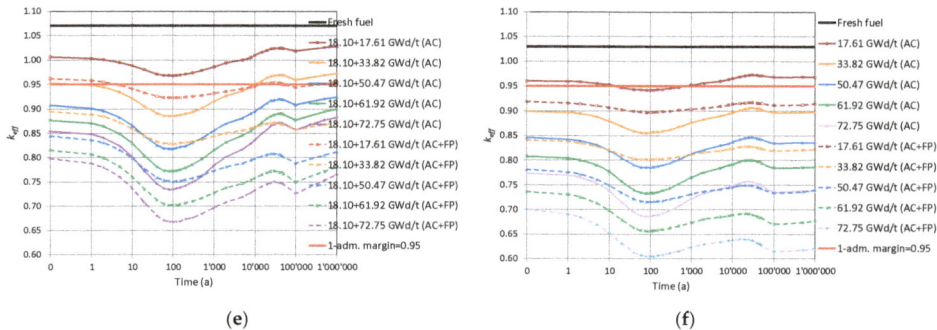

Figure 4. The evolution of k_{eff} for the intact canister loaded with (**a**) spent UO$_2$ fuel; (**b**) mixed burnup UO$_2$ fuel (the first burnup value for one position, the second value for the three remaining); (**c**) ERU fuel; (**d**) MOX fuel; (**e**) one low burnup MOX (18 GWd/tHM) and three UO$_2$ fuel at different burnups; and (**f**) 3 UO$_2$ assemblies and one empty position.

The results obtained are summarised below in Sections 3.2.1–3.2.4. It should be mentioned that the observed behaviour is in line with already published results for similar situations analysed and discussed in detail in Reference [28]. In brief, it can be recalled here that the decrease of k_{eff} within the first approximate 100 years is mainly related to the decay of ^{241}Pu (T$_{1/2}$ ≈ 14.4 years), as well as the build-up of ^{241}Am (in the case of AC+FP, also the build-up of ^{155}Gd as a result of the beta-decay of ^{155}Eu is important). Later, the increase of k_{eff} up to around 30,000 years is explained by the decay of ^{241}Am (T$_{1/2}$ ≈ 432 years) and ^{240}Pu (T$_{1/2}$ ≈ 6,560 years). At a later time, k_{eff} starts to decrease again mainly due to ^{239}Pu decay (T$_{1/2}$ ≈ 24,100 years). The given explanations and the half-life (T$_{1/2}$) data are borrowed from Reference [28].

3.2.1. UO$_2$ Fuel Assemblies

Figure 4a shows the k_{eff} results as a function of the decay time obtained for the disposal canister loaded with the UO$_2$ FAs of the same burnup for both the AC and the AC+FP approaches. It was assessed using quadratic interpolation that the spent fuel with a burnup of less than 24 GWd/tHM (For the bounding fuel case assessments reported in this section, a quadratic interpolation technique was applied to estimate the limiting burnup values based on the data illustrated in Figure 4. A more robust method explained in Section 6 was finally applied for the derivation of the limiting burnup values for the loading curves.) (for the AC+FP case) does not meet even the limit of k_{eff} = 0.95 for loading unless a mixed configuration with higher burnup fuel was to be considered. If an AC approach was considered, a burnup higher than approximately 38 GWd/tHM needs to be reached. The k_{eff} of the system after 10,000 years could reach a value above that of the initial discharged fuel for the AC approach.

3.2.2. UO$_2$ Fuel Assemblies of Mixed Burnups

Next, a mixed burnup configuration was investigated, as reported in Figure 4b. One FA at low burnup (17.61 GWd/tHM) is considered in the mixed loading with the other three FAs of higher burnups. The results show that high burnup fuels are needed (above 50 GWd/tHM) when taking credit only for actinides.

3.2.3. ERU Fuel Assemblies

The ERU FA corresponds to fuel enriched up to 4.6 w/o ^{235}U$_{eq}$ and operated until 61.72 GWd/tHM. The evolution of k_{eff} through time, shown in Figure 4c for both the AC and AC+FP cases, is very similar to that of UO$_2$ fuel.

3.2.4. MOX Fuel Assemblies

The MOX FA corresponds to fuel enriched to 4.8 w/o Pu_{fiss} and operated until 51.72 GWd/tHM. The evolution of k_{eff} through time in Figure 4d has stronger dip and peak values around 100 and 30,000 years (case AC+FP) or 45,000 years (case AC), respectively. Notable is that the k_{eff} peak after thousands of years would be higher than any previous value calculated for the AC case, meaning that using the discharge compositions without decay would not be a bounding assumption for the whole disposal period (moreover, taking only actinide changes into account leads to a remarkable case where k_{eff} of partly burned fuel can go above the k_{eff} of fresh fuel; see the Figure 4d case for 18.1 GWd/tHM (AC)).

On the other hand, considering additional minor actinides and fission products in the fuel composition also has a stronger impact on the k_{eff} than for UO_2 fuel, and this impact is more important at later periods, thus reducing the k_{eff} peak below former values in time. Therefore, the use of the discharge compositions would be bounding for the AC+FP approach.

3.2.5. One MOX and Three UO_2 Fuel Assemblies

In this model, the canister is loaded with one MOX and three UO_2 FAs. Figure 4e shows the k_{eff} evolution for the canister loaded with low burnup MOX (18 GWd/tHM) together with the three UO_2 assemblies at different burnup levels. The main findings from these graphs are as follows:

- For the AC case, k_{eff} is increasing from 100,000 years and keeps growing at the end of the period considered of 1 million years. At the cooling time of 1 million years, the k_{eff} values are greater than at the zero cooling time.
- For the AC+FP case, the bounding value of k_{eff} corresponds to zero cooling time.

The AC approach would require UO_2 fuel burnt to around 40 GWd/tHM, and the AC+FP approach would require burnup of approximately 20 GWd/tHM.

3.2.6. Empty Position and Three UO_2 Fuel Assemblies

The empty position is assumed to be also flooded with water. The results plotted in Figure 4f show that fuel with a limiting burnup of 19 GWd/tHM can be considered if using the AC approach and 10 GWd/tHM for the AC+FP approach.

3.3. Findings from the Bounding Fuel Case Analysis

The bounding fuel case analyses consisted of assessing the multiplication factor variation as a function of the discharge burnup and decay time, ranging between fresh fuel conditions and best-estimate burnt fuel configurations. The main findings of the analysis are that UO_2 fuel could be problematic if featuring low burnups, especially for the AC case where all fuels suffer a rise in k_{eff} in later time periods after disposal, which may violate the USL value. ERU fuel has a similar behaviour to the UO_2 fuel. The mixing of UO_2 and MOX fuel in the canisters could be a good compromise to keep k_{eff} below the safety margin while avoiding the thermal limitations more easily violated by MOX fuel filled canisters.

Figure 5 summarises the minimum burnup credit which would be required for every type of loaded nuclear fuel considered in this study for both the AC and AC+FP cases.

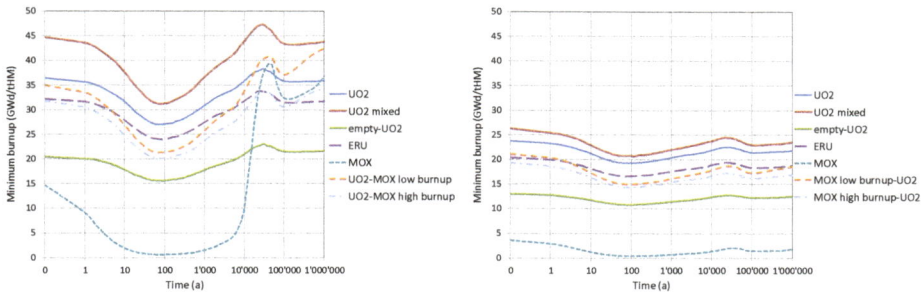

Figure 5. The evolution of minimum burnup credit required to comply with a k_{eff} value below 0.95 within the geological disposal timeframe: AC case (**left**) and AC+FP case (**right**).

For the AC case, the loadings of UO_2 fuel operated for just one cycle (17.61 GWd/tHM) could be allowed only if mixed with 3 other UO_2 assemblies with burnups above 45 GWd/tHM (the second k_{eff} peak is decisive for the AC approach). It can also be noted from Figure 5 for the AC case that the minimum burnup required for a homogeneous UO_2 fuel loading is not always above the minimum burnup required for the canisters filled with MOX fuel (mixed UO_2/MOX case or full MOX case) since the second k_{eff} peak in the MOX cases is very high.

For the AC+FP case, the mixed burnup loading (1 MOX + 3 UO_2) requires a minimum burnup of 20 GWd/tHM for the UO_2 fuel (fuel operated at least for two operating cycles). In this case, the burnup credit required for canisters loaded with MOX fuel would be dominated by the k_{eff} at discharge and it will not be higher than the corresponding UO_2 case.

Finally, it should be borne in mind that all these results imply that the fuel matrix is still intact, maintaining the actinide and fission product mixture even after 100,000 years which is far from the normal FA design target, which ensures FA integrity during irradiation in core but not for the geological timeframe.

The main driving parameter for criticality in the current canister design is the distance between the FAs, due to the compact configuration (and also because neutronic poisons are not present in the canister basket material). Fresh fuel calculations indicate a difference of 2–5% in k_{eff} values between nominal and displaced configurations, which is very important.

The results obtained are in line with the analyses performed for other canister designs [28] and indicate that the inclusion of the fission products in the burnup credit could allow 2-cycle-operated FAs to be safely loaded into disposal canisters. However, the margin is very close to the limit, and once the biases, uncertainties and further bounding assumptions are introduced for the loading curves generation, it will be further reduced, as will be discussed in the following sections.

Fuel with low burnup (corresponding to one cycle of operation) cannot be loaded to fill all the positions of the same canister. A dedicated study with mixed UO_2-MOX models demonstrated that the fuel with low burnup could meet the requirements in some mixed configurations or in the empty position configuration.

Finally, based on the reported findings for the currently considered canister design, the canister model loaded with 4 similar PWR UO_2 spent FAs (i.e., the type of fuel employed at KKG) will be used for the preliminary loading curve derivation as it is the most problematic configuration among those studied.

4. Methodology for Preliminary Loading Curve Derivation

As a main outcome of the study outlined to this point, the application of burnup credit to the criticality calculations for disposal canisters appears to be necessary for the safe disposal of the PWR spent FAs operated in the Swiss reactors according to the current design concept for the Nagra disposal canister.

This section presents the results of the final stage of the preliminary loading curve derivation. The application of a best-estimate computational route is now complemented with conservative coverage in the form of a stochastic analysis of the main uncertainty components in combination with the application of bounding burnup profiles, as illustrated below. Furthermore, the USL value considered in this final stage of the work is now based on the comprehensive validation study for LWR fuel performed at PSI for the MCNP code in combination with the ENDF/B-VII.1 library, which has been selected for the routine criticality calculations [10,11].

Note that the general practice in the burnup credit applications is based on choosing a set of bounding parameters for the burnup calculations in terms of power density, fuel and coolant temperatures, coolant densities, etc. so that the k_{eff} at discharge for such conservative assumptions will be higher than the k_{eff} obtained with any possible real irradiation history. This path, however, was not adopted for the BUCSS-R project. In fact, the approach in the BUCSS-R project is different because real operational data are employed (using SIMULATE-3 for accurate core-follow calculations) for the FAs of different enrichments in order to estimate the loading curves on the basis of best-estimate assessments integrated with a conservative but rational treatment of the uncertainties. Therefore, at present, only some representative FAs operated at KKG could be considered and explicitly analysed. It is foreseen that, in the future, the studies performed should be repeated for a statistically significant number of FAs, at least for the most reactive design/enrichment types, to allow for statistically confident verification/updating of the presently evaluated preliminary loading curves.

4.1. Chosen Criterion for Criticality Safety Accounting for Burnup Credit

The criticality safety criteria applied for derivation of the loading curves can be presented with the following relations (adapted from Reference [29]):

$$k_{eff}\Big|_{Bounding\ FA\ pos}^{Canister}(BU) + \Delta k_{eff}^{Ax}(BU) + \Delta k_{eff}^{Rad}(BU) + 2\sigma_{tot}(BU) < USL = K_{eff}^{LTB}\Big|_{AOA} - \Delta k_{eff}^{AM}, \quad (1)$$

where

$$\sigma_{tot}(BU) = \sqrt{\sigma_{ND}^2(BU) + \sigma_{BU-eff}^2(BU) + \sigma_{OC}^2(BU) + \sigma_{TP}^2 + \sigma_{T1/2}^2 + \sigma_{MC}^2}; \quad (2)$$

$$\sigma_{ND}(BU) = \sigma_{ND-SNF}^{CASMO}(BU) + \sigma_{ND-Keff}^{MCNP}(BU); \quad (3)$$

$k_{eff}\Big|_{Bounding\ FA\ pos}^{Canister}$ is the neutron multiplication factor corresponding to the disposal canister loaded with SNF placed in the most penalising positions considering the canister technological tolerances (see Section 3.1 for details); Δk_{eff}^{Ax} and Δk_{eff}^{Rad} are the k_{eff} penalties to cover the bounding axial and radial burnup profiles, respectively; USL is the Upper Subcritical Limit (see Reference [10] for details); and σ_{ND}, σ_{BU-eff}, σ_{OP}, σ_{TP}, $\sigma_{T1/2}$ and σ_{MC} are the uncertainties at one standard deviation level respectively for the nuclear data (ND), radiation/burnup-induced changes/effects (BU-eff), operating conditions (OC), technological parameters components (TP), decay constants ($T_{1/2}$) and the Monte Carlo statistical uncertainty of the employed MCNP code for the criticality calculations. The listed components of the $\sigma_{tot}(BU)$ uncertainty are assumed to be random (not systematic) and uncorrelated. The resulting $\sigma_{tot}(BU)$ is further assumed to be normally distributed. Under these conditions, the term $2\sigma_{tot}(BU)$ in Equation (1) is assumed to represent the 95% confidence interval for k_{eff}, which is, for instance, in line with the recommendations provided, e.g., in References [30,31].

The σ_{ND-SNF}^{CASMO} nuclear data-related component is responsible for the k_{eff} uncertainty associated with the SFC (due to the propagation of nuclear data uncertainties during depletion calculations), and the $\sigma_{ND-Keff}^{MCNP}$ component is the k_{eff} uncertainty due to the nuclear data uncertainties themselves (see Section 5.3.3. for further details).

The $K_{eff}^{LTB}\Big|_{AOA}$ term stands for the Lower Tolerance Bound for the particular Area of Applicability (AOA; here, this is limited to Swiss LWR fuel), and its value was reported in Reference [10] as 0.99339 (following the Gaussian-based derivation; see Reference [10] for details) for the PSI CSE methodology using the MCNP code in conjunction with the ENDF/B-VII.1 library. As outlined in Section 2.4, Δk_{eff}^{AM} is the "administrative margin" normally imposed to cover the unknown uncertainties to ensure subcriticality, which is assumed here to be 0.05000 (The administrative margin to criticality is normally set at 5000 pcm, i.e., the k_{eff} of the system plus the calculation bias and uncertainty in the bias should not exceed 0.95. More recently, an administrative margin of 2000 pcm was suggested for very unlikely accident conditions [32].) (5000 pcm). Thus, for the final loading curve derivations, USL = 0.99339 − 0.05000 = 0.94339 is employed. As for the above listed components of the σ_{tot} uncertainty, they are presented in more detail in Section 5.4.

4.2. Spatial Burnup Distribution Assumptions

This section comprises most of the modelling and simulation assumptions employed. First of all, it addresses the derivation of bounding burnup profiles and the assessment of their impact on the canister k_{eff} value, to derive the Δk_{eff}^{Ax} and Δk_{eff}^{Rad} penalties for Equation (1). The impact of the cooling time is also addressed.

4.2.1. Axial Burnup Distribution

Axial burnup profiles of the spent fuel operated at KKG and irradiated to different average burnups were retrieved from the PSI CMSYS database, which includes all the burnup values per fuel assembly at every axial node of the SIMULATE-3 calculations, as described in Section 2. The burnup profiles were normalised with these average values and separated into two families corresponding to the models with 40 axial nodes in SIMULATE-3 with active fuel regions between 358 and 352 cm height and the models with 38 axial nodes in SIMULATE-3 with an active region of 340 cm height. The reason for having SIMULATE-3 models with different numbers of axial layers of nodes with fuel is that the active fuel length of KKG FAs was changed from earlier cycles to the later ones. The older FAs have reflector segments at the 2 bottom nodes instead of fuel nodes. It can also be noted that the older and shorter FAs had lower fuel enrichment compared to the later and longer FAs. Therefore, it is important to take into account an actual axial burnup profile for every specific enrichment for the correct derivation of the loading curves.

Following standard practice [33], the approach was to choose the lowest normalised burnup values of all the profiles for the first and the last 9 nodes and the highest normalised burnup values of the profiles for the remaining central nodes. This resulted in the renormalised profiles shown later in Figure 6 (the average node burnup fraction is normalised to unity in both the cases of 38 and 40 axial nodes).

The change in the bounding axial profiles with the average assembly burnup was not considered and could be a way of reducing conservatism if needed. Also, a mass calculation with all the profiles in the database could be a different approach to deriving bounding profiles.

4.2.2. Radial Burnup Distribution

For the radial burnup profiles within the FAs, there were no operational or CMSYS data available at the time of the study (such data could be obtained by upgrading the reference CMSYS models and also can be obtained in line with the discussions provided later in Section 7). Therefore, as an alternative solution, the publicly open information on the bounding horizontal burnup profile reported in Reference [34] was employed (and this is one of the items requiring further verification). The bounding profile is expressed with Equations (4) and (5), which were derived from real measurements (see Reference [34] and references therein for details) to generate a radial burnup tilt varying for each assembly row:

$$B_{rel} = \frac{B_H - B_{av}}{B_{av}} = 0.33 - \frac{0.08}{15} \cdot (B_{av} - 10), \tag{4}$$

$$B(n) = \left[B_{rel} + 1 - \frac{4}{N} \cdot B_{rel} \cdot \left(n - \frac{N+2}{4} \right) \right] \cdot B_{av} \tag{5}$$

where N is the number of rows in the square assembly, 15 in our case; B_{rel} is the relative difference between the horizontally averaged burnup value for the half of the assembly (B_H) with the highest burnup and the horizontally averaged assembly burnup B_{av}; and n is the row number in the assembly to which the computed burnup B(n) corresponds (The second formula includes a correction of a sign from the one in the original report [34].).

It must be underlined that, in fact, CASMO5 does not allow the specification of the burnup value desired for each row of the FA. To overcome this difficulty, a surrogate approach was utilised: the CASMO5 input file was modified such that the fuel composition is printed at the 15 burnup steps which correspond to the desired burnups of each of the fuel rod rows, as obtained with Equations (4) and (5). After this, the fuel compositions from every burnup step were transferred to the SERPENT and later to the MCNP6 models row by row, providing required intra-assembly fuel burnup horizontal distributions. In this sense, the approach differs slightly from the one described in Reference [34], where it was assumed that "all the fuel rods belonging to one and the same row have one and the same burnup". In the present approach, each fuel rod has its own composition, but the horizontally averaged burnup of the entire row is preserved as defined by the above procedure. However, an examination of the typical ratios between the burnup value of each pin and the average assembly burnup showed that, to avoid burning the pins in the regions of higher power above the desired value for the row, a factor of 0.93 should be applied to each B(n). This implies that the assembly burnup is lowered by 7%, which introduces an additional conservatism in the sequence as peripheral pins typically have already lower burnup than average. In addition, the lowest burnup regions of the assemblies are later faced in the canister so as to produce the highest k_{eff}.

To illustrate the outcome of the employed methodology, Figure 6 below shows an example of the radial (horizontal) U-235 concentration distribution on a pin-by-pin basis within a FA axial node.

4.3. Impact of Cooling Time

The cooling time between cycles was explicitly considered in the burnup calculations with CASMO5. As it was illustrated in Section 3, in the case of the actinides only credit, the impact of cooling time after discharge on the system k_{eff} is characterised by an initial decrease in k_{eff} in the first approximate 100 years and then a steady increase which reaches its maximum at around 30,000 years after discharge. The important point is that the k_{eff} value at that time for intact canister configurations could be higher than the initial k_{eff} value just after discharge, so taking this initial value cannot be considered bounding in all cases and that decay calculations up to 100,000 years also need to be considered to generate the loading curves. Beyond that time, the flooded intact canister approximation would be totally unrealistic for a canister with a lifetime of approximately 10,000 years, and degraded models should start to be considered in that range.

The time positions where decay compositions have been used to compute k_{eff} values were 0, 5, 20,000, 30,000, 40,000 and 50,000 years.

4.4. Canister Modelling

At the stage of the loading curve analyses, the canister model was updated based on the most detailed and actual design information received by PSI from Nagra. The modelling included all the details provided in Reference [12] as well as the detailed structure of the FAs, including heads, grids and rods from the internal documentation available at PSI. The illustration of the final MCNP model with the indications of the spatial burnup distributions employed in the loading curve derivation analysis is shown in Figure 6.

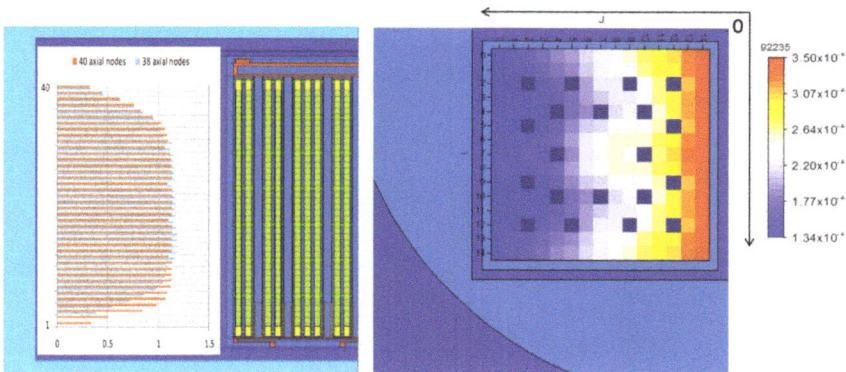

Figure 6. The illustrative MCNP model (1/4th symmetry sector; schematic and not to scale): the axial (**left**) and radial (**right**) view of the refined canister model. The bounding axial burnup profiles (relative units) for the models with 38 and 40 axial fuel nodes and radial burnup profiles (here, the U-235 atomic density is $*10^{24}$ at/cm^3) are illustrated.

As outlined in Section 3, the fuel assemblies were conservatively placed towards the centre of the canister at the storage positions as this was found to be the most reactive configuration.

5. Quantification of the Bounding Effects and Random Uncertainties Components

The results for criticality calculations of the canister loaded with the same fuel assembly in the four positions were compiled for the different enrichments covering the values employed from the initial to the latest fuel cycles of KKG. In the following sections, the Δk_{eff} effects resulting from substituting the nominal burnup profiles by the penalising (conservative) ones are quantified and reported.

5.1. Axial Burnup Effect

The results of substituting the original burnup profiles by the penalising profiles while keeping the average assembly burnup are illustrated in Table 2 for the highest considered fuel enrichment of 4.94 w/o.

Table 2. The Δk_{eff} penalty due to the bounding axial burnup profiles for the case of 4.94 w/o, pcm.

Time (a)	Discharge Burnup (GWd/tHM)									
	17.61 *		33.82		50.47		61.92		72.75	
	AC	AC+FP	AC	AC+FP	AC	AC+FP	AC	AC+FP	AC	AC+FP
0	-	-	983	1792	2359	3390	3604	4880	4737	6223
5	-	-	1273	2203	2752	4042	4141	5642	5322	7144
20,000	-	-	1209	2445	2886	4946	4571	7113	6210	9078
30,000	-	-	1197	2445	2930	4968	4675	7218	6414	9236
40,000	-	-	1225	2544	2950	5064	4729	7307	6538	9493
50,000	-	-	1213	2533	2993	5154	4901	7381	6693	9609

* At the first discharge burnup of 17.61 GWd/tHM, the nominal axial burnup profile is actually more reactive than the penalising one.

At first glance, it can be observed that the impact on k_{eff} is stronger for the following cases:

- AC+FP credit
- Longer decay periods
- Increasing burnup

Table 3 shows similar information for the lower enrichment of 3.5 w/o. In this case, the proposed profile is conservative even for the lowest burnup, so the impact of the conservative axial burnup profile is apparently also stronger with lower enrichments. As in the previous case, the added k_{eff} is notably larger in the AC+FP approach.

Table 3. The Δk_{eff} penalty due to the bounding axial burnup profiles for the case of 3.5 w/o, pcm.

	Discharge Burnup (GWd/tHM)							
	18.9		33.66		45.25		56.15	
Time (a)	AC	AC+FP	AC	AC+FP	AC	AC+FP	AC	AC+FP
0	401	1037	1703	2306	3711	4705	4910	6211
5	592	1387	1931	2860	4129	5575	5484	7252
20,000	587	1769	2143	3603	4904	7184	6808	9472
30,000	577	1905	2248	3698	4974	7258	6982	9647
40,000	674	1919	2274	3760	5064	7411	7271	9944
50,000	634	2026	2350	3826	5294	7593	7401	10,144

Finally, regarding the effect for the lowest enrichments of 1.9 w/o and 2.5 w/o, the impact of the proposed profiles is not conservative and the k_{eff} from the nominal profile is higher and will be maintained for the final loading curves. Other profiles could be proposed to create a unique bounding axial burnup profile for these enrichments.

5.2. Radial Burnup Effect

In this section, the impact is considered of the intra-assembly burnup profiles obtained in accordance with the descriptions given in Section 4.2.2, in such a way that the lowest radial burnup regions are facing the centre of the canister to raise k_{eff}. Results for the cases of 4.94 w/o and of 3.5 w/o are given respectively in Tables 4 and 5.

Table 4. The Δk_{eff} penalty due to the bounding radial burnup profiles for the case of 4.94 w/o, pcm.

	Discharge Burnup (GWd/tHM)									
	17.61		33.82		50.47		61.92		72.75	
Time (a)	AC	AC+FP	AC	AC+FP	AC	AC+FP	AC	AC+FP	AC	AC+FP
0	1380	1729	2008	2230	2100	2656	2449	2829	2375	2820
5	1464	1947	2187	2475	2364	3016	2584	3268	2548	3103
20,000	1246	1832	2194	2743	2553	3550	2917	3959	2926	3858
30,000	1202	1831	2185	2821	2601	3652	3009	3952	3040	3883
40,000	1208	1860	2225	2815	2630	3663	3028	3976	3121	3973
50,000	1260	1860	2219	2888	2670	3751	3073	3997	3102	4045

Table 5. The Δk_{eff} penalty due to the bounding radial burnup profiles for the case of 3.5 w/o, pcm.

	Discharge Burnup (GWd/tHM)							
	18.9		33.66		45.25		56.15	
Time (a)	AC	AC+FP	AC	AC+FP	AC	AC+FP	AC	AC+FP
0	1966	2339	2278	2511	2866	3210	2780	3163
5	2140	2575	2248	2975	2928	3710	2972	3617
20,000	2150	2761	2633	3553	3567	4429	3642	4601
30,000	2107	2799	2603	3657	3577	4572	3701	4617
40,000	2151	2770	2666	3677	3586	4624	3851	4750
50,000	2198	2911	2754	3733	3754	4638	3879	4826

In general, the k_{eff} impact is

- stronger for the AC+FP credit;
- increasing from lower to higher burnups; and
- mainly increasing during the decay period up to 20,000 years and then stabilising.

As with the axial burnup profiles for the lowest enrichments of 1.9 and 2.5 w/o, the proposed radial profiles are not conservative in any case and so the nominal profiles are kept.

5.3. Assessment of Uncertainties

In the following subsections, quantitative assessments are given for the uncertainties components of σ_{tot} (BU) from Equations (1)–(3) listed in Section 4.1. It must be outlined that, at present, some of the values provided are rather preliminary and may require verification depending on the availability of related information.

5.3.1. Reactor Operating Conditions and Radiation-Induced Changes

The impact of the reactor operational parameter variations was estimated in a dedicated study with the help of the CASMO-5 code (the results have been partly presented in Reference [8]). As mentioned, proprietary information on the PIE data for various burned fuel rods from Swiss reactors is available at PSI together with detailed information on the fuel operation and fuel design parameters. This allows for the validation of the fuel burnup calculations together with an assessment of the calculational vs. experimental uncertainty components of the obtained C/E results. Such studies were conducted for a KKG fuel rod sample (from a 15×15 fuel assembly irradiated during 3 cycles up to the sample final burnup above 50 GWd/tHM), and two types of uncertainties were assessed within this analysis:

- The uncertainties related to operating conditions, including boron concentration, moderator temperature, reactor power, etc. (the power and the moderator density were assumed fully correlated with the fuel and moderator temperatures in the underlying work described in Reference [8])
- The radiation (BU-)induced changes in the geometry (i.e., fuel pin position shift, moderator pin position shift, fuel pellet diameter increase, etc.)

As can be seen, these uncertainties represent the terms σ_{OC} and σ_{BU-eff} of Equation (2). The final assessments accepted for the given study are shown below in Table 6 (only the first figures are deemed significant). It shall be noticed that at present stage, the uncertainty components of Equation (2) are assumed to be uncorrelated. In reality, there might be some correlations between them, although they are not expected to be strong a priori, to the best knowledge of the authors. Stronger correlations would occur if one imposed into analysis certain reactor operational constraints, e.g., $k_{eff} = 1$, at the reactor core follow simulations. However, in such cases the posterior uncertainties of the affected parameters are normally significantly reduced. Some representative illustrations on the example of the nuclear data evaluations can be found in References [35–37]. Furthermore, these types of uncertainties are most difficult for quantification and are already quite conservative by themselves, and therefore, questions on the uncertainties of these uncertainties or their correlations go far beyond the current work on the preliminary loading curves development.

Table 6. The available uncertainty data on the operating conditions and the BU-induced geometry changes, pcm.

Burnup (GWd/tHM)	0.0–17.6	17.6–33.8	33.8–50.5
Operating conditions	100	400	500
BU-induced changes	200	200	700

5.3.2. Technological Tolerances Impact

The impact of the PWR fuel technological and manufacturing parameter tolerances on the criticality calculations was analysed in Reference [38] with another PSI in-house tool "MTUQ" (Manufacturing and Technological Uncertainty Quantification). Taking into account only the fuel assembly-related uncertainties from the list of parameters used, the total σ_{TP} uncertainty component is assessed as only 10 pcm. In particular, the uncertainty components from all parameters considered in Figure 9 of Reference [38], except parameters 11 (Absorber Box—Inner boundary) and 13 (FA rack—Centre-to-centre distance), should be summed as random uncorrelated uncertainties, thus leading to a total uncertainty value limited by 10 pcm.

5.3.3. Nuclear Data Uncertainty Impact

The uncertainties in the nuclear data employed in the calculations contribute to the uncertainty in the computed canister k_{eff} values. Their impact was considered in the CASMO5 burnup calculations using SHARK-X methodology (see References [39,40] and references therein), providing the $\sigma_{ND-SNF}^{CASMO}(BU)$ estimation (uncertainties of the cross sections and the fission yields were taken into account), and in the MCNP6 criticality calculations using the Nuclear data Uncertainty Stochastic Sampling (NUSS) methodology to assess the $\sigma_{ND-Keff}^{MCNP}(BU)$ component (see References [39] and [29]; uncertainties of the thermal scattering (S(α,β)) data were ignored (see details and discussions in Reference [41]). The approach to estimate $\sigma_{ND-SNF}^{CASMO}(BU)$ was explained also in Reference [29]: "The nuclear data uncertainty propagation in CASMO depletion calculations, resulting in the spread of the SNF composition, was done using SHARK-X tool. The obtained set of the different SNF compositions was further translated to the SERPENT decay module for the decay simulation and finally provided to the MCNP6 models of the disposal canister to compute the spread of the k_{eff} values due to the spread of the SNF compositions, using the nominal ENDF/B-VII.1 ND files." Interested readers can find alternative ways of assessing the uncertainties associated with depletion calculations, for example, in References [42,43].

For an additional illustration, Figure 7 shows the scheme of the ND-related uncertainties (given as covariance matrices (CM)) propagation in compliance with the flowchart in Figure 3.

The Monte Carlo sampling method employed to obtain the estimated uncertainty in k_{eff} requires a large number of calculations and has thus been realised only for the 4.94 w/o fuel enrichment case so far.

Figure 8 shows the estimated $\sigma_{ND-SNF}^{CASMO}(BU)$ and $\sigma_{ND-Keff}^{MCNP}$ values for the fuel just after discharge and after 50,000 years of decay; uncertainties from all AC and FP are considered. The direct effect from the nuclear data in the MCNP6 calculation, $\sigma_{ND-Keff}^{MCNP}$, is similar in both periods and decreases slightly with the burnup level attained. The indirect effect of the nuclear data contained in the isotopic uncertainties, $\sigma_{ND-SNF}^{CASMO}(BU)$, increases with decay time and grows with burnup. These observations are valid for UO_2 fuel and for the employed ENDF/B-VII.1 CM. Details of the calculations performed are given in Reference [39].

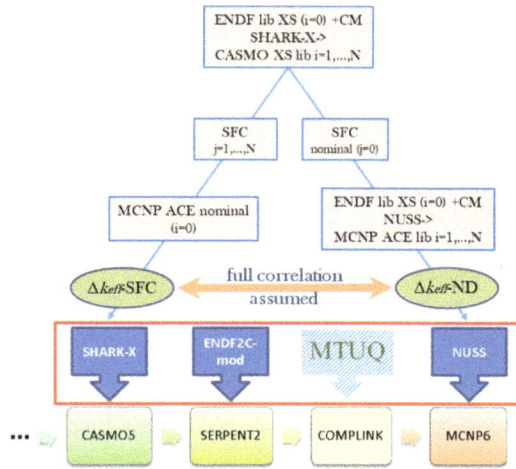

Figure 7. The presently employed nuclear data (ND) stochastic sampling methodology ("XS" is cross sections).

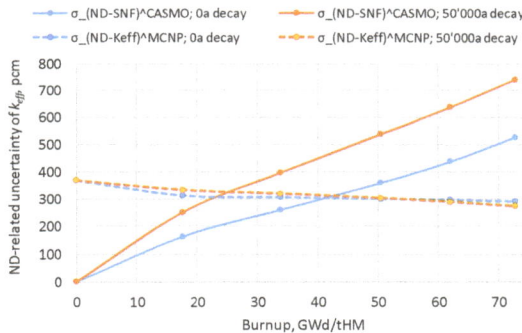

Figure 8. The estimated nuclear data-related uncertainties of k_{eff}.

The decrease of the $\sigma_{ND-Keff}^{MCNP}$ uncertainty component with burnup is consistent with the results presented earlier in Reference [44], where even the net effect (uncertainty propagations through both the depletion and criticality calculations) of the ND uncertainties originating from the ENDF/B-VII.1 library was also found to decrease with burnup for UO$_2$ fuel (see [44] and note that the impact of the fission yields uncertainties was not accounted for in [44], while in the given work the fission yields' contribution is taken into account [39]). The decreasing uncertainty can potentially be explained by the decrease in the contribution of U-235 cross section uncertainties with burnup and probably with some spectrum-related effects (see also comments provided in Reference [29] for Figure 6 of Reference [29]).

It should be noted that the uncertainty components σ_{ND-SNF}^{CASMO} and $\sigma_{ND-Keff}^{MCNP}$ must be correlated since the underlying nuclear data are the same for the independent estimations performed for both components. However, the correlation level has not been assessed. In the ideal case, all calculations should be done in a single set using the same original perturbation factors for the nuclear data in both the depletion and the criticality calculations; however, this will require significant additional computational burdens. Therefore, it will be conservative to assume a full correlation between both components and thus to estimate the total ND-related k_{eff} component according to Equation (3). More advanced assessments are planned for the near future with the latest release of the

ENDF/B library (ENDF/B-VIII), including the analysis of the ND-related correlations, as proposed in References [10,29].

Next, to be on the conservative side, the total ND-related uncertainty will be composed of the σ_{ND-SNF}^{CASMO} component corresponding to 50,000 years of cooling and the $\sigma_{ND-Keff}^{MCNP}$ corresponding to zero cooling time (Figure 8).

5.3.4. Long-Term Nuclide Evolution

The performance of the decay module of the SERPENT2 code and the employed nuclear data library was investigated and benchmarked in Reference [18]. Later, the impact of the nuclear data uncertainty on the decay calculations realised with the SERPENT2 code was studied by perturbing the decay data with a modified version of the ENDF2C tool [39,45]. The main outcome shows an impact of $\sigma_{T_{1/2}} \approx 15$ pcm on the k_{eff} for the studied load.

5.3.5. MCNP Monte Carlo Uncertainty

The Monte Carlo statistical uncertainty σ_{MC} of the calculations for the loading curve analysis was approximately $\pm$ 25 pcm, which is sufficiently low.

5.4. Summary of Bounding Burnup Distributions and Random Uncertainties

The analysed random uncertainty components are summarised in Table 7 (adapted from Reference [29]) together with the total sum of the uncertainties, σ_{tot}, which will be used in Equation (1). It is important to note that the total uncertainty is burnup-dependent due to the burnup dependency of the components σ_{ND}, σ_{OP} and σ_{BU-eff}.

Table 7. A summary of all the random uncertainty components, pcm.

Burnup (GWd/tHM)	σ_{ND}	σ_{OP}	σ_{BU-eff}	σ_{TP}	$\sigma_{T_{1/2}}$	σ_{MC}	$1\sigma_{tot}$	$2\sigma_{tot}$
0	367	0	0	10	15	25	368	737
17.61	560	100	200	10	15	25	604	1208
33.82	700	400	200	10	15	25	831	1662
50.47	834	500	700	10	15	25	1199	2397
61.92	930	500	700	10	15	25	1267	2534
72.75	1026	500	700	10	15	25	1339	2679

To better illustrate the impact of the considered burnup profile penalties and the uncertainty components, Figure 9 shows the results obtained for the case of AC+FP based on the data reported in Table 6, Figure 8 and Table 7 (the red line is the direct sum of the axial (Table 2) and radial (Table 4) effects; adapted from Reference [29]).

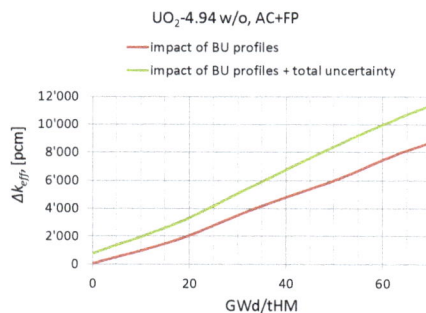

Figure 9. The impact of the burnup profiles and the total uncertainty on the canister k_{eff} value.

6. Loading Curves with Combined Uncertainties Effects

The final target of this work was to assess a minimum average burnup for individual fuel assemblies required for a full loading of the disposal canister without exceeding the defined upper subcritical limit. This goal is accomplished by the development of specific loading curves for discharged spent fuels, where the initial enrichment and final burnup of a fuel bundle will function as the acceptance criteria for the loading of the disposal canister.

The development of the curve is done as follows: the left part of Equation (1) is represented by a curve depending on the burnup, while the right part of Equation (1) is a constant corresponding to the given USL value. If the burnup-dependent curve of Equation (1) and the burnup-independent USL line intersect, the burnup at the point of the intersection becomes the point on the loading curve corresponding to the given fuel enrichment. If the burnup-dependent curve of Equation (1) is always below the USL value, then for the given enrichment, the minimum burnup equals zero on the loading curve. As presented in Section 4.1, for the PSI CSE methodology using the MCNP code in conjunction with ENDF/B-VII.1 and assuming the "administrative margin" equals 5000 pcm, the USL value is defined according to Equation (1) as $0.99339 - 0.05000 = 0.94339$.

To give an outlook on the general behaviour of the curve over the burnup, Figure 10 shows the examples for the case of AC (left) and AC+FP (right) corresponding to the highest of all the considered enrichments, 4.94 w/o.

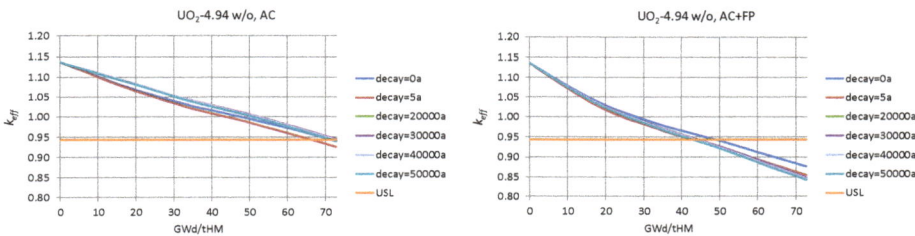

Figure 10. An illustration of the determination of the minimum burnup required for fuel to meet the Upper Subcritical Limit (USL) criticality safety criteria for AC (**left**) and AC+FP (**right**).

It can be seen again that the most conservative case for the credit of AC+FP is zero cooling time after discharge, while in the case of only AC, the most conservative results generally correspond to the cooling time of about 30,000 years (in particular, for the AC cases with 2.50 w/o and 3.50 w/o enrichments, the highest k_{eff} results took place at zero cooling time, and in the case of 3.50 w/o enrichment, at 20,000 years cooling time; in all other cases, the most conservative cooling time was 30,000 years).

The computed minimum required burnups for each particular enrichment are shown in Table 8.

Table 8. The minimum burnup required for meeting the criticality safety criteria.

Enrichment w/o	AC	AC+FP
1.90	0	0
2.50	~6.0 *	3.7
3.20	21.5	13.8
3.50	33.9	21.4
4.10	~51.0 *	31.5
4.30	~51.7 *	~36.4 *
4.94	~73.0 *	49.1

* These values are based on extrapolations or involve certain (minor) interpolations.

The loading curves are then derived by applying the data from Table 8 as shown in Figure 11 (adapted from Reference [29]).

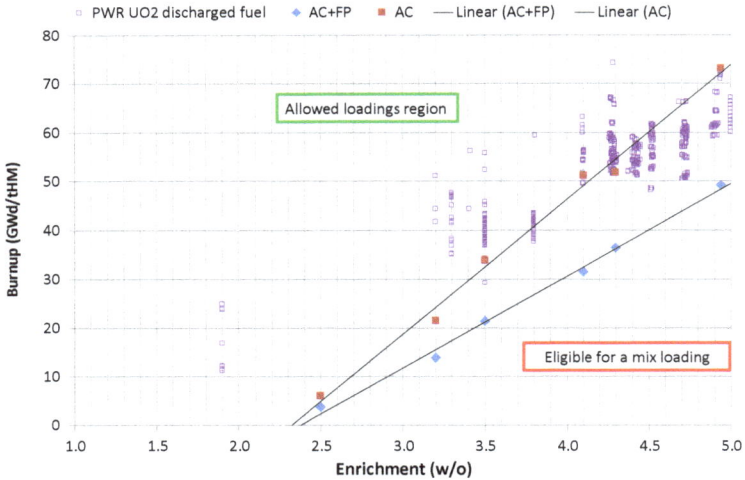

Figure 11. The preliminary loading curves with all the conservative effects for discharged spent fuel (valid only for the criticality safety criteria, and subject to discussed important assumptions).

Some examples on the KKG discharged fuel assemblies' characteristics (viz., initial enrichment and discharged burnup) are also demonstrated in Figure 11. From these results, it becomes clear that the given canister design meets the criticality safety criteria if the burnup credit is based on the AC+FP approach. A potential exception could be, for instance, the 5 w/o enriched FAs if they were irradiated for a lower number of cycles than designed, e.g., discharged once-burned from a last operated core. In reality, the last cycles can be loaded with a lower enriched fuel to avoid the inefficient use of fissile material.

It can be seen from Figure 11 that, for the given USL value and for the case of UO_2, 4.94 w/o fuel, the saving in terms of the minimum required burnup as compared between the cases of AC+FP vs. AC is about 24 GWd/tHM (this has been illustrated in more detail in Reference [29]). This result was obtained by comparing the cases which take into account the bounding burnup profiles and also all uncertainties listed in Table 7. Basically, the same effect (23 GWd/tHM) can be observed for the case when only bounding burnup profiles are taken into account with no uncertainties contribution. A much weaker effect would be obtained (16 GWd/tHM) if the calculations were done with nominal burnup profiles instead of the bounding ones. Finally, a comparison between two values of the administrative margin was also shown in Reference [29]: for the conventional value of 5000 pcm and for a reduced one of 2000 pcm. This illustration allowed the prediction of what saving in the minimum burnup requirement could be achieved, provided the administrative margin could hypothetically be relaxed to 2000 pcm. The extra saving for such a case was estimated as approximately 12 GWd/tHM.

7. Outlook and Discussion

The classical practice for criticality safety calculations for applications out of the reactor has relied on the fresh fuel assumption. The advantages of taking credit for changes in the actinide compositions after irradiation and the build-up of fission products in the spent fuel, with a net effect of decreasing the k_{eff} of the systems formed by burned fuel assemblies, have been given major attention during the last decades. Initially applied to spent fuel pools and fuel dissolution facilities and later to transport and dry storage, burnup credit has also been used in waste disposal applications in the USA, Sweden and Finland. The waste management applications in these three countries have been reviewed by the regulators, are under review or are under development respectively.

In line with this, a preliminary criticality assessment study has been performed taking into account the burnup credit within the PSI/Nagra BUCSS-R research project recently conducted at PSI. Ways for further methodological improvements and the verification of the present results are discussed below.

7.1. Ways to Improve the Reference BUCSS-R Methodology

Based on the analyses performed so far, it can be concluded that the most significant uncertainty components at the current stage are

- the nuclear data (ND) combined impact (σ_{ND}),
- the operating conditions (σ_{OC}), and
- the radiation-/burnup-induced geometry changes (σ_{BU-eff}).

Obviously, it can be recommended at first to verify, in particular, the most significant contributors to the loading curve burnup penalties in future work (although future studies should not be limited to only the listed components since other parameters can appear significant in more comprehensive simulations). Also, it would be relevant to assess the SIMULATE-3/BOHR/CASMO5 methodology component approximations, namely the infinite lattice calculations of the pin-wise SNF compositions with CASMO5 code. With respect to this limitation, the transition to the SIMULATE5 code in combination with another Studsvik code "SNF" (in quotation marks to distinguish from the earlier spent nuclear fuel abbreviation) could provide intra-assembly pin-wise isotopic distributions from the real core-follow calculations, as discussed in the next section. Such work is currently ongoing at PSI outside the present research collaboration with NAGRA.

Furthermore, a comparison of the current methodologies, which can be classified as a combination of the best estimate plus uncertainties (BEPU) approach with certain conservative assumptions, against more traditional methodologies can be recommended. In particular, the comparison of the BEPU SNF compositions versus analogue data but with application of the so-called isotopic correction factors [13] obtained from the validation studies with the post-irradiation examination (PIE) data should be considered.

Finally, the inclusion of the correlation analysis in the CSE methodology, as discussed in Reference [29], should be considered, since it would lead to more accurate and probably less penalising loading curves.

7.2. Alternative Advanced Way to Derive the Loading Curves under Development and Verification

The presently developed BUCSS-R calculation scheme is considered at PSI to be sufficiently detailed and accurate for the construction of the BEPU-type loading curves, provided that statistically representative samples can be obtained for FAs with all considered fuel types and enrichments. A drawback of this system is that it is rather demanding in terms of calculation resources. Important to mention is that the SIMULATE/BOHR/CASMO calculations required for the BUCSS-R scheme are basically not needed for any other type of simulation, so they are practically solely dedicated to the BUCSS-R methodology for the loading curve derivations (As outlined before, the BOHR methodology is also useful for the PIE data analysis, but in that case, only important FA segments can be calculated, which drastically reduces the computational demands.). Since the CASMO calculations generally have to be redone every time an updated/new version of the CASMO code or/and its cross section library are released, such a method for the derivation of the loading curves becomes costly and inefficient.

However, an alternative way to derive the loading curves recently became available based on the exploitation of the advanced features of the additional CMS code "SNF", which can be seamlessly integrated into the PSI CMSYS system. A description of the new methodology developed at PSI outside the BUCSS-R project, which is based on the CASMO/SIMULATE/SNF/COMPLINK/MCNP system of codes and which was named the CS_2M method has been presented in Reference [46], together with examples of its trial application. Thus, the concept of the new and more advanced approach for the loading curve derivation is to rely on the nominal and validated CMSYS CASMO/SIMULATE core-follow models and calculations, with no need for complementary CASMO calculations to derive the SFC. All required information can now be obtained directly from the CASMO/SIMULATE results with the help of the "SNF" code, as illustrated schematically in Figure 12 (see Reference [47] for details on the relative perturbation factors).

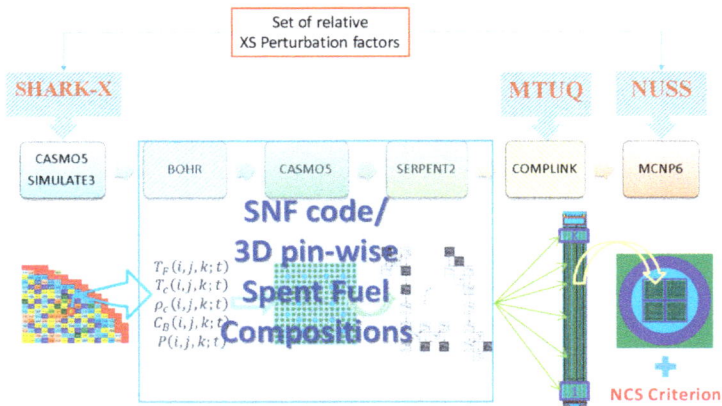

Figure 12. An alternative computational scheme based on the "SNF" code integration into the CMSYS system.

Furthermore, as already mentioned, it is recommended in the future to use the SIMULATE5 code, as it has a more accurate inter-assembly SFC characterisation capability. In doing so, it will be in principle possible to extract the SFC for every individual fuel assembly ever operated in Switzerland and thus to avoid all undesirable approximations to realise the BEPU calculations. The uncertainty propagation can be done with the same tools as used before in the BUCSS-R scheme. This is illustrated with the shaded boxes in the top part of Figure 12.

Therefore, although the new CS2M approach is also time-consuming, it does not require any complementary and unnecessary simplified assessments, and for that reason, it is more efficient and transparent for verification and validation.

Finally, since both approaches being developed at PSI to derive the loading curves for the Swiss reactor spent fuel assemblies differ from the more conventional "conservative" approaches (see Section 4 "Group Discussions" of Reference [48]), it is definitely an advantageous situation at PSI that two different methods can be compared, and in such way, the resulting loading curves can be verified with high confidence. In line with this, the results of the work in Reference [46] and of the BUCSS-R assessments presented in this paper have been compared, as demonstrated in Figure 13. For a consistent comparison, the BUCSS-R results were reevaluated for this particular graph, using the same conditions as were applied in Reference [46]: no uncertainty components and no bounding burnup profiles are taken into account in this test exercise, i.e., only the reference/nominal burnup profiles are considered. As can be seen, the results are in good agreement. Future work will be focused on the extended and more representative verification studies using both BUCSS-R and CS2M calculation schemes.

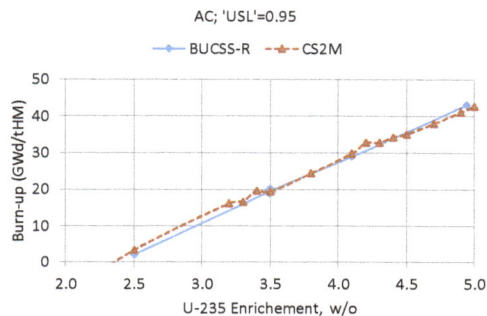

Figure 13. A comparison of the test loading curves obtained with the reference BUCSS-R and the advanced CS2M schemes (for the same simplified conditions).

8. Conclusions

This work was oriented towards deriving the preliminary loading curves for the SNF disposal canisters to be used in a Swiss deep geological repository currently being planned by Nagra. This paper contains a description of the applied methodology and presents the main findings and results.

In the initial stage of the study, bounding fuel type analyses using representative fuel assemblies operated in the Swiss PWR plant KKG were performed. The UO_2, MOX and ERU fuels were analysed using the highest enrichments used to date and operated to the highest burnups to properly assess their behaviour. As detailed as possible, intra-assembly SNF compositions have been used in the MCNP criticality calculations based on the results of calculations corresponding to realistic cycle operating conditions extracted from the validated PSI core management models. The case of PWR UO_2 has been confirmed to be the most limiting, and consequently, the loading curve analysis was done for this fuel.

The methodology applied for the loading curve derivation integrated the outcome of the standard PSI criticality safety validation procedure with the estimated penalties on the computed due to the uncertainties in the nuclear data, fuel assembly design parameters and operating conditions as well as radiation-induced changes in the fuel assembly geometry. Furthermore, bounding axial and radial burnup profiles and the most reactive fuel loading configuration in the canister, in terms of penalising radial tilt, were taken into account accordingly.

The loading curves obtained for the reference disposal canister (as illustrated in Figure 11) show what minimum average fuel assembly burnup is required for the given original fuel enrichment of fresh fuel assemblies so that the k_{eff} of the canister would comply with the imposed criticality safety criterion.

The loading curves presented were obtained for a reference disposal canister design provided by Nagra in the course of the project. However, Nagra is exploring various options for the selection of the materials and design concepts for the canister, which may require a reevaluation of the loading curves. A preliminary study demonstrating the plausibility of the alternative canister designs can be found in Reference [2].

The loading curves presented in this work show that the AC credit approach would not be sufficient to meet the USL criticality safety criterion for a non-mixed loading with fuel with an initial enrichment above approximately 3.5 w/o, while the AC+FP approach justifies the applicability of the canister design considered for the safe disposal of spent nuclear fuel with all the existing enrichments with required minimum burnups.

A postulated hypothetical case consisting of FAs with 5 w/o initial enrichment and relatively low burnup would be the only exception to fitting the loading criteria; however, this case belongs only to a theoretical last core discharge, where, in reality, a lower enriched fuel should be employed. The option of some canisters being not fully loaded could be an alternative approach but would be challenging from another point of view (logistics and cost optimisation). Theoretically, mixed configurations of SNF with different burnup levels could also be a solution for the most problematic cases.

The loading curves must be treated as preliminary since there is still room for improvement in the assessment of different components of Equations (1)–(3). However, valuable findings have already been obtained concerning the identification of the most problematic scenarios and loading schemes, which will serve as guiding information for the next phases of the research and development at PSI and Nagra, in particular for the optimisation of the canister design and the mixed burnup/fuel assembly design loading schemes.

Among the topics for further improvement of the BUCSS-R methodology (with respect to only the criticality safety and burnup credit assessments), other aspects can be proposed for consideration, such as the extension of the analysed FA sample to yield 95%/95% tolerance bounds for the loading curves (i.e., to safely cover the potential uncertainties from the operating condition variations); a refinement of the treatment of the uncertainties with respect to the burnup axial and radial profiles, allowing a consistent "Total Monte Carlo" assessment ("seamless" calculations of both depletion/decay and criticality models using the same ND perturbation factors, as outlined in Figure 12); and the assessment of long-term scenarios with canister and fuel evolution and degradation. In addition, an improvement of the nuclear criticality safety criterion based on a detailed analysis of the nuclear data-related uncertainties and correlations, which can be done, e.g., with NUSS [10,49], has been proposed in References [10,29] and should be further explored.

In parallel, assessments of an alternative option for the currently used BUCSS-R methodology are ongoing independently at PSI, for which Studsvik's code "SNF" is integrated into the CS_2M scheme [46] to substitute the BOHR component of the BUCSS-R scheme. At present, the new approach is basically used for the verification of the base BOHR/BUCSS-R results; however, if confirmed to be more efficient, this new calculation scheme can replace the original one in the future studies at PSI.

Author Contributions: Conceptualization, A.V., H.F. and S.C.; data curation, H.F. and S.C.; formal analysis, A.V., J.H. and D.R.; funding acquisition, A.V., H.F. and S.C.; investigation, J.H. and D.R.; methodology, A.V., J.H., M.P., D.R. and H.F.; project administration, A.V., H.F. and S.C.; resources, H.F.; software, J.H., M.P., D.R. and H.F.; supervision, A.V. and H.F.; validation, A.V., J.H., M.P. and D.R.; visualization, A.V., J.H., M.P. and D.R.; writing—original draft, A.V., J.H., M.P., D.R. and S.C.; writing—review and editing, A.V., J.H., M.P., D.R., H.F. and S.C.

Funding: This research was partially funded by PSI and Nagra within the framework of bilateral collaboration on the "BUCSS-R" project.

Acknowledgments: The authors express their gratitude to former Nagra collaborator L. Johnson and former LRS/NES/PSI head M. Zimmermann for their contributions to initiating the joint PSI/Nagra "BUCSS-R" research project. The authors are also grateful to L. McKinley/Nagra for the kindly provided comments and English grammar check.

Conflicts of Interest: The authors declare no conflict of interest.

References

1. Kühl, H.; Johnson, L.H.; McGinnes, D.F. *Criticality and Canister Shielding Calculations for Spent Fuel Disposal in a Repository in Opalinus Clay*; Nagra Technical Report NAB 12-42; Nationale Genossenschaft für die Lagerung radioaktiver Abfälle: Wettingen, Switzerland, 2012.
2. Gutierrez, M.; Caruso, S.; Diomidis, N. Criticality and Shielding Assessment of Canister Concepts for the Disposal of Spent Nuclear Fuel. In Proceedings of the International High-Level Radioactive Waste Management Conference, Charlotte, NC, USA, 9–13 April 2017.
3. Scaglione, J.M.; Wagner, J.C. Burnup Credit Approach Used in the Yucca Mountain License Application. In Proceedings of the International Workshop on Advances in Applications of Burnup Credit for Spent Fuel Storage, Transport, Reprocessing, and Disposition, Córdoba, Spain, 27–30 October 2009.
4. Ewing, R.C.; Weber, W.J. Nuclear-waste management and disposal. In *Fundamentals of Materials for Energy and Environmental Sustainability*; Ginley, D.S., Cahen, D., Eds.; Cambridge University Press: Cambridge, UK, 2011; pp. 178–193.

5.   Gmal, B.; Kilger, R.; Thiel, J. Issues and future plans of burnup credit application for final disposal. In Proceedings of the Technical Meeting "Advances in Applications of Burnup Credit to Enhance Spent Fuel Transportation, Storage, Reprocessing and Disposition", Storage, UK, 29 August–2 September 2005; pp. 129–138.

6.   Leray, O.; Rochman, D.; Grimm, P.; Ferroukhi, H.; Vasiliev, A.; Hursin, M.; Perret, G.; Pautz, A. Nuclear data uncertainty propagation on spent fuel nuclide compositions. *Ann. Nucl. Energy* **2016**, *94*, 603–611. [CrossRef]

7.   Grimm, P.; Hursin, M.; Perret, G.; Siefman, D.; Ferroukhi, H. Analysis of reactivity worths of burnt PWR fuel samples measured in LWR-PROTEUS Phase II using a CASMO-5 reflected-assembly model. *Prog. Nucl. Energy* **2017**, *101*, 280–287. [CrossRef]

8.   Rochman, D.; Vasiliev, A.; Ferroukhi, H.; Janin, D.; Seidl, N. Best Estimate Plus Uncertainty Analysis for the ^{244}CM Prediction in Spent Fuel Characterization. In Proceedings of the Best Estimate Plus Uncertainty International Conference, Lucca, Italy, 13–18 May 2018.

9.   Leray, O.; Ferroukhi, H.; Hursin, M.; Vasiliev, A.; Rochman, D. Methodology for core analyses with nuclear data uncertainty quantification and application to Swiss PWR operated cycles. *Ann. Nucl. Energy* **2017**, *110*, 547–559. [CrossRef]

10.   Vasiliev, A.; Rochman, D.; Pecchia, M.; Ferroukhi, H. On the options for incorporating nuclear data uncertainties in criticality safety assessments for LWR fuel. *Ann. Nucl. Energy* **2018**, *116*, 57–68. [CrossRef]

11.   Pecchia, M.; Vasiliev, A.; Ferroukhi, H.; Pautz, A. Updated Validation of the PSI Criticality Safety Evaluation Methodology using MCNPX2.7 and ENDF/B-VII.1. In Proceedings of the ANS Topical Meeting on Reactor Physics PHYSOR-2014, Kyoto, Japan, 28 September–3 October 2014.

12.   Patel, R.; Punshon, C.; Nicholas, J. *Canister Design Concepts for Disposal of Spent Fuel and High Level Waste*; Nagra Technical Report NTB 12-06; Nationale Genossenschaft für die Lagerung radioaktiver Abfälle: Wettingen, Switzerland, 2012.

13.   Agrenius, L. Criticality Safety Calculations of Disposal Canisters. SKB Public Report–1193244 4.0. 2010. Available online: http://www.mkg.se/uploads/Arende_SSM2011_37/SSM201137_011_bilaga2_Criticality_safety_calculations_of_disposal_canisters_2.pdf (accessed on 1 February 2019).

14.   Rhodes, J.; Smith, K.; Lee, D. CASMO-5 Development and Applications. In Proceedings of the ANS Topical Meeting on Reactor Physics PHYSOR-2006, Vancouver, BC, Canada, 10–14 September 2006.

15.   Scandpower, S. SIMULATE-3. Advanced Three-Dimensional Two-Group Reactor Analysis Code, SSP-95/15–Rev 4. 1995. Available online: https://www.studsvik.com/ (accessed on 1 February 2019).

16.   Herrero, J.J.; Pecchia, M.; Ferroukhi, H.; Canepa, S.; Vasiliev, A.; Caruso, S. Computational Scheme for Burnup Credit applied to Long Term Waste Disposal. In Proceedings of the Nuclear Criticality Safety International Conference, ICNC 2015, Charlotte, NC, USA, 13–17 September 2015.

17.   Leppänen, J. Serpent—A Continuous-energy Monte Carlo Reactor Physics Burnup Calculation Code. In *User's Manual*; VTT Technical Research Centre of Finland: Espoo, Finland, 18 June 2015.

18.   Herrero, J.J.; Vasiliev, A.; Pecchia, M.; Ferroukhi, H.; Caruso, S. Review calculations for the OECD/NEA Burn-up Credit Criticality Safety Benchmark. *Ann. Nucl. Energy* **2016**, *87*, 48–57. [CrossRef]

19.   Pelowitz, D.B. *MCNP6 User's Manual*; Code Version 6.1.1beta; Los Alamos National Laboratory: Los Alamos, NM, USA, 2014.

20.   Ferroukhi, H.; Hofer, K.; Hollard, J.-M.; Vasiliev, A.; Zimmermann, M. Core Modelling and Analysis of the Swiss Nuclear Power Plants for Qualified R & D Applications. In Proceedings of the International Conference on the Physics of Reactors, PHYSOR'08, Interlaken, Switzerland, 14–19 September 2008.

21.   Pecchia, M.; Canepa, S.; Ferroukhi, H.; Herrero, J.; Vasiliev, A.; Pautz, A. COMPLINK: A Versatile Tool for Automatizing the Representation of Fuel Compositions in MCNP Models. In Proceedings of the International Conference on Nuclear Criticality Safety ICNC-2015, Charlotte, NC, USA, 13–17 September 2015.

22.   Chadwick, M.B.; Obložinský, P.; Herman, M.; Greene, N.M.; McKnight, R.D.; Smith, D.L.; Young, P.G.; MacFarlane, R.E.; Hale, G.M.; Frankle, S.C.; et al. ENDF/B-VII.0 next generation evaluated nuclear data library for nuclear science and technology. *Nucl. Data Sheets* **2006**, *107*, 2931–3060. [CrossRef]

23.   Chadwick, M.B. ENDF/B-VII.1 nuclear data for science and technology: Cross sections, covariances, fission product yields and decay data. *Nucl. Data Sheets* **2011**, *112*, 2887–2996. [CrossRef]

24.   Nuclear Energy Agency (OECD/NEA). *Burn-Up Credit Criticality Safety Benchmark—Phase VII*; UO2 Fuel: Study of Spent Fuel Compositions for Long-Term Disposal; NEA No. 6998; Nuclear Energy Agency: Paris, France, 2012; ISBN 978-92-64-99172-9.

25. Brown, F. A Review of Best Practices for Monte Carlo Criticality Calculations. In Proceedings of the American Nuclear Society Nuclear Criticality Safety Topical Meeting, Richland, WA, USA, 13–17 September 2009.

26. American Nuclear Society (ANS). ANSI/ANS-8.17-2004: American National Standard. In *Criticality Safety Criteria for the Handling, Storage, and Transportation of LWR Fuel Outside Reactors*; American Nuclear Society: La Grange Park, IL, USA, 2004.

27. Nuclear Energy Agency (OECD/NEA/CSNI). *Proceedings of the Workshop "Operational and Regulatory Aspects of Criticality Safety" held in Albuquerque, NM, USA, 19–21 May 2015*; Nuclear Energy Agency: Paris, France, 2016.

28. U.S. Nuclear Regulatory Commission (U.S. NRC). *Recommendations on the credit for cooling time in PWR burnup credit analysis*; NUREG/CR-6781; U.S. Nuclear Regulatory Commission: Washington, DC, USA, 2003.

29. Vasiliev, A.; Herrero, J.; Rochman, D.; Pecchia, M.; Ferroukhi, H.; Caruso, S. Criticality Safety Evaluations for the Concept of Swiss PWR Spent Fuel Geological Repository. In Proceedings of the Best Estimate Plus Uncertainty International Conference, BEPU 2018, Lucca, Italy, 13–18 May 2018.

30. U.S. Nuclear Regulatory Commission (U.S. NRC). *Guide for Validation of Nuclear Criticality Safety Calculational Methodology*; NUREG/CR-6698; U.S. Nuclear Regulatory Commission: Washington, DC, USA, 2001.

31. U.S. Department of Energy (DOE). Criticality Safety Good Practices Program. In *Guide for DOE Nonreactor Nuclear Facilities*; DOE G 421.1-1; US Department of Energy: Washington, DC, USA, 2009.

32. Mennerdahl, D. *Review of the Nuclear Criticality Safety of SKB's Licensing Application for a Spent Nuclear Fuel Repository in Sweden*; Technical Note-2012:65; Swedish Radiation Safety Authority: Stockholm, Sweden, 2012.

33. U.S. Nuclear Regulatory Commission (U.S. NRC). *Recommendations for Addressing Axial Burnup in PWR Burnup Credit Analysis*; NUREG/CR-6801; U.S. Nuclear Regulatory Commission: Washington, DC, USA, 2003.

34. Neuber, J.C. Evaluation of Axial and Horizontal Burnup Profiles. In Proceedings of the Technical Committee Meeting, Implementation of Burnup Credit in Spent Fuel Management Systems, Vienna, Austria, 10–14 July 2000.

35. Rochman, D.; Bauge, E.; Vasiliev, A.; Ferroukhi, H. Correlation ν_p-σ-χ in the fast neutron range via integral information. *EPJ Nucl. Sci. Technol.* **2017**, *3*, 1–8. [CrossRef]

36. Rochman, D.; Bauge, E.; Vasiliev, A.; Ferroukhi, H.; Perret, G. Nuclear data correlation between different isotopes via integral information. *EPJ Nucl. Sci. Technol.* **2018**, *4*, 1–10. [CrossRef]

37. Rochman, D.; Bauge, E.; Vasiliev, A.; Ferroukhi, H.; Pelloni, S.; Koning, A.J.; Sublet, J.C. Monte Carlo nuclear data adjustment via integral information. *Eur. Phys. J. Plus* **2018**, *133*, 1–23. [CrossRef]

38. Pecchia, M.; Vasiliev, A.; Ferroukhi, H.; Pautz, A. Criticality Safety Evaluation of a Swiss wet storage pool using a global uncertainty analysis methodology. *Ann. Nucl. Energy* **2015**, *83*, 226–235. [CrossRef]

39. Herrero, J.J.; Rochman, D.; Leray, O.; Vasiliev, A.; Pecchia, M.; Ferroukhi, H.; Caruso, S. Impact of nuclear data uncertainty on safety calculations for spent nuclear fuel geological disposal. In Proceedings of the Nuclear Data 2016 Conference, Bruges, Belgium, 11–16 September 2016.

40. Leray, O.; Grimm, P.; Hursin, M.; Ferroukhi, H.; Pautz, A. Uncertainty Quantification of Spent Fuel Nuclide Compositions due to Cross Sections, Decay Constants and Fission Yields. In Proceedings of the ANS Topical Meeting on Reactor Physics PHYSOR-2014, Kyoto, Japan, 28 September–3 October 2014.

41. Vasiliev, A.; Rochman, D.; Zhu, T.; Pecchia, M.; Ferroukhi, H.; Pautz, A. Towards Application of Neutron Cross-Section Uncertainty Propagation Capability in the Criticality Safety Methodology. In Proceedings of the International Conference on Nuclear Criticality Safety ICNC-2015, Charlotte, NC, USA, 13–17 September 2015.

42. Radulescu, G.; Gauld, I.C.; Ilas, G.; Wagner, J.C. An Approach for Validating Actinide and Fission Product Burnup Credit Criticality Safety Analyses-Isotopic Composition Predictions. *Nucl. Technol.* **2014**, *188*, 154. [CrossRef]

43. Yun, H.; Park, K.; Choi, W.; Hong, S.G. An efficient evaluation of depletion uncertainty for a GBC-32 dry storage cask with PLUS7 fuel assemblies using the Monte Carlo uncertainty sampling method. *Ann. Nucl. Energy* **2017**, *110*, 679–691. [CrossRef]

44. Rochman, D.; Leray, O.; Hursin, M.; Ferroukhi, H.; Vasiliev, A.; Aures, A.; Bostelmann, F.; Zwermann, W.; Cabellos, O.; Diez, C.J.; et al. Nuclear Data Uncertainties for Typical LWR Fuel Assemblies and a Simple Reactor Core. *Nucl. Data Sheets* **2017**, *139*, 1–76. [CrossRef]

Materials **2019**, *12*, 494

45. Cullen, D.E. *Program ENDF2C: Convert ENDF Data to Standard Fortran, C and C++ Format*; IAEA-NDS-217; The International Atomic Energy Agency: Vienna, Austria, May 2015.

46. Rochman, D.; Vasiliev, A.; Ferroukhi, H.; Pecchia, M. Consistent criticality and radiation studies of Swiss spent nuclear fuel: The CS2M approach. *J. Hazard. Mater.* **2018**, *357*, 384–392. [CrossRef]

47. Zhu, T.; Vasiliev, A.; Ferroukhi, H.; Rochman, D.; Pautz, A. Testing the sampling-based NUSS-RF tool for the nuclear data–related global sensitivity analysis with Monte Carlo neutronics calculations. *Nucl. Sci. Eng.* **2016**, *184*, 69–83. [CrossRef]

48. The International Atomic Energy Agency (IAEA). Advances in Applications of Burnup Credit to Enhance Spent Fuel Transportation, Storage, Reprocessing and Disposition. Proceedings of a Technical Meeting, London, UK, 29 August–2 September 2005.

49. Vasiliev, A.; Rochman, D.; Pecchia, M.; Ferroukhi, H. Exploring stochastic sampling in nuclear data uncertainties assessment for reactor physics applications and validation studies. *Energies* **2016**, *9*, 1039. [CrossRef]

![materials logo]

Article

Investigating the Durability of Iodine Waste Forms in Dilute Conditions

R. Matthew Asmussen *, Joseph V. Ryan, Josef Matyas, Jarrod V. Crum, Joelle T. Reiser,
Nancy Avalos, Erin M. McElroy, Amanda R. Lawter and Nathan C. Canfield

Energy and Environment Directorate, Pacific Northwest National Laboratory, Richland, WA 99352, USA;
joe.ryan@pnnl.gov (J.V.R.); josef.matyas@pnnl.gov (J.M.); jarrod.crum@pnnl.gov (J.V.C.);
Joelle.t.reiser@pnnl.gov (J.T.R.); Nancy.avalos@pnnl.gov (N.A.); erin.mcelroy@wsu.edu (E.M.M.);
amanda.lawter@pnnl.gov (A.R.L.); nathan.canfield@pnnl.gov (N.C.C.)
* Correspondence: matthew.asmussen@pnnl.gov; Tel.: +1-150-9371-7223

Received: 27 December 2018; Accepted: 14 February 2019; Published: 26 February 2019

Abstract: To prevent the release of radioiodine during the reprocessing of used nuclear fuel or in the management of other wastes, many technologies have been developed for iodine capture. The capture is only part of the challenge as a durable waste form is required to ensure safe disposal of the radioiodine. This work presents the first durability studies in dilute conditions of two AgI-containing waste forms: hot-isostatically pressed silver mordenite (AgZ) and spark plasma sintered silver-functionalized silica aerogel (SFA) iodine waste forms (IWF). Using the single-pass flow-through (SPFT) test method, the dissolution rates respective to Si, Al, Ag and I were measured for variants of the IWFs. By combining solution and solid analysis information on the corrosion mechanism neutral-to-alkaline conditions was elucidated. The AgZ samples were observed to have corrosion preferentially occur at secondary phases with higher Al and alkali content. These phases contained a lower proportion of I compared with the matrix. The SFA samples experienced a higher extent of corrosion at Si-rich particles, but an increased addition of Si to the waste led to an improvement in corrosion resistance. The dissolution rates for the IWF types are of similar magnitude to other Si-based waste form materials measured using SPFT.

Keywords: iodine; waste form; corrosion; microscopy; silver iodide

---

1. Introduction

In the reprocessing of used nuclear fuel, radioiodine will be released, primarily during the dissolution of the fuel [1]. A portion of this iodine is ^{129}I with a half-life of 15.7 million years; to prevent discharge of this long-lived radionuclide, the released iodine needs to be captured in the off-gas management system of the reprocessing facility. Multiple approaches to removing the iodine from the off-gas system (which contains large amounts of water and NO_X) and can be grouped into: (A) wet scrubbing methods such as Mercurex, Iodox, electrolytic scrubbing, and alkaline scrubbing [2,3]; and (B) solid sorbent capture including resins [4], carbon-based materials [5–7], metal organic frameworks [8,9], zeolites [10–12], silica [13] and aerogels [14–16]. The wet scrubbing processes would all require a secondary process(es) for the iodine-loaded product to be converted to a waste form such as grouting or vitrification. One of the primary advantages of solid sorbents is their potential to be readily transformed into a final waste form, through either direct post-processing in a canister or densification.

The presence of silver (Ag) in solid sorbents can enhance iodine capture through the generation of silver iodide (AgI) in the material. AgI is widely considered of as a desirable form of iodine for disposal because it has a low solubility (AgI $K_{sp} = 8 \times 10^{-17}$) [17]. However, the stability of AgI can be impacted by its local environment, as its dissolution can be highly affected by redox conditions [18], increasing pH or the presence of sulfide [19]. Placing the AgI within a durable matrix can ensure further protection in long-term disposal.

While many solid sorbents with and without Ag have been developed for iodine capture, the consolidated waste form development for the materials and their associated durabilities have been sparsely investigated. The durabilities of these candidate waste forms need to be understood to facilitate predictions of their behavior over long disposal time frames in a repository.

Two of the most technologically mature iodine waste forms (IWF) developed to date are silver-exchanged zeolites, such as silver mordenite (AgZ) [20–22], and silver-functionalized silica aerogels (SFA) [14,23,24]. Silver exchanged zeolites, specifically the mordenite form that has higher Si:Al ratio compared with faujasite zeolites [25], have been researched for iodine capture in the US. Reduced Ag (Ag^0) is present in the AgZ to reduce iodine (I_2) to iodide (I^-) and the eventual formation of AgI while in use [20]. The loaded AgZ can then be converted to a final waste form through post-processing in a canister [22,26]. These demonstrations have been performed using hot isostatic pressing (HIP) and hot uniaxial pressing (HUP) to create a consolidated AgZ in steel canisters.

SFAs have been developed in the last decade as a moderate specific surface area (~150 m^2/g) material for iodine capture. An aerogel backbone can be thiolated and functionalized with Ag^0 nanoparticles to create the SFAs [23]. The SFAs are capable of high iodine loadings (up to 40 wt %), are stable in expected off-gas operating conditions (e.g., high humidity and NO_x) and can be directly densified to a final waste form using HIP or spark plasma sintering (SPS). In both processes, the application of heat collapses the aerogel backbone, reducing its volume and eliminating void spaces to create a final, high density waste form. This densification process has been demonstrated previously [23].

This work presents the first study of the corrosion behavior of AgZ and SFA based IWF in aqueous environments using the single-pass flow-through (SPFT) technique [27]. The consolidated AgZ and densified SFA samples are comprised of multi-component microstructures, which may lead to heterogeneous dissolution of the waste form. To assess such behavior, monolithic samples of each material were evaluated to track corrosion using solution and solid analyses. The materials were evaluated pre- and post-corrosion with optical microscopy, optical profilometry, electron microscopy, and X-ray diffraction (XRD). This study will help inform further development of IWF to improve durability and the data within can be used to develop long-term predictive models for iodine releases from candidate waste forms. It should be made clear that the IWF samples used in this study have not yet been optimized for durability, but can be used as comparisons for any future assessments of IWF durability.

2. Materials and Methods

2.1. Materials

The AgZ samples were prepared at Oak Ridge National Laboratory [26]. The base zeolite used was Ionex Type Ag 900 E16 from Molecular Products and had a chemical composition of $Ca_8(Al_8Si_{40}O_{96})\cdot24H_2O$ [22] with ~9 wt % Ag content. A steel cylindrical canister (25 mm" diameter, 75 mm" tall) was filled with AgZ and HIPed for 3 h at 175 MPa. Three samples were included in this testing: AgZ 1-3 (HIPed at 700 °C at 175 MPa for 3 h, no iodine loading), AgZ 1-7 (HIPed at 525 °C at 175 MPa for 3 h, loaded with iodine) and AgZ 1-8 (HIPed at 700 °C at 175 MPa for 3 h, loaded with iodine); the numeration sequence corresponds to the previously reported sample preparation [28]. From the canisters, two horizontal pucks were sectioned from the ingot (2 mm thick) to produce a flat surface encapsulated in a steel ring. Visual images of the samples are shown in Figure 1a–c. The chemical makeup of the AgZ samples was determined using energy dispersive X-ray spectroscopy (EDS, Bruker Quantax 6 | 60; Bruker Nano GmbH, Berlin, Germany) by taking the average composition of a minimum of four 250 μm × 350 μm areas on the sample and these are listed in Table 1. Because there were only two small AgZ samples, and to keep the integrity of the samples, no digestion for chemical composition was conducted.

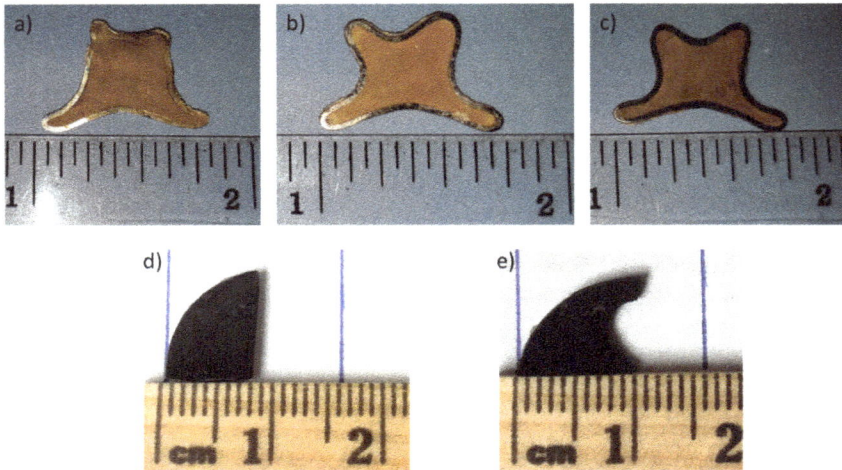

Figure 1. Photographs showing the samples used in this study: (**a**) AgZ 1-3; (**b**) AgZ 1-7; (**c**) AgZ 1-8; (**d**) SPS-1; and (**e**) SPS-2.

Table 1. Composition of the AgZ samples used in this study determined using the average area scan from multiple EDS maps. The standard deviation is representative of the multiple areas imaged to determine the average composition.

Sample	AgZ 1-3		AgZ 1-7		AgZ 1-8	
Element	wt %	St.Dev	wt %	St.Dev	wt %	St.Dev
Ag	8.70	2.21	10.94	1.74	9.78	1.57
I	0.00	0.00	6.02	1.34	4.79	1.52
O	43.10	2.02	37.84	2.96	39.85	2.23
Na	0.30	0.10	0.27	0.19	0.22	0.21
Mg	0.62	0.11	0.53	0.11	0.38	0.24
Al	6.48	0.75	6.18	0.90	6.02	0.59
Si	34.64	2.89	33.73	3.59	33.85	3.07
K	0.66	0.17	0.70	0.25	0.53	0.24
Ca	0.87	0.08	0.86	0.15	0.74	0.23
Fe	1.20	0.86	1.35	1.05	1.45	0.96
Others	3.42	-	1.57	-	2.39	-

The spark plasma sintered (SPS) SFA samples were prepared using SFA fabricated at Pacific Northwest National Laboratory using a commercially available silica aerogel from United Nuclear (Laingbrugh, MI, USA). The as-received SFA materials were functionalized in-house using the method reported previously [29]. Two samples were used in the study and given the designations of SPS-1 and SPS-2. SPS-1 was densified without alterations to the materials while the SPS-2 sample included an additional 20 wt % of raw SFA added prior to the SPS process. The samples (~5 g) were sintered in a graphite die set and heated to 1200 °C (ramp of 100 °C/min) under Ar atmosphere. The temperature was held at 1200 °C for 30 min at 70 MPa and allowed to cool to room temperature under Ar atmosphere until below 400 °C. The final waste form samples to be used in the testing are shown in Figure 1d,e. Because there was only a single sample for each condition, no digestion for chemical composition was possible. The compositions of the SPS-SFA samples were determined from the original SFA material prior to densification and are given in Table 2. The densified SFA materials are highly sensitive to electron beam exposure and lose I with increased exposure time. A difference between the composition measured with EDS for SFA and the actual composition has been observed in a previous study [23]. Thus, EDS compositions were not used for the SFA in this work.

Table 2. Composition of the SFA SPS samples based on the initial composition of the SFA material. The "Others" is comprised primarily of oxygen and minor species (e.g., Fe).

Sample	SPS-1	SPS-2
Element	wt %	wt %
Ag	24.9	19.9
I	30.0	24.0
Si	16.8	33.0
S	0.7	0.5
Others	27.6	22.1

One face of the each sample was polished prior to being exposed to the SPFT test. For the AgZ samples, the faces polished for each of the samples were adjacent to each other when cut. The samples were placed on a rotating polishing unit at 15 μm SiC for 20 min with 20 lbs of force at 240 rpm, followed by successive 5 min sets at 9 μm and 3 μm. Following these steps, the samples were finalized on a vibratory polisher with 1-μm SiC followed by 0.05-μm colloidal silica for 4 h each. A final ethanol rinse was used to remove any remaining debris. The opposite face of the samples was then masked using room temperature vulcanizing (RTV) silicone (Locktite®). The masking was done to limit damage to the samples as there were only a few unique samples.

2.2. Single-Pass Flow-Through Testing

Corrosion testing of the samples was performed with the SPFT technique following ASTM Method C1662-17 [27]. In general, the SPFT test utilizes a flowing solution through a saturated, sealed vessel containing a sample and the effluent solution from the reactor was monitored. The monolithic samples were placed on a cage within a 60 mL reaction vessel made of high-density polyethylene (HDPE, Savillex) with the polished face directed upward. The flow rate was provided by a syringe pump (Nordgren-Khloen, V6 syringe drive pump, Las Vegas, NV, USA) and was targeted at 20 mL/day to provide dilute conditions (<5 mg/L for species in solution) yet keep the concentrations within a measurable range (above instrument detection limits). All experiments were performed in an oven, in open atmosphere (sealed reactor), at 90 °C. Effluent samples were collected in polytetrafluorethylene (PTFE) bottles and flow rates were determined gravimetrically. Only a single test was performed at pH 7 and pH 11 to preserve the unique samples due to the loss of material experienced during the test and the post-test polishing procedure. A repeat experiment at shorter duration (17 days compared to 36 days) was performed on the AgZ samples at pH 9 to ensure reproducibility of the SPFT technique.

At the conclusion of the test, flow to the reactors was stopped and the samples were removed and rinsed three times with double deionized water (18.2 MΩ·cm) and three times with anhydrous ethanol (98%, Fisher Scientific). Solutions buffered at pH (at room temperature, RT) 7 and pH 9 were made with 0.05 M tris(hydroxylmethyl)aminomethane (TRIS, Fisher Scientific) adjusted to the desired pH using HNO_3, while solutions at pH (RT) 11 were a 0.001 M LiOH + 0.01 M LiCl solution.

2.3. Post Analysis

Concentrations of the analytes in the collected effluents were measured using inductively coupled plasma (ICP) mass spectroscopy (Thermo X-Series 2, Waltham, MA, USA) for total I (detection limit of 1.26 μg/L) and ICP optical emission spectroscopy (Perkin Elmer Optima 8300 DV, Perkin Elmer, Shelton, CT, USA) for Si, Ag and Al (with detection limits of 54.6 μg/L, 17.9 μg/L and 15.6 μg/L, respectively).

The sample surfaces were imaged using scanning electron microscopy (SEM) and elemental distributions were determined using EDS. Attempts were made to correlate the same area on the sample surface both before and after corrosion. Images were collected at 70× and 250× magnifications. SEM analyses were performed with a JSM-7001F microscope (JEOL USA, Inc., Peabody, MA, USA) with an XFlash 6|60 EDS Si-drift detector (Bruker) for elemental mapping and spot analysis.

The samples were also characterized post-corrosion for any changes in their structure using X-ray diffraction (XRD). The samples were not altered prior to the XRD measurements; they were analyzed as intact coupons. The XRD patterns were collected with a Bruker D8 Advance XRD system (Bruker AXS, Tuscon, AZ, USA) equipped with a Cu target (Kα1 = 0.15406 nm) over a scan range of 5° 2θ to 75° 2θ using a step size of 0.015° 2θ and a hold time of 4 s per step. The scans were analyzed with TOPAS (v4.2) whole pattern fitting software according to the fundamental parameters approach [30]. Structure patterns were selected from the Inorganic Crystal Structure Database (release 2013) with unit cell dimensions refined in the fitting process of each pattern.

The topographical evolution of the surface following corrosion was observed using optical profilometry (OP) on a Bruker GTK profilometer with a 5× or 50× lens before and after corrosion.

3. Results

3.1. Pre-Corrosion Characterization

The initial microstructures of the AgZ samples prior to corrosion were observed with SEM and EDS. The multiphase structure of AgZ 1-3 is shown in the SEM micrograph in Figure 2a. The microstructure contained a continuous matrix and large isolated secondary phases (the lighter grey regions in the SEM image) within the matrix. The Ag particles (Figure 2b) were present in both the matrix and in the secondary phases. No iodine was present in AgZ 1-3 (Figure 2c). The secondary phases present within the matrix contained high amounts of Al and K (i.e., Figure 2d,e, respectively). Si comprised the matrix phase (Figure 2f). The larger white inclusion in the center of Figure 2a was composed primarily of Zr and S, the origin of which is unknown. An example of a commonly observed secondary phase is highlighted with a white box in Figure 2a. EDS analysis of this location, shown in Table 3, revealed higher amounts of K (1.5 wt %) than the matrix as a whole. This phase also contained Ag inclusions.

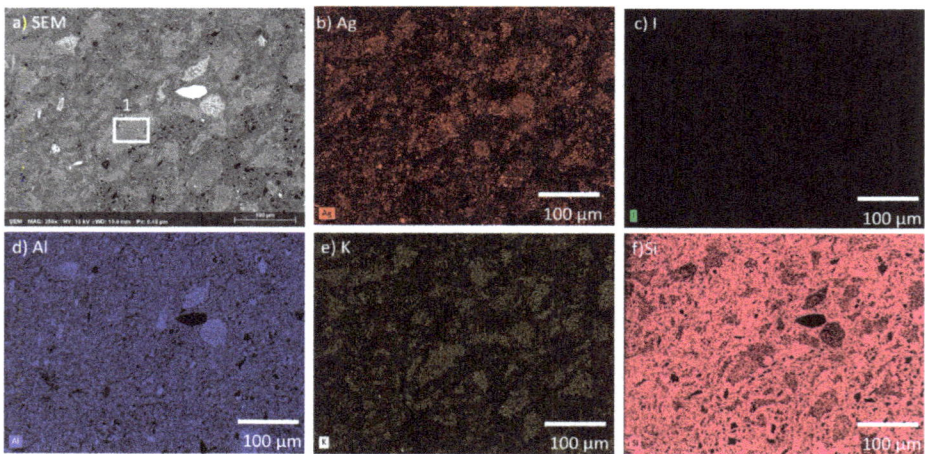

Figure 2. Microstructure of the AgZ 1-3 sample shown by: (**a**) SEM micrograph; and the corresponding EDS maps of: (**b**) Ag; (**c**) I; (**d**) Al; (**e**) K; and (**f**) Si. The large white inclusion in (**a**) is made of Zr and S whose origin are not known. The white box marked "1" is the location where an EDS spot analysis was performed and listed in Table 3.

Table 3. Composition of the features highlighted in the SEM images in Figures 2a and 3a determined with EDS spot analysis.

Image	Figure 2a	Figure 3a	Figure 3a
Location	1	2	3
Element	wt %	wt %	wt %
Ag	5.7	1.8	4.5
I	0	0.2	0.4
O	46.3	41.6	43.0
Na	0.5	3.4	0.2
Mg	0.3	0.08	0.9
Al	6.9	10.9	8.0
Si	37.2	35.5	36.9
K	1.5	5.1	0.9
Ca	0.6	0.2	3.4
Fe	0.3	0.3	0.2
Others	0.7	0.92	1.6

The AgZ 1-7 sample is shown in Figure 3a. AgZ 1-7 had similar features to AgZ 1-3. The two main differences between AgZ 1-3 and AgZ 1-7 were the presence of I (Figure 3c), and more even distribution of Ag in the AgZ 1-7 sample. The I and Ag distribution in the AgZ 1-7 sample were observed to be even with one another and with few discrete Ag particles, such as those observed in the AgZ 1-3 sample. These changes were possibly due to the higher temperature used in the HIP process of the AgZ 1-3 sample. Similar secondary phases of Al and K (Figure 3d,e, respectively) were observed within the widespread Si matrix (Figure 3f). A different type of inclusion was observed in this image being comprised of Fe and Mn, the origin of these species is not known. (EDS not shown). Two common microstructural features are highlighted in the SEM micrograph (Figure 3a). Area #2 was measured to be comprised of higher levels of Na (3.4 wt %), Al (10.9 wt %), and K (5.1 wt %) with lower Ag (1.8 wt %) and I (0.2 wt %) compared to the bulk composition (see Table 3). Area #3 contained higher amounts of Ca (3.4 wt %) and Al (8 wt %) than the bulk.

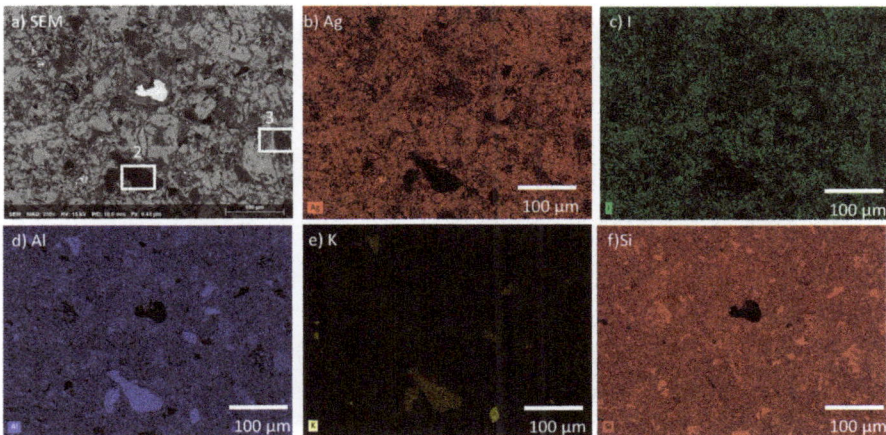

Figure 3. Microstructure of the AgZ 1-7 sample shown by: (**a**) SEM micrograph; and the corresponding EDS maps of: (**b**) Ag; (**c**) I; (**d**) Al; (**e**) K; and (**f**) Si. The large white inclusion in (**a**) is comprised of Fe and Mn whose sources are not known. The white boxes marked "2 and 3" are the locations where an EDS spot analysis was performed and listed in Table 3.

Figure 4 displays the microstructure of the AgZ 1-8 sample. The elemental distributions across the microstructure were similar to AgZ 1-3 including small isolations of Ag that were associated with

I in the AgZ 1-8. Such a distribution can be expected as the AgZ 1-3 and 1-8 sample had identical processing parameters.

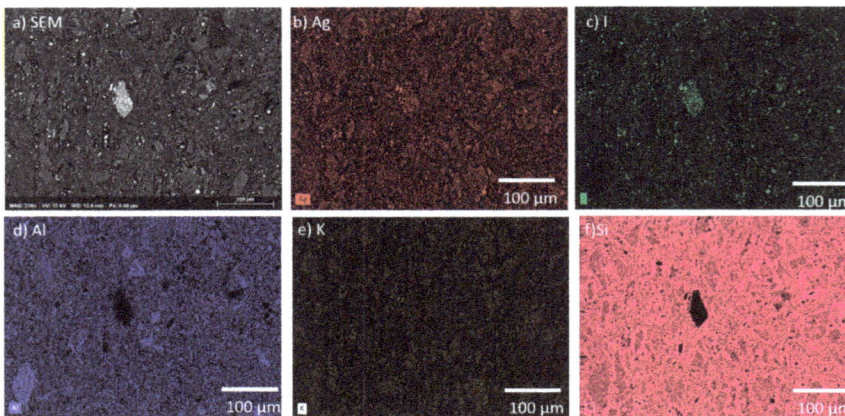

Figure 4. Microstructure of the AgZ 1-8 sample shown by by: (**a**) SEM micrograph; and the corresponding EDS maps of: (**b**) Ag; (**c**) I; (**d**) Al; (**e**) K; and (**f**) Si. The large white inclusion in (**a**) is Fe (source not known) with AgI particles within the structure.

The SFA samples were also comprised of a multiphase microstructure. The SPS-1 sample can be seen in the SEM micrograph in Figure 5a and large features were observed in the image. The Ag was observed to be sitting on the edges of the large particles and in smaller discrete isolations (see Figure 5b). The I was generally located throughout the sample but not as intimately associated with areas of high Ag (see Figure 5c). The SFA samples contained S from the thiol backbone of the original aerogel and the S was distributed evenly (Figure 5d), except for areas of high Si observed in Figure 5e. The SPS-2 sample, with 20 wt % additional Si added had a similar microstructure to the SPS-1 with a larger coverage of Si-rich particles (see Figure 6) and a more widespread distribution of Ag compared with the SPS-1 sample.

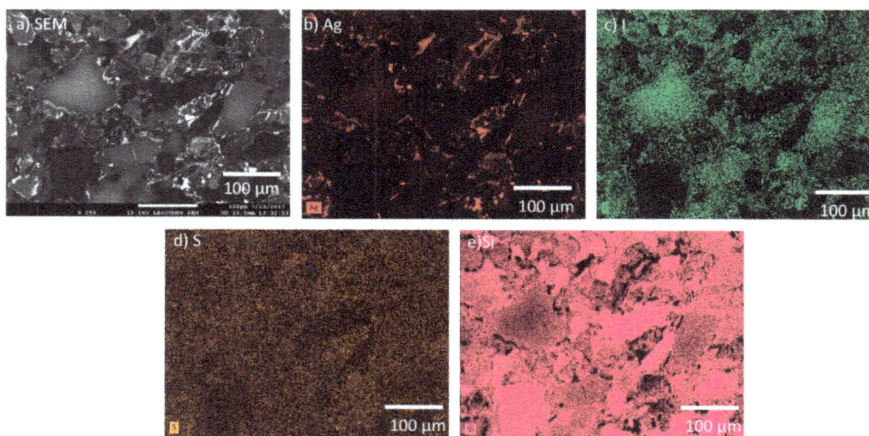

Figure 5. Microstructure of the SPS-1 sample shown by: (**a**) SEM micrograph; and the corresponding EDS maps of: (**b**) Ag; (**c**) I; (**d**) S; and (**e**) Si.

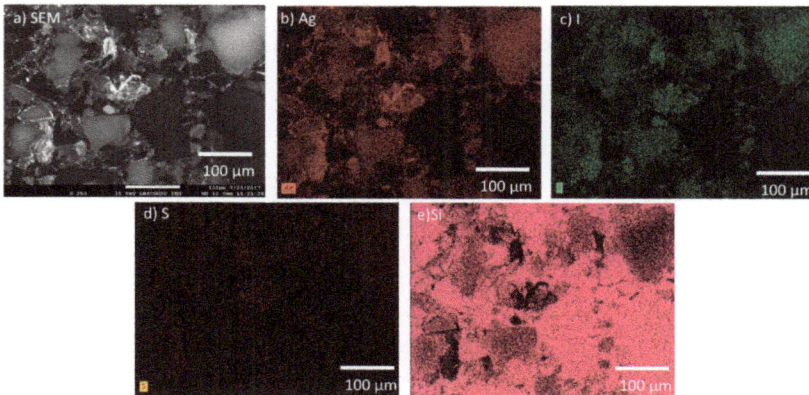

Figure 6. Microstructure of the SFA SPS-2 sample shown by: (**a**) SEM micrograph; and the corresponding EDS maps of: (**b**) Ag; (**c**) I; (**d**) S; and (**e**) Si.

3.2. Corrosion Testing of HIPed Ag Mordenite

SPFT testing was performed on the AgZ samples with inlet solutions at pH 7, pH 9, and pH 11. The errors presented represent the standard deviation of the individual rates measured during the test. All dissolution rates in this work were normalized to the individual sample compositions (Tables 1 and 2) and the dissolution rates of the samples were determined using the following equation:

$$rate = \frac{[X] * V}{SA * t * f_i}$$

where

X is the	concentration of the analyte in the effluent, g/L;
V is the	volume of the collected effluent during the interval, L;
SA is the	surface area of the sample, m^2;
t is the	duration of the interval, day; and
f_i is the	normalization factor based on the mass % of analyte, unitless.

Figure 7 displays the normalized dissolution rates measured for the three AgZ samples in pH 7 solution. For the iodine-free AgZ 1-3 sample (Figure 7a), the dissolution rates were fairly constant over the duration of the test. The decreases observed at 63 days were due to a pump failure. The Ag dissolution rate was higher (0.65 ± 0.07 g/m^2/day average) compared with the Si dissolution rate (0.17 ± 0.01 g/m^2/day average). There was only detectable Al in four samples throughout the duration of the test, the rest falling below the instrument detection limit. Using the instrument detection limit for Al as an input, a maximum rate of 0.06 g/m^2/day can be presumed for the Al dissolution rate. The AgZ 1-7 sample (Figure 7b) showed an average Si dissolution rate of 0.066 ± 0.009 g/m^2/day while the I dissolution rate was lower at 0.015 ± 0.008 g/m^2/day. Neither the Ag nor the Al had measurable concentrations in the effluent and maximum rates of 0.04 g/m^2/day and 0.07 g/m^2/day, respectively, can be assumed using the associated instrument detection limit. The AgZ 1-8 sample, having identical processing parameters to the AgZ 1-3 but with I (Figure 7c), behaved similarly to the iodine-free sample. The Ag dissolution rate was again higher (0.30 ± 0.11 g/m^2/day) than the Si dissolution rate (0.08 ± 0.01 g/m^2/day). The Al was measurable for this sample for the majority of the test with an average dissolution rate of 0.09 ± 0.02 g/m^2/day being measured. The I dissolution rate was measured at 0.005 ± 0.001 g/m^2/day, which was lower than the AgZ 1-7. Near the conclusion of the test, the I dissolution rate increased with time and is possibly due to the corrosion of the surface exposing more AgI that could dissolve. The difference between the AgZ 1-3 and Ag 1-8 Ag and Si dissolution rates compared with the AgZ 1-7 sample may have arisen from Ag particles being present outside of the Si matrix in the 1-3 and 1-8 samples (Figures 2 and 4) and thus more readily attacked.

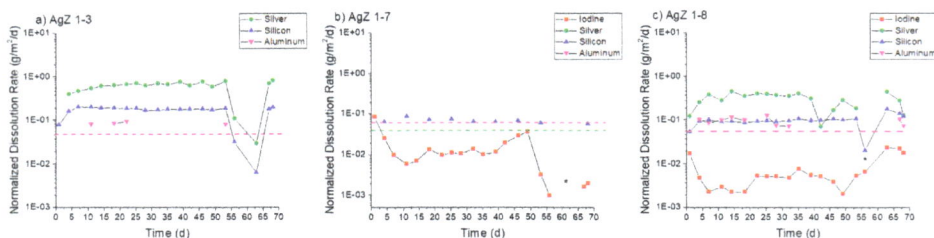

Figure 7. Normalized dissolution rates calculated from SPFT experiments using the AgZ samples at pH 7 for: (**a**) AgZ 1-3 (iodine-free sample); (**b**) AgZ 1-7; and (**c**) AgZ 1-8. The dashed lines, when present, represent the maximum rate for samplings where the analyte concentration was below the detection limit of the instrument and a dissolution rate calculated using the detection limit value. The asterisks (*) mark samplings where the flow rate deviated by >10% from the average flow of the test.

Figure 8 presents the AgZ normalized dissolution rates in pH 9 solution. With the increased alkalinity of the test solution measurable analyte concentrates were present in all effluent samples. AgZ 1-3 (Figure 8a) showed a higher Ag dissolution rate (1.16 ± 0.49 g/m^2/day in the 36-day test and 1.01 ± 0.37 g/m^2/day in the 17-day test) compared with the Si dissolution rate (0.34 ± 0.12 g/m^2/day in the 36-day test and 0.19 ± 0.05 g/m^2/day in the 17-day test). The Al dissolution rates were measured to be 0.30 ± 0.13 g/m^2/day (36 day) and 0.30 ± 0.13 g/m^2/day (17 day). The AgZ 1-7 (Figure 8b) showed I dissolution rates of 0.25 ± 0.09 g/m^2/day (36 day) and 0.27 ± 0.08 g/m^2/day (17 day), Ag dissolution rates of 0.14 ± 0.05 g/m^2/day (36 day) and 0.31 ± 0.08 g/m^2/day (17 day), Si dissolution rates of 0.31 ± 0.23 g/m^2/day (36 day) and 0.15 ± 0.04 g/m^2/day (17 day) and Al dissolution rates of 0.14 ± 0.06 g/m^2/day (36 day) and 0.13 ± 0.10 g/m^2/day (17 day). The AgZ 1-8 (Figure 8c) showed I dissolution rates of 0.14 ± 0.06 g/m^2/day (36-day test) and 0.30 ± 0.14 g/m^2/day (17-day test), Ag dissolution rates of 1.32 ± 0.53 g/m^2/day (36 day) and 1.01 ± 0.36 g/m^2/day (17 day), Si dissolution rates of 0.49 ± 0.24 g/m^2/day (36 day) and 0.20 ± 0.08 g/m^2/day (17 day) and Al dissolution rates of 0.39 ± 0.30 g/m^2/day (36 day) and 0.41 ± 0.13 g/m^2/day (17 day). The measured dissolution rates at pH 9 in the 36-day tests and 17-day tests highlight the reproducibility using the SPFT technique. Similar to the pH 7 tests, the AgZ 1-3 and AgZ 1-8 samples showed similar dissolution rates with the rates for Ag being larger than the Si and Al. The AgZ 1-7 sample showed dissolution rates that tracked with one another for all four analytes. The I dissolution rates for AgZ 1-7 and AgZ 1-8 were similar despite the higher Ag dissolution rate for the AgZ 1-8. This would suggest some free Ag is generated at the higher HIP temperature of the AgZ 1-8 sample.

Moving to pH 11 (see Figure 9), an expected increase in overall dissolution of the samples was observed with new trends in the elemental releases. The AgZ 1-3 (Figure 9a) displayed an increase in dissolution rate until >7 days and the values measured beyond this were used to determine the average rates. The AgZ 1-3 sample showed higher Si dissolution rates (1.05 ± 0.21 g/m^2/day) than Ag (0.14 ± 0.05 g/m^2/day), which were different than rates for pH 7 and pH 9. The inversion of the two rates may have been due to the increased solubility of Si and potential decrease in Ag solubility (through formation of Ag$_2$O) with increased alkalinity [31]. The Al dissolution rate was measured to be 0.49 ± 0.18 g/m^2/day. The AgZ 1-7 (Figure 9b) showed a similar trend with a Si dissolution rate of 1.20 ± 0.32 g/m^2/day and an Ag dissolution rate of 0.09 ± 0.02 g/m^2/day. The I dissolution rate was measured to be 0.22 ± 0.02 g/m^2/day and the Al dissolution rate was 0.39 ± 0.18 g/m^2/day. AgZ 1-8 (Figure 9c) showed a Si dissolution rate of 0.99 ± 0.46 g/m^2/day, an Ag dissolution rate of 0.81 ± 0.19 g/m^2/day, an I dissolution rate of 0.06 ± 0.02 g/m^2/day, and an Al dissolution rate of 0.52 ± 0.38 g/m^2/day. The last three sampling of the pH 11 test had a lower flow rate through the reactor and conditions within the reactor may have changed, leading to the stark decreases observed after 14 days.

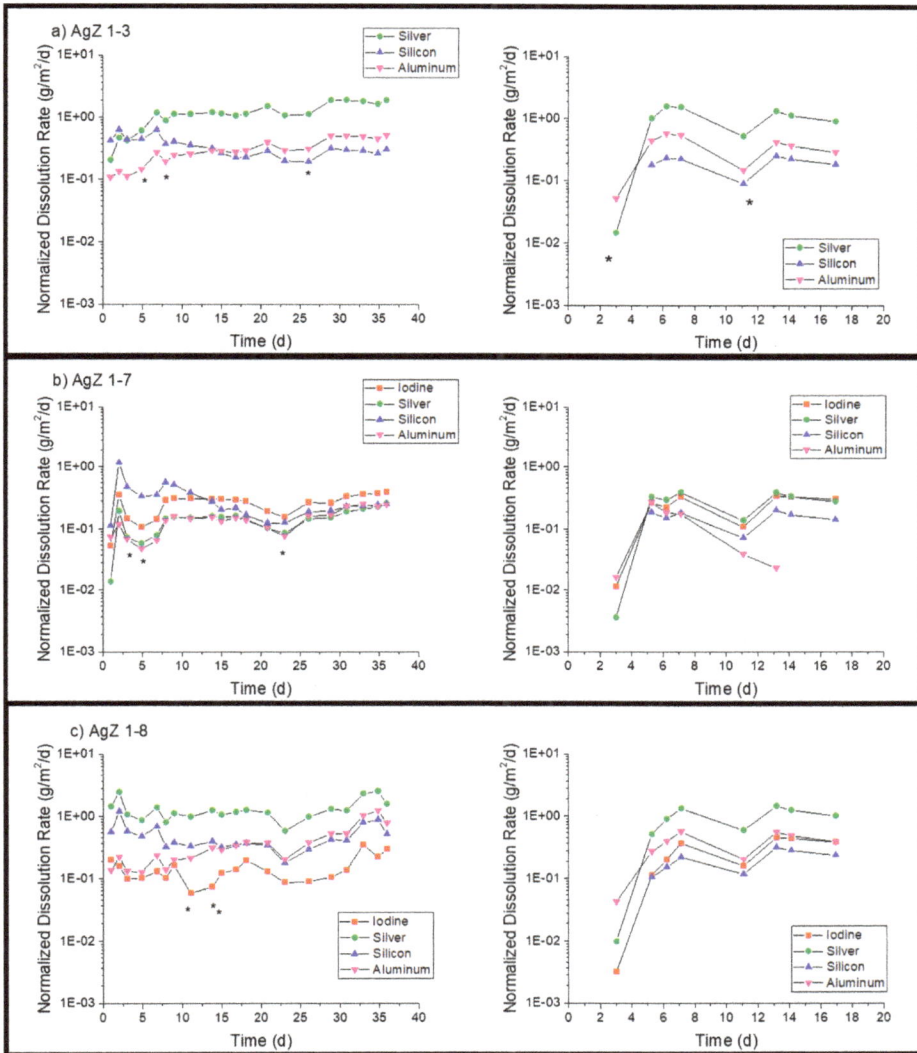

Figure 8. Normalized dissolution rates calculated from SPFT experiments using the AgZ samples at pH 9 for: (**a**) AgZ 1-3 (iodine-free sample); (**b**) AgZ 1-7; and (**c**) AgZ 1-8. The asterisks (*) mark samplings where the flow rate deviated by >10% from the average flow of the test.

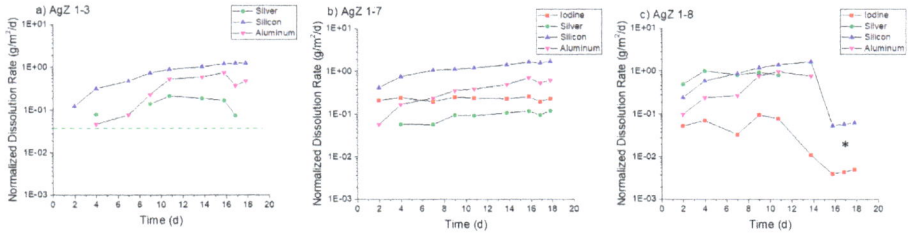

Figure 9. Normalized dissolution rates calculated from SPFT experiments using the AgZ samples at pH 11 for: (**a**) AgZ 1-3 (iodine-free sample); (**b**) AgZ 1-7; and (**c**) AgZ 1-8. The dashed lines, when present, represent the maximum rate for samplings where the analyte concentration was below the detection limit of the instrument and a dissolution rate calculated using the detection limit value. The asterisks (*) mark samplings where the flow rate deviated by >10% from the average flow of the test.

Based on the solution data presented above, an incongruent dissolution of the sample surface is likely occurring. The different phases of the heterogeneous microstructure shown in Figures 1–3 can each corrode independently of one another. The monolithic samples were imaged following SPFT testing to observe any physical changes on the sample surface. Using SEM, no observable changes were present on the AgZ samples following the pH 7 and pH 9 tests. Following the pH 11 tests, noticeable changes were present on the AgZ samples. Figure 10 shows the AgZ sample surfaces before and after the pH 11 exposure. The AgZ 1-3 sample (Figure 10a) appears to have corroded at the secondary phases and not the continuous Si matrix. The Ag particles (bright spots) appeared larger following corrosion as the higher alkalinity environment may increase their stability while the rest of the material corrodes. Based on the Pourbaix diagram for Ag, above pH 9 AgO becomes a stable phase for Ag and such a process may be occurring in the pH 11 tests [32]. For the AgZ 1-7 sample (Figure 10b), the secondary phases also appeared to have corroded. This observation is best exemplified by the rhomboid-shaped particle in the left center of the image, which was a K-rich particle. Following corrosion, the sharp edges of this phase had disappeared. The AgZ 1-8 sample also showed attack of the secondary phases and, similar to the AgZ 1-3 sample, an apparent growth of the Ag-containing particles (Figure 10c). The SEM micrographs suggest that corrosion preferentially occurred at the secondary phases, yet this was only observed from a two-dimensional view.

Optical profilometry was used to observe the three-dimensional (3D) profile of the AgZ samples following the SPFT experiments (Figure 11). At pH 7, all three samples showed only minor surface topography. In fact, the surface had retained enough of its polished nature to make it difficult to create the proper reflection to image at higher resolution, and as a result, a lower magnification image is shown. At pH 9, the surface morphology resembled what was suggested by the SEM images in Figure 10. Here, the lowest points on the surface were found to be the secondary phases for all three samples. The shapes and distributions of the phases suggest that these are the alkali- and alkaline-earth-rich phases shown in Figures 2–4. At pH 11, more extensive damage was observed and the AgZ 1-7 sample could not be fully resolved to generate a 3D image.

a) AgZ 1-3

b) AgZ 1-7

c) AgZ 1-7

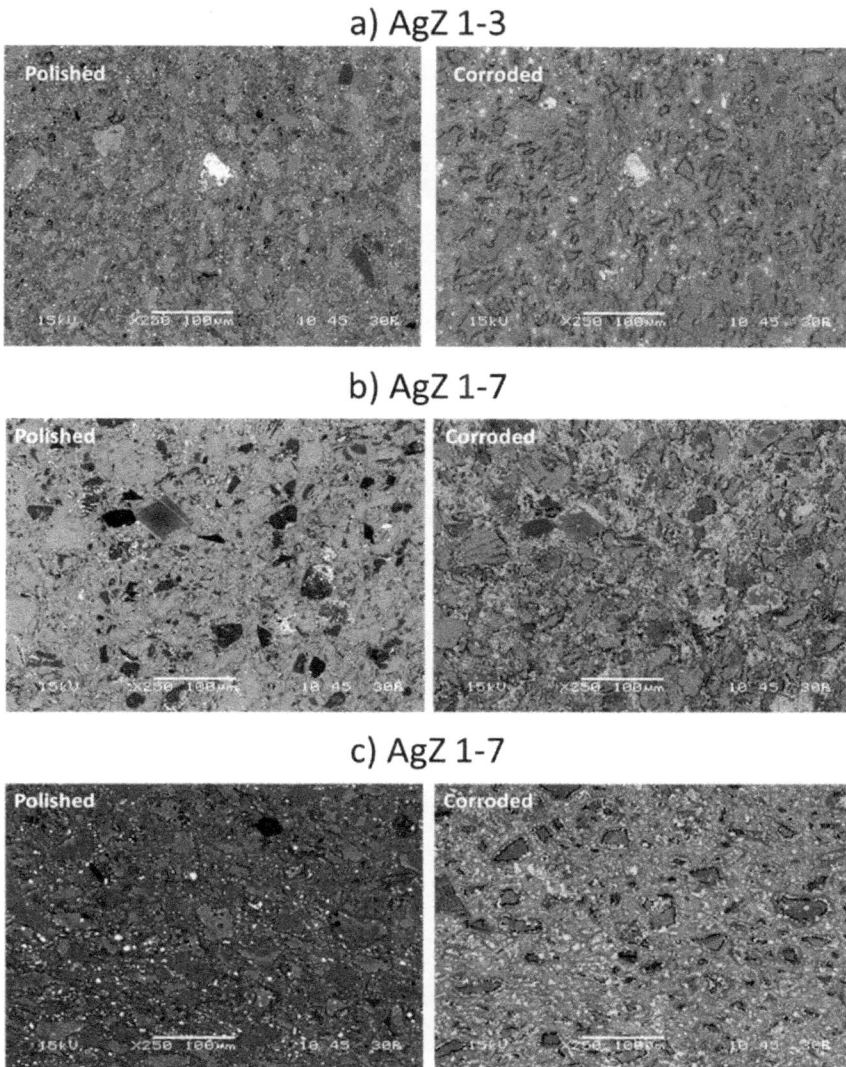

Figure 10. SEM micrographs of the samples both before and after corrosion in pH 11 SPFT experiments: (**a**) AgZ 1-3; (**b**) AgZ 1-7; and (**c**) AgZ 1-8.

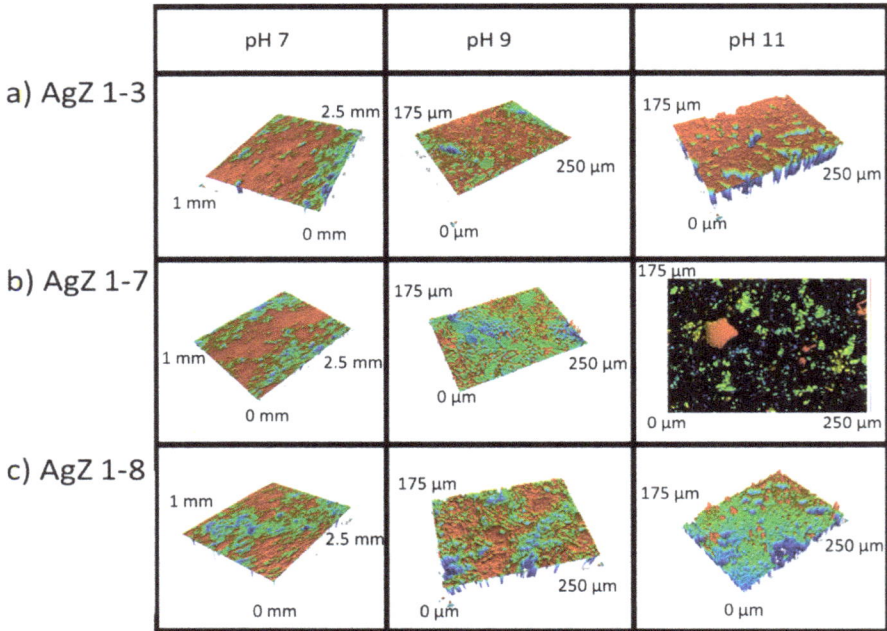

Figure 11. Optical profilometry images of the samples following the SPFT experiments: (**a**) AgZ 1-3;
(**b**) AgZ 1-7; and (**c**) AgZ 1-8. At pH 7, the surface could not be resolved at the higher magnification
(50×) so a lower magnification (5×) was used.

XRD analysis of the AgZ samples following testing at pH 9 and pH 11 showed no substantial
difference (spectra not shown) to the starting material [26]. Following the test, the sample surface was
composed of a mixture of silicon oxides, Ag metal, and AgI (Table 4). It should be noted that the XRD
mode used generated excitation volumes between 5 μm and 50 μm and the information within this
table includes the signal from the surface and inner sample in the excitation volume.

Table 4. Summary of the crystalline phases measured with XRD following SPFT AgZ tests at pH 9 and
pH 11.

Sample	pH 9	pH 11
AgZ 1-3	Ag metal, aluminum silicon oxide, silicon oxide, anorthite	Ag metal, aluminum silicon oxide, silicon oxide
AgZ 1-7	Ag metal, silicone oxide, anorthite, Ag iodide, aluminum silicate, cristobalite	Ag metal, silicone oxide, anorthite, Ag iodide, aluminum silicate, cristobalite
AgZ 1-8	Ag metal, silicone oxide, anorthite, Ag iodide, aluminum silicate, cristobalite	Ag metal, silicone oxide, anorthite, Ag iodide, aluminum silicate, cristobalite

3.3. Corrosion Testing of Spark Plasma Sintered Silver-Functionalized Silica Aerogels

The densified SFA materials were tested using the SPFT method in a similar fashion to the
AgZ. In pH 7 solution (Figure 12a), the SPS-1 sample experienced consistent dissolution, with an
average Si dissolution rate of 4.49 ± 1.52 g/m^2/day and an I dissolution rate of 0.12 ± 0.05 g/m^2/day.
The SPS-2 sample (Figure 12b), with higher Si content, measured lower Si dissolution rates averaging
0.65 ± 0.16 g/m^2/day and I dissolution rates measuring 0.06 ± 0.02 g/m^2/day. No Ag release was
measured at pH 7 for either sample.

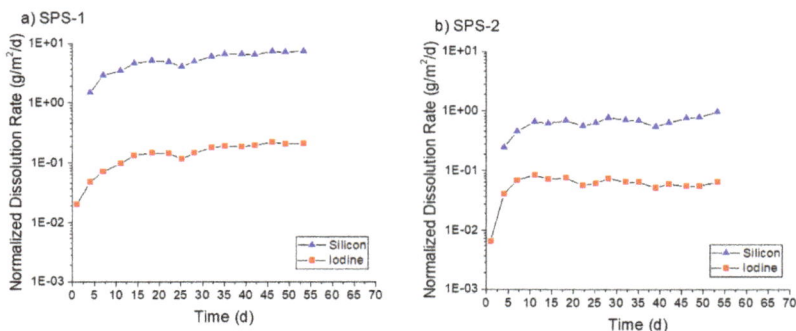

Figure 12. Normalized dissolution rates calculated from SPFT experiments using the SFA samples at pH 7 for: (**a**) SPS-1; and (**b**) SPS-2 (+20 wt % SFA). The asterisks (*) mark samplings where the flow rate deviated by >10% from the average flow of the test.

In pH 9 solution (Figure 13), both samples showed a continual increase in dissolution rate with time. At the conclusion of the test on the SPS-1 sample (Figure 13a), the measured Si dissolution rate was 4.67 g/m^2/day and the measured I dissolution rate was 0.37 g/m^2/day. The SPS-2 sample (Figure 13b) showed a Si dissolution rate of 1.26 g/m^2/day and an I dissolution rate of 0.56 g/m^2/day. Only at the conclusion of the test was Ag measurable for the SPS-1 sample, corresponding to an Ag dissolution rate of 0.02 g/m^2/day. No Ag release was measurable for the SPS-2 sample.

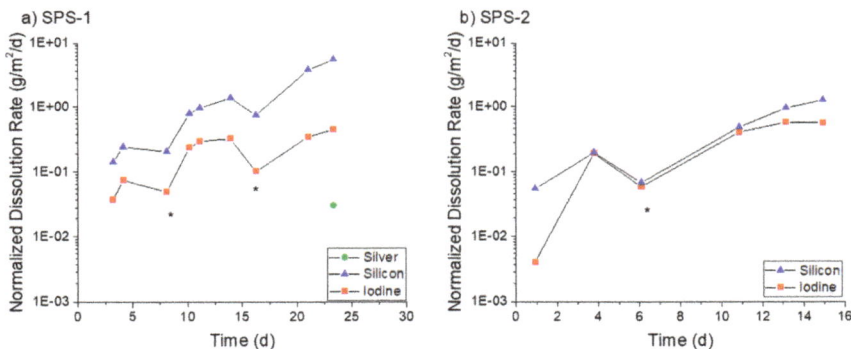

Figure 13. Normalized dissolution rates calculated from SPFT experiments using the SFA samples at pH 9 for: (**a**) SPS-1; and (**b**) SPS-2 (+20 wt % SFA). The asterisks (*) mark samplings where the flow rate deviated by >10% from the average flow of the test.

At pH 11 (Figure 14a), the SPS-1 sample measured an average Si dissolution rate of 33.3 ± 5.6 g/m^2/day and an I dissolution rate of 1.04 ± 0.56 g/m^2/day prior to the decrease at the final interval. The SPS-2 sample (Figure 14b) showed a Si dissolution rate of 10.21 ± 1.73 g/m^2/day. The I release was initially low before increasing past seven days. After this increase, the average I dissolution was measured at 0.54 ± 0.16 g/m^2/day. Ag was measured in the effluent at two time points for the SPS-1 sample equaling an Ag dissolution rate of 0.02 g/m^2/day.

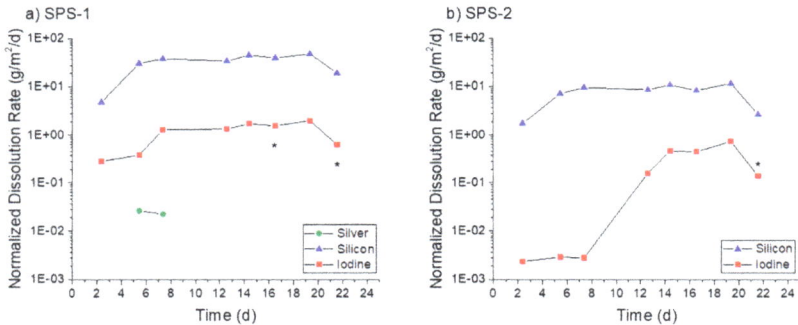

Figure 14. Normalized dissolution rates calculated from SPFT experiments using the SFA samples at pH 11 for: (**a**) SPS-1; and (**b**) SPS-2 (+20 wt % SFA). The asterisks (*) mark samplings where the flow rate deviated by >10% from the average flow of the test.

In all cases, the Si dissolution rates measured for SPS-2 were lower than those for SPS-1. The SPS-2 sample had additional Si added (as raw SFA) prior to sintering to improve durability and this methodology appeared to be successful. With the exception of the SPS-2 at pH 9, the I dissolution rates were also lower than the Si dissolution rates for the SFA samples. The microstructure of the SFA samples had areas of higher Si without any I present. Dissolution of those particles may have caused the higher Si dissolution rates, and more-so if the Si isolations were less durable than the matrix. The minimal release of Ag observed may be a result of the S presence in the SFA. AgS is a very insoluble compound and it is possible that dissolved Ag can become associated with S and be retained on the surface.

As with the AgZ samples, there were observable changes on the SFA surfaces following pH 11 exposure. SEM-EDS analysis performed on SPS-1, shown in Figure 15, provided some insight as to the retention of Ag during the testing of the SFA samples. In the SEM micrographs (Figure 15a), the large Si particles in the uncorroded image (those depleted in Ag and I) were heavily corroded, the large particle in the upper right being a perfect example. The attack appeared to have moved from the outer edge of the particles inward. The Ag remained evenly distributed following corrosion while some new Ag particles also appeared (Figure 15b). The I-rich particles in the uncorroded image near the large particles had disappeared in the corroded image (Figure 15c). The dissolution of the large particles likely drove the I-release. The most prominent change, however, is the increased definition of S in the image following corrosion (Figure 15d). After corrosion, a large particle has been exposed or generated that also contained Ag (Figure 15b) and I (Figure 15c). The appearance of the Ag-S-containing particles coordinated with I (see the large particle in the center of the corroded images) suggests that S may be responsible for the low Ag release and present a possible mechanism for improving I-retention in the sintered SFA. Previous work has shown that S behaves as a redox control agent over the Ag [33]. More work is planned to pursue understanding of this possible mechanism.

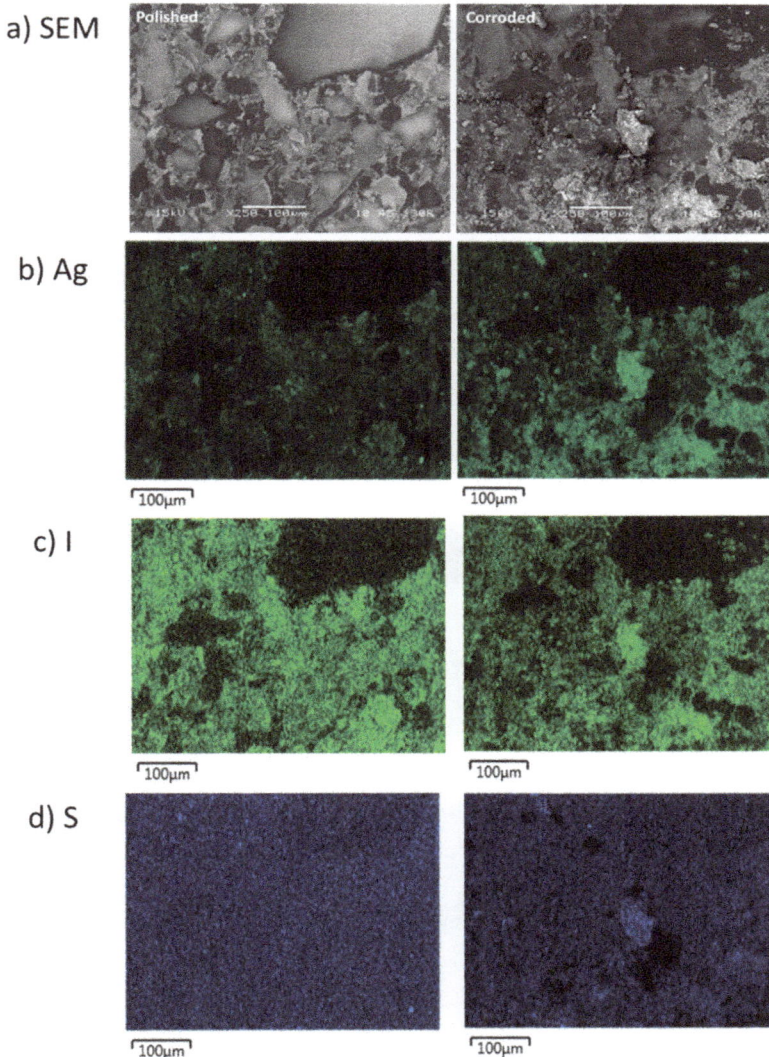

Figure 15. (a) SEM micrographs of the SPS-1 sample (left) before and (right) following SPFT testing at pH 11; and the corresponding EDS maps of: (b) Ag; (c) I; and (d) S.

For the SPS-2 sample, clear corrosion attack of the Si-rich particles was observed (Figure 16a), with the large particle in the center of the image being almost fully removed. An AgI particle was observed in the center of the non-corroded image (Figure 16b,c). Following corrosion, this particle was more visible as a result of the Si matrix removal around the particle. Other large AgI isolations behaved similarly in the images. Compared with the small AgI particles at the boundaries of the Si-particles in SPS-1 that were removed, large AgI isolations appeared to be retained better on the surface of the SPS-2.

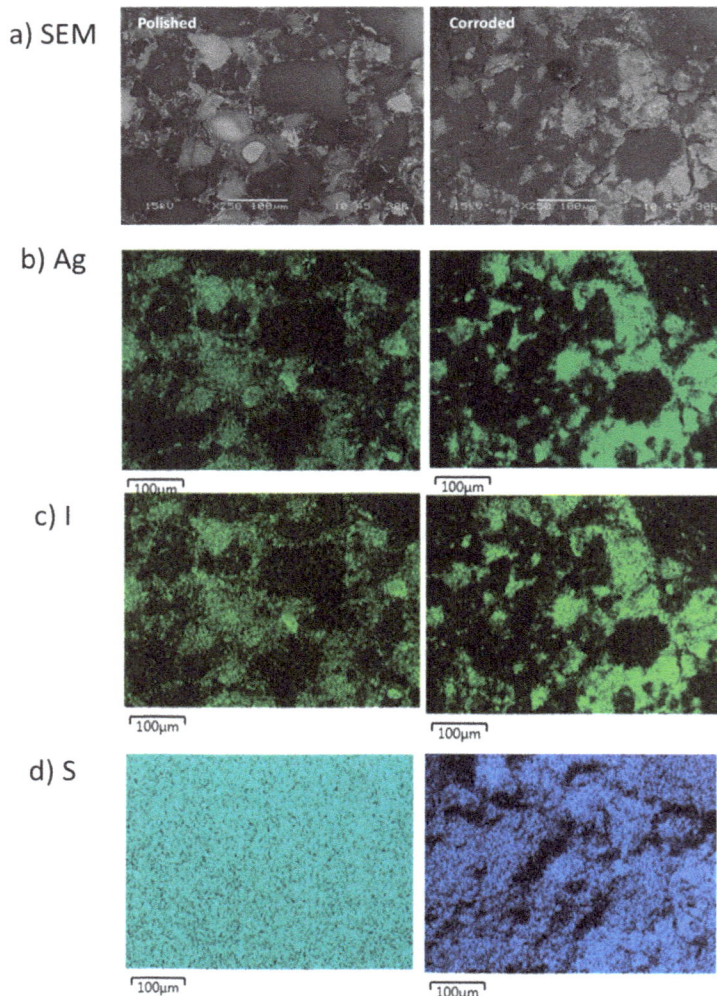

Figure 16. (**a**) SEM micrographs of the SPS-2 sample before (left) and following (right) SPFT testing at pH 11; and the corresponding EDS maps of: (**b**) Ag; (**c**) I; and (**d**) S.

The change in overall surface roughness of the SFA samples based on exposure pH can be observed in the optical profilometry images shown in Figure 17. For both the SPS-1 (Figure 17a) and SPS-2 (Figure 17b), following exposure at pH 7, the surface was notably roughened compared with the polished surface. The suppressed regions of the sample following corrosion appeared to be isolated and would suggest a similar dissolution pathway targeting the Si particles, as was observed at pH 11 in the SEM images. The surface was heavily corroded at pH 11.

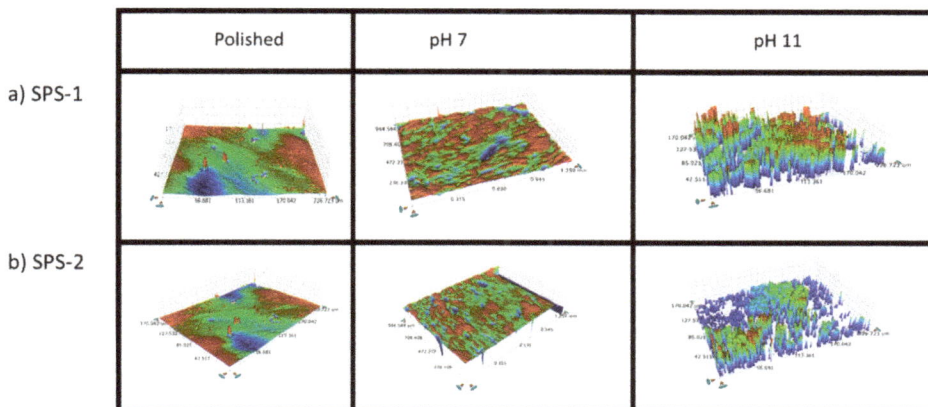

	Polished	pH 7	pH 11
a) SPS-1			
b) SPS-2			

Figure 17. Optical profilometry images before and after the SPFT experiments at pH 7 and pH 11 of: (**a**) SPS-1; and (**b**) SPS-2.

3.4. Comparison to Other Materials

Dissolution rates for Si-based materials in SPFT testing are highly dependent on the flow to surface area (q/S) ratio. This can limit direct comparisons between the dissolution rates measured for materials in different SPFT testing efforts. However, any comparative assessment of the overall durability of IWFs should be made against other material types under investigation for the long-term disposal of nuclear wastes. Other iodine-containing waste forms have been tested with SPFT but the tests were performed with differing conditions. Neeway et al. performed SPFT on iodine-containing fluidized bed steam reforming (FBSR) material at 40 °C and at far lower q/S (largest being 3×10^{-4} m/day) than this work [34]. Higher temperature dissolution data on FBSR material (without iodine) has been reported but the q/S used in the testing was not included [35]. Mowry et al. used a small-volume SPFT design to assess the durability of low-temperature Bi-Si-Zn oxide glass-composite materials (GCM) that contained AgZ [36]. The experiments focused on solutions with pH < 7, a maximum temperature of 60 °C and a q/S of 2×10^{-4} m/day. An iodine-containing glass (BNDL-A-S98) was investigated with SPFT at 90 °C but the raw data is not available in the report to compare the q/S values [37]. The best available comparisons are works on the dissolution of high-level nuclear waste glasses [38] and glass ceramic waste forms [39] where SPFT tests were performed (on powdered samples) at 90 °C in pH 9 and pH 11 solutions with similar q/S values to this work. The glass ceramic waste forms were multi-phase, borosilicate-based materials comprised of a borosilicate glass matrix with crystalline powellite and oxyapatite phases within. A summary of the normalized dissolution rates determined in this study as well as the comparative examples is given in Table 5. The three high-level waste glasses AFCI, ISG, and SON68 had Si dissolution rates of 0.350 g/m^2/day, 0.154 g/m^2/day, and 0.369 g/m^2/day, respectively, while the glass-ceramic waste form had a Si dissolution rate of 3.39 g/m^2/day in pH 9 tests. The highest pH 9 dissolution rates in the current study were 0.20 g/m^2/day for AgZ 1-8 and 4.67 g/m^2/day for SPS-1. This comparison suggests that the Si-matrices of the IWFs in the study are as durable as other Si-based waste form materials at pH 9. At pH 11, ISG had the highest Si dissolution rate at 3.44 g/m^2/day while AgZ 1-7 had a Si dissolution rate of 1.20 g/m^2/day and SPS-1 was higher at 33.3 g/m^2/day. The limited number of available datasets to directly compare IWF durability highlights the need for a standardized test to be defined to assess IWFs on an even playing field and to provide data to be used in long-term modelling predictions of IWF durability upon disposal.

Table 5. Summary of the normalized dissolution rates measured with SPFT testing in this report and comparison to dissolution rates measured for other materials in similar test conditions.

Sample	Test pH (Room Temp)	Length (d)	q/S (m/day)	I Dissolution Rate (g/m²/day)	Ag Dissolution Rate (g/m²/day)	Si Dissolution Rate (g/m²/day)	Al Dissolution Rate (g/m²/day)
AgZ 1-3	7	68	0.21	*N/A*	0.65 ± 0.07	0.17 ± 0.01	*< 0.06*
	9	17	0.18	*N/A*	1.01 ± 0.37	0.19 ± 0.05	0.35 ± 0.14
		36	0.17	*N/A*	1.16 ± 0.49	0.34 ± 0.12	0.30 ± 0.13
	11	18	0.21	*N/A*	0.14 ± 0.05	1.05 ± 0.21	0.49 ± 0.18
AgZ 1-7	7	68	0.24	0.015 ± 0.008	*<0.04*	0.066 ± 0.009	*< 0.07*
	9	17	0.21	0.27 ± 0.08	0.31 ± 0.08	0.15 ± 0.04	0.13 ± 0.10
		36	0.17	0.25 ± 0.09	0.14 ± 0.06	0.32 ± 0.23	0.14 ± 0.06
	11	18	0.23	0.22 ± 0.02	0.09 ± 0.02	1.20 ± 0.32	0.39 ± 0.18
AgZ 1-8	7	68	0.24	0.005 ± 0.001	0.30 ± 0.11	0.08 ± 0.01	0.09 ± 0.02
	9	17	0.18	0.30 ± 0.14	1.01 ± 0.36	0.20 ± 0.08	0.41 ± 0.13
		36	0.17	0.14 ± 0.07	1.32 ± 0.53	0.49 ± 0.24	0.39 ± 0.30
	11	18	0.23	0.06 ± 0.02	0.81 ± 0.19	0.99 ± 0.46	0.52 ± 0.38
SPS-1	7	68	0.41	0.12 ± 0.05	ND	4.49 ± 1.52	NA
	9	17	0.36	0.37	0.02	4.67	NA
	11	18	0.31	1.04 ± 0.56	0.02	33.3 ± 5.6	NA
SPS-2	7	68	0.37	0.06 ± 0.02	ND	0.65 ± 0.16	NA
	9	17	0.32	0.56	ND	1.26	NA
	11	18	0.28	0.54 ± 0.16	ND	10.21 ± 1.73	NA
AFCI (€)	9	21	0.35	NA	NA	0.350	NA
	11	21	0.35	NA	NA	3.36	NA
ISG (€)	9	21	0.35	NA	NA	0.154	NA
	11	21	0.35	NA	NA	3.44	NA
SON68 (€)	9	21	0.35	NA	NA	0.369	NA
	11	21	0.35	NA	NA	2.11	NA
Glass Ceramic (¥)	9	21	4.1E-01	NA	NA	3.39	NA

NA, data not available or analyte not present in sample; ND, analyte below detection limit, thus no rate calculations; € Ref [38]; ¥, Ref [39].

4. Conclusions

In summary, the dilute-condition chemical durability of two IWF types, HIPed AgZ and SPS-SFA, were investigated using the SPFT method. For the AgZ samples, the following trends were observed: (1) at pH 7 and 9, the releases of Ag were larger than that of Si for samples HIPed at higher temperature; (2) at pH 7, the release of I was much slower compared to the other analytes; (3) at pH 11, the release of Si was higher than Ag; and (4) preferential corrosion attack was observed on secondary phases that contained higher amounts of Al and alkali species but were lower in overall I content. The following observations were made for the SFA samples: (1) lowered I release compared with Si; (2) corrosion attack was preferential at Si-rich particles making small AgI isolations near these particle boundaries susceptible to dissolution; (3) an increased addition of Si (in the form of 20 wt % SFA) to the SFA waste form improved chemical durability; (4) minimal Ag release was observed; and (5) Ag- and S-containing isolations were found to contain I after corrosion testing. Both IWF types had similar dissolution rates to other Si-based waste forms at pH 9 and pH 11. The information collected here can be used in the development of long-term predictive models for disposal of IWFs, help direct improved chemical durability of the IWFs, and highlighted a need for a standardized test to be used for the durability of IWF.

Author Contributions: Conceptualization: R.M.A. and J.V.R. Methodology: R.M.A., J.V.R. and A.R.L. Investigation: R.M.A., J.V.C., J.M., J.T.R., N.A., E.M.M., N.C.C. Data Curation: R.M.A. and A.R.L. Writing: R.M.A.

Funding: Funding was received from the US Department of Energy—Office of Nuclear Energy.

Acknowledgments: The authors would like to acknowledge the US Department of Energy—Office of Nuclear Energy for project funding that was carried out under the Material Recovery and Waste Form Development Program. We would like to thank Kimberly Gray, Terry Todd and John Vienna for project support and program leadership. Robert Jubin and Stephanie Bruffey (ORNL) provided the AgZ samples and the authors appreciate their collaboration. Ian Leavy, Steven Baum and Keith Geiszler of the Environmental Sciences Laboratory at PNNL are thanked for their analysis and data reviews.

Conflicts of Interest: The authors declare no conflict of interest.

References

1. Jubin, R.T.; Strachan, D.M.; Soelberg, N.R. *Iodine Pathways and Off-Gas Stream Characteristics for Aqueous Reprocessing Plants—A Literature Survey and Assessment*; INL/EXT-13-30119 United States 10.2172/1111056 INL English; Idaho National Laboratory (INL): Idaho Falls, ID, USA, 2013.

2. Soelberg, N.R.; Garn, T.G.; Greenhalgh, M.R.; Law, J.D.; Jubin, R.; Strachan, D.M.; Thallapally, P.K. Radioactive Iodine and Krypton Control for Nuclear Fuel Reprocessing Facilities. *Sci. Technol. Nucl. Install.* **2013**, *2013*, 12. [CrossRef]

3. Riley, B.J.; Vienna, J.D.; Strachan, D.M.; McCloy, J.S.; Jerden, J.L. Materials and processes for the effective capture and immobilization of radioiodine: A review. *J. Nucl. Mater.* **2016**, *470*, 307–326. [CrossRef]

4. Decamp, C.; Happel, S. Utilization of a mixed-bed column for the removal of iodine from radioactive process waste solutions. *J. Radioanal. Nucl. Chem.* **2013**, *298*, 763–767. [CrossRef]

5. Sun, H.; La, P.; Zhu, Z.; Liang, W.; Yang, B.; Li, A. Capture and reversible storage of volatile iodine by porous carbon with high capacity. *J. Mater. Sci.* **2015**, *50*, 7326–7332. [CrossRef]

6. Yu, F.; Li, D.-D.; Cheng, L.; Yin, Z.; Zeng, M.-H.; Kurmoo, M. Porous Supramolecular Networks Constructed of One-Dimensional Metal–Organic Chains: Carbon Dioxide and Iodine Capture. *Inorg. Chem.* **2015**, *54*, 1655–1660. [CrossRef] [PubMed]

7. Scott, S.M.; Hu, T.; Yao, T.; Xin, G.; Lian, J. Graphene-based sorbents for iodine-129 capture and sequestration. *Carbon* **2015**, *90*, 1–8. [CrossRef]

8. Katsoulidis, A.P.; He, J.; Kanatzidis, M.G. Functional Monolithic Polymeric Organic Framework Aerogel as Reducing and Hosting Media for Ag nanoparticles and Application in Capturing of Iodine Vapors. *Chem. Mater.* **2012**, *24*, 1937–1943. [CrossRef]

9. Yao, R.-X.; Cui, X.; Jia, X.-X.; Zhang, F.-Q.; Zhang, X.-M. A Luminescent Zinc(II) Metal–Organic Framework (MOF) with Conjugated π-Electron Ligand for High Iodine Capture and Nitro-Explosive Detection. *Inorg. Chem.* **2016**, *55*, 9270–9275. [CrossRef] [PubMed]

10. Chapman, K.W.; Chupas, P.J.; Nenoff, T.M. Radioactive Iodine Capture in Silver-Containing Mordenites through Nanoscale Silver Iodide Formation. *J. Am. Chem. Soc.* **2010**, *132*, 8897–8899. [CrossRef] [PubMed]

11. Bennett, T.D.; Saines, P.J.; Keen, D.A.; Tan, J.-C.; Cheetham, A.K. Ball-Milling-Induced Amorphization of Zeolitic Imidazolate Frameworks (ZIFs) for the Irreversible Trapping of Iodine. *Chem. A Eur. J.* **2013**, *19*, 7049–7055. [CrossRef] [PubMed]

12. Nenoff, T.M.; Rodriguez, M.A.; Soelberg, N.R.; Chapman, K.W. Silver-mordenite for radiologic gas capture from complex streams: Dual catalytic CH3I decomposition and I confinement. *Microporous Mesoporous Mater.* **2014**, *200*, 297–303. [CrossRef]

13. Yang, J.H.; Cho, Y.-J.; Shin, J.M.; Yim, M.-S. Bismuth-embedded SBA-15 mesoporous silica for radioactive iodine capture and stable storage. *J. Nucl. Mater.* **2015**, *465*, 556–564. [CrossRef]

14. Matyas, J.; Fryxell, G.; Busche, B.; Wallace, K.; Fifield, L. Functionalised silica aerogels: Advanced materials to capture and immobilise radioactive iodine. In *Proceedings of Ceramic Engineering and Science Proceedings*; American Ceramic Society, Inc.: Columbus, OH, USA, 2011; pp. 23–32.

15. Riley, B.J.; Chun, J.; Ryan, J.V.; Matyáš, J.; Li, X.S.; Matson, D.W.; Sundaram, S.K.; Strachan, D.M.; Vienna, J.D. Chalcogen-based aerogels as a multifunctional platform for remediation of radioactive iodine. *RSC Adv.* **2011**, *1*, 1704–1715. [CrossRef]

16. Subrahmanyam, K.S.; Sarma, D.; Malliakas, C.D.; Polychronopoulou, K.; Riley, B.J.; Pierce, D.A.; Chun, J.; Kanatzidis, M.G. Chalcogenide Aerogels as Sorbents for Radioactive Iodine. *Chem. Mater.* **2015**, *27*, 2619–2626. [CrossRef]

17. Haynes, W.M. *CRC Handbook of Chemistry and Physics*; CRC Press: Boca Raton, FL, USA, 2014.

18. Tanabe, H.; Sakuragi, T.; Yamaguchi, K.; Sato, T.; Owada, H. Development of new waste forms to immobilize iodine-129 released from a spent fuel reprocessing plant. In *Proceedings of the Advances in Science and Technology*; Trans Tech Publications: Stafa-Zurich, Switzerland, 2010; pp. 158–170.

19. Asmussen, R.M.; Pearce, C.I.; Lawter, A.R.; Miller, B.W.; Neeway, J.J.; Lawler, B.; Smith, G.; Serne, J.; Swanberg, D.J.; Qafoku, N. Preparation, Performance and Mechanism of Tc and I Getters in Cementitious Waste Forms. In Proceedings of the Waste Management Symposium, Phoenix, AZ, USA, 5–9 March 2017; p. 17124.

20. Bruffey, S.H.; Jubin, R.T.; Jordan, J.A. Capture of Elemental and Organic Iodine from Dilute Gas Streams by Silver-exchanged Mordenite. *Procedia Chem.* **2016**, *21*, 293–299. [CrossRef]

21. Jubin, R.; Ramey, D.; Spencer, B.; Anderson, K.; Robinson, S. Impact of Pretreatment and Aging on the Iodine Capture Performance of Silver-Exchanged Mordenite. In Proceedings of the Waste Management Symposium, Phoenix, AZ, USA, 27 February–3 March 2011; p. 12314.

22. Bruffey, S.H.; Jubin, R.T. *Recommend HIP Conditions for AgZ*; FCRD-MRWFD-2015-000423; Oak Ridge National Laboratory: Oak Ridge, TN, USA, 2015; p. 23.

23. Matyas, J.; Canfield, N.; Silaiman, S.; Zumhoff, M. Silica-based waste form for immobilization of iodine from reprocessing plant off-gas streams. *J. Nucl. Mater.* **2016**, *476*, 255–261. [CrossRef]

24. Asmussen, R.M.; Matyáš, J.; Qafoku, N.P.; Kruger, A.A. Silver-functionalized silica aerogels and their application in the removal of iodine from aqueous environments. *J. Hazard. Mater.* **2018**. [CrossRef] [PubMed]

25. Chibani, S.; Chebbi, M.; Lebègue, S.; Cantrel, L.; Badawi, M. Impact of the Si/Al ratio on the selective capture of iodine compounds in silver-mordenite: A periodic DFT study. *Phys. Chem. Chem. Phys.* **2016**, *18*, 25574–25581. [CrossRef] [PubMed]

26. Jubin, R.T.; Bruffey, S.H. High-Temperature Pressing of Silver-Exchanged Mordenite into a Potential Iodine Waste Form–14096. In Proceedings of the Waste Management 2014, WM Symposia, Tempe, AZ, USA, 2–6 March 2014.

27. ASTM. *Standard Practice for Measurement of the Glass Dissolution Rate Using the Single-Pass Flow-Through Test Method*; ASTM C1662-17 2017; ASTM International: West Conshohocken, PA, USA, 2017.

28. Jubin, R.T.; Bruffey, S.H.; Patton, K.K. *Expanded Analysis of Hot Isostatic Pressed Iodine-Loaded Silver Exchanged Mordenite*; FCRD-SWF-2014-000278; Oak Ridge National Laboratory: Oak Ridge, TN, USA, 2014; p. 43.

29. Riley, B.J.; Kroll, J.O.; Peterson, J.A.; Matyáš, J.; Olszta, M.J.; Li, X.; Vienna, J.D. Silver-Loaded Aluminosilicate Aerogels as Iodine Sorbents. *ACS Appl. Mater. Interfaces* **2017**, *9*, 32907–32919. [CrossRef] [PubMed]

30. Cheary, R.W.; Coelho, A.A.; Cline, J.P. Fundamental Parameters Line Profile Fitting in Laboratory Diffractometers. *J. Res. Natl. Inst. Stand. Technol.* **2004**, *109*, 1–25. [CrossRef] [PubMed]

31. Pourbaix, M. *Atlas of Electrochemical Equilibria in Aqueous Solutions*, 2d ed.; National Association of Corrosion Engineers: Houston, TX, USA, 1974.

32. Delahay, P.; Pourbaix, M.; Van Rysselberghe, P. Potential-pH Diagram of Silver Construction of the Diagram—Its Applications to the Study of the Properties of the Metal, its Compounds, and its Corrosion. *J. Electrochem. Soc.* **1951**, *98*, 65–67. [CrossRef]

33. Matyáš, J.; Ilton, E.S.; Kovařík, L. Silver-functionalized silica aerogel: Towards an understanding of aging on iodine sorption performance. *RSC Adv.* **2018**, *8*, 31843–31852. [CrossRef]

34. Neeway, J.J.; Qafoku, N.P.; Williams, B.D.; Snyder, M.M.V.; Brown, C.F.; Pierce, E.M. Evidence of technetium and iodine release from a sodalite-bearing ceramic waste form. *Appl. Geochem.* **2016**, *66*, 210–218. [CrossRef]

35. Jantzen, C.M.; Lorier, T.H.; Marra, J.C.; Pareizs, J. Durability Testing of Fluidized Bed Steam Reforming (FBSR) Waste Forms. In Proceedings of the Waste Management Symposium 2006, Tucson, AZ, USA, 26 February–2 March 2006.

36. Mowry, C.D.; Brady, P.V.; Garino, T.J.; Nenoff, T.M. Development and Durability Testing of a Low-Temperature Sintering Bi–Si–Zn Oxide Glass Composite Material (GCM) 129I Waste Form. *J. Am. Ceram. Soc.* **2015**, *98*, 3094–3104. [CrossRef]

37. McGrail, B.P.; Schaef, H.T.; Martin, P.F.; Bacon, D.H.; Rodriguez, E.A.; McCready, D.E.; Primak, A.N.; Orr, R.D. *Initial Evaluation of Steam Reformed Low Activity Waste for Direct Land Disposal*; Pacific Northwest National Laboratory: Richland, WA, USA, 2003.

38. Neeway, J.J.; Rieke, P.C.; Parruzot, B.P.; Ryan, J.V.; Asmussen, R.M. The dissolution behavior of borosilicate glasses in far-from equilibrium conditions. *Geochim. Cosmochim. Acta* **2018**, *226*, 132–148. [CrossRef]

39. Asmussen, R.M.; Neeway, J.J.; Kaspar, T.C.; Crum, J.V. Corrosion Behavior and Microstructure Influence of Glass-Ceramic Nuclear Waste Forms. *Corrosion* **2017**, *73*, 1306–1319. [CrossRef]

materials

MDPI

Article

An Assessment of Initial Leaching Characteristics of Alkali-Borosilicate Glasses for Nuclear Waste Immobilization

Osama M. Farid [1], Michael I. Ojovan [2,3], A. Massoud [4] and R.O. Abdel Rahman [5,*]

[1] Reactors Department, Nuclear Research Center, Atomic Energy Authority of Egypt, Cairo 13759, Egypt; usamafa98@hotmail.co.uk
[2] Department of Materials Science and Engineering, The University of Sheffield, Sheffield S1 3JD, UK; m.ojovan@sheffield.ac.uk
[3] Department of Radiochemistry, Lomonosov Moscow State University, 119991 Moscow, Russia; m.i.ojovan@gmail.com
[4] Chemistry Unit of Cyclotron, Nuclear Research Center, Atomic Energy Authority of Egypt, Cairo 13759, Egypt; ayman_mass@yahoo.com
[5] Hot Laboratory Center, Atomic Energy Authority of Egypt, Cairo 13759, Egypt
*   Correspondence: alaarehab@yahoo.com; Tel.: +20-01061404462

Received: 2 April 2019; Accepted: 2 May 2019; Published: 6 May 2019

Abstract: Initial leaching characteristics of simulated nuclear waste immobilized in three alkali-borosilicate glasses (ABS-waste) were studied. The effects of matrix composition on the containment performance and degradation resistance measures were evaluated. Normalized release rates are in conformance with data reported in the literature. High Li and Mg loadings lead to the highest initial de-polymerization of sample ABS-waste (17) and contributed to its thermodynamic instability. Ca stabilizes non-bridging oxygen (NBO) and reduces the thermodynamic instability of the modified matrix. An exponential temporal change in the alteration thickness was noted for samples ABS-waste (17) and Modified Alkali-Borosilicate (MABS)-waste (20), whereas a linear temporal change was noted for sample ABS-waste (25). Leaching processes that contribute to the fractional release of all studied elements within the initial stage of glass corrosion were quantified and the main controlling leach process for each element was identified. As the waste loading increases, the contribution of the dissolution process to the overall fractional release of structural elements decreases by 43.44, 5.05, 38.07, and 52.99% for Si, B, Na, and Li respectively, and the presence of modifiers reduces this contribution for all the studied metalloids. The dissolution process plays an important role in controlling the release of Li and Cs, and this role is reduced by increasing the waste loading.

Keywords: fractional release; alkali borosilicate glass; leaching processes; modeling

1. Introduction

Radioactive waste disposal is considered to be the last step (end point) in radioactive waste management systems [1–3]. The design of both geological and near-surface disposal facilities relies on the application of passive safety functions to ensure the containment and confinement of the radiological hazards of these wastes, where the wastes are isolated for periods sufficient to allow for radioactive decay of the short-lived radionuclides and limit the release of long-lived radionuclides [1,2,4,5]. To ensure safe performance of these facilities throughout their life cycles, assessment studies have to be conducted to support the decision-making process. In these assessments, temporal evolution of engineering barriers and the dynamic nature of hydrological and biological subsystems in the host environment are considered by applying a modular approach [3,5,6]. In this approach, the disposal system is divided into near- and far-field subsystems that are subsequently divided into their main components [3].

The waste-immobilizing matrix is the main component of the near-field subsystem. Its main safety functions are to ensure structural stability, resist degradation, and limit water ingress and radio-contaminant releases. Several waste matrices have been proposed to stabilize the radioactive/nuclear wastes, including, cement-, bitumen- and polymer-, glass-, and ceramic-based matrices [4,6–14]. The main safety function of glass waste matrices is to slow down radionuclide releases from a geological disposal facility [15]. In this respect, two performance indicators are used to assess the quality, reliability, and efficiency of the waste matrices, namely the glass–water reaction rate and the radionuclide leach rate that ensure the degradation resistance and containment ability of the matrices, respectively. These indicators are evaluated by conducting leaching experiments that simulate leaching conditions under conservative disposal conditions.

Generally, leaching characteristics of radioactive/nuclear waste matrices are highly dependent on the chemical compositions of the waste matrices and leaching experimental conditions [6–9,11–17]. A huge research effort was directed at studying the leaching characteristics of glass-based waste matrices using static and dynamic leaching experiments, i.e., PCT (product consistence test), MCC (Material Characterization Center), and single pass flow through tests, by investigating different waste matrices and leachant compositions at varying pH and temperature values and leachant-to-waste volumes [10–25]. These studies identified hydrolysis, ion exchange, diffusion, dissolution, and re-precipitation as the main corrosion processes for glass structural elements that led to glass degradation [10–25]. The overall temporal evolution of the glass waste matrix was attributed to these processes and their interactions and is conventionally divided into four [11–14] or three [16,17,24] basic stages, namely initial/forward (inter-diffusion and hydrolysis), residual/final, and resumption of alteration.

Safety assessment studies for the glass waste matrices are based on kinetic models to predict temporal variation in radio-contaminant releases and glass degradation [17,23]. Long-term assessment studies are challenged by the quantification of potential formation of zeolites and their roles in enhancing long-term glass degradation, whereas short-term assessments are challenged by the dynamic changes in the leachant chemical composition and glass surface area [11–14,16,17,23,24]. In addition, the initial leaching stage is characterized by the fastest leaching rates that result from contributions of different leaching processes [11–14,17,24,26]. An understanding of the leaching characteristics of all the matrix elements at this stage and an assessment of initiating leaching processes can help in predicting and controlling the releases at subsequent stages of the degradation process.

Borosilicate glasses (BSs) were proposed as nuclear-waste-immobilizing matrices because of their ability to incorporate a wide variety of metal oxides, high waste loading, physical and radiological stability, and simplicity of production [10–14,27]. Alkali modifiers can affect the durability of borosilicate matrices as a result of a boron anomaly and formation of non-bridging oxygen (NBO) [10,11,28]. Table 1 summarizes normalized release rates for different contaminants and structural elements for different alkali-borosilicate waste glass (ABS) matrices [29–33]. In this work, the short-term temporal evolution of glass-waste matrices will be investigated by assessing the initial glass leaching characteristics for all the matrix constituents in three borosilicate waste glasses. The aim is to identify the effects of waste loading and matrix modification on the containment performance and degradation resistance and vindicate the controlling leaching mechanism for each metal group. In this context, we investigate short-term MCC1 leaching characteristics of three borosilicate waste glass matrices that represent modified/unmodified vitreous waste forms of varying metal oxide loading. Temporal changes in the leaching solutions' composition will be presented for all the matrices constituents, glasses composition evolution will be traced by calculating the non-bridging oxygen (NBO), and the associated degradation will be evaluated by calculating the corresponding altered glass fraction ($\delta AGF(t)$) and alteration thickness (ET). The hydration free energies of the glasses will be calculated to have insights into the effect of the chemical composition on the glass stability and identify the role of the structural elements, modifiers, and different waste constituents on the initial thermodynamic stability of the matrices. The leaching mechanisms of all the studied elements will be identified, and corresponding leaching parameters will be estimated. Finally, the contribution of each leaching process to the short-term releases will be presented and linked to the structure of the glasses. The main text is divided into two sections; the first (Section 2) presents the glass preparation, leaching test, free energy of hydration calculation, and leaching mechanism evaluation procedures and the second (Section 3) presents the results and discussions of the experimental and theoretical investigations.

Table 1. The normalized release rate ($mg \cdot m^{-2} \cdot d^{-1}$) of different elements from different alkali-borosilicate glasses matrices (ABS), including calcined Prototype Fast Reactor-Raffinate (PFR), Reactor Bolshoy Moshchnosty Kanalny-concentrate (RBMK), Water-Water Energetic Reactor-concentrate (WWER), RBMK-evaporator concentrate (K-26), High Level Waste Simulant (BS-5), and PyroGreen salt waste (PG).

ABS Glass Waste			PFR	RBMK	WWER	K-26	BS-5	PG
Test Type			PCT	ISO-6961	PCT	PCT	PCT	Field Data
Alkali	Na		16.9–21.7	10^1–10^2	10^2	59.3–90.9	378	1.42–8.57
	Li		-	-	-	-	-	5.7–37.14
	Cs		-	10^1–10^2	10^2	-	-	-
Alkaline earth metal	Ca		3.62–5.89	-	-	-	-	-
	Sr		-	10^0–10^1	10^1	-	-	-
Post-transition	Al		0.29	-	-	-	-	-
Transition	Mo		4.44–6.38			-	-	-
	Ba		1.47–4.43	10^0–10^{-1}	10^0	-	-	-
	Cr		0.16–0.35			-	-	-
Metalloid	Si		7.18–8.4	-	-	28.1–29.3	174	4.28–17.1
	B		32.4–33.3	$<10^{-1}$	$<10^{-1}$	31.3–40.5	435	1.42–18.57
Rare earth elements			-	10^{-1}	10^{-1}	-	7.11	-
Reference			[29]	[30]	[30]	[31]	[32]	[33]

2. Materials and Methods

2.1. Glasses Preparation

Alkali-borosilicate glasses were prepared using the melt quenching technique, where powders were mixed, as indicated in Tables 2–4, and milled to obtain homogeneous batches. These samples simulate the performance of ABS-17% Magnox (ABS-waste (17)), Modified ABS-20% Magnox (MABS-Waste (20)), and ABS-25% Mixed oxide (ABS-Waste (25)). The powder mixes were melted in a platinum crucible at 1060 °C for 1 h and stirred for 4 h before casting into blocks using a preheated stainless steel mould. Glasses were allowed to cool before being placed into an annealing furnace at 500 °C for 1 h then to cool to room temperature at a rate of 1 °C/min. The glasses were kindly supplied by Dr. Cassingham, N.C. and Prof. Hyatt, N.C., Immobilization Science Laboratory, The University of Sheffield, Sheffield, UK.

Table 2. Chemical composition of the studied glasses (structural elements and modifiers).

Compound	SiO_2	B_2O_3	Na_2O	Li_2O	CaO	ZnO	Total
ABS-Waste (17)	50.200	15.400	8.800	8.700	–	–	83.100
MABS-Waste (20)	44.260	17.950	9.010	2.110	1.390	4.430	79.150
ABS-Waste (25)	46.280	16.430	8.330	3.980	–	–	75.020

Table 3. Chemical composition of the studied glasses (waste components: alkali, alkaline, post-transitions, and metalloids).

Compound	Alkali		Alkaline Earth Metals			Post-Transitions and Metalloids		
	Cs_2O	BaO	MgO	SrO	Total	Al_2O_3	TeO_2	Total
ABS-Waste (17)	0.300	0.200	8.200	0.200	8.60	3.100	0.100	3.200
MABS-Waste (20)	0.890	0.40	4.100	0.240	4.740	4.110	0.150	4.260
ABS-Waste (25)	1.590	0.470	1.610	0.410	2.490	1.910	0.280	2.190

Table 4. Chemical composition of the studied glasses (waste components: transitions and rare earth elements).

Compound	Transition Metals*							Rare Earth *			
	Cr_2O_3	Fe_2O_3	MoO_3	RuO_2	ZrO_2	Y_2O_3	Total	CeO_2	La_2O_3	Nd_2O_3	Total
ABS-Waste (17)	0.300	1.300	0.700	0.200	0.800	0.100	3.400	0.500	0.100	0.400	1.000
MABS-Waste (20)	0.630	2.790	1.320	0.520	1.240	0.160	6.660	0.960	0.520	1.530	3.010
ABS-Waste (25)	0.510	2.060	2.490	0.550	2.820	0.310	8.740	1.450	0.730	2.170	4.350

* Ni, Pr, and Gd oxides were neglected in this study.

2.2. Leaching Test

Glass leaching was assessed by conducting an MCC1 (ASTM C1220-10) static leaching test [11], where glass coupons of $1 \times 1 \times 0.5$ cm^3 were immersed in deionized water in Perfluoroalkoxy (PFA) vessels. The test was performed at 90 °C using a constant surface area to volume ratio (S/V) 10 m^{-1} for all samples studied. The spectroscopic analyses of the leachants as a function of time were conducted using inductively coupled plasma optical emission spectroscopy (ICP-OES). The experimental data (average of triplicates) were used to calculate four performance measures that represent temporal changes in the leaching solution composition and glass waste matrices compositions, i.e., normalized release rates (NR$_i$, mg·m^{-2}·d^{-1}) and non-bridging oxygen (NBO), and its corresponding degradation, i.e., altered glass fraction (δAGF(t)) and altered thickness (ET(t), µm) [6,10,19,22,23,33,34]:

$$NR_i = \frac{C_i V}{f_i S \Delta t} \tag{1}$$

$$NBO = 2(R_2O + RO) + 6R_2O_3 - 2(Al_2O_3 + Fe_2O_3) + 4RO_2 \tag{2}$$

$$\delta AGF(t) = \left(C_{B_t} - C_{B_{t-1}}\right)\left(\frac{V}{m_B}\right) \tag{3}$$

$$ET(t) = \left(1 - (1 - AGF(T))^{\frac{1}{3}}\right)\left(\frac{3}{\rho \times SA}\right) \tag{4}$$

where C_i is the measured element (i) concentration in leachant released at a specified time t (g/m^3), V and S are the leachant volume (m^3) and sample surface area (m^2), respectively, f_i is the fraction of the element in the sample, Δt is the time change, R_xO_y is the metal oxide amount, m_B is the mass of boron (g), ρ is the glass density (g/cm^3), and SA is the specific surface area (m^2/g).

2.3. Free Energy of Hydration

Leaching behavior can be viewed as a combination of two subsequent reactions. The first is the waste matrix hydration followed by elemental transport through the matrix and interaction with the leachant solution. Subsequently, the tendency to undergo a hydration reaction could be seen as an indication of the waste matrix instability. The hydration free energy (ΔG) for glass waste matrices was correlated to the thickness of the altered glass, pH, Eh, and former normalized release rates [29,30,35]. The free energy of hydration reaction is expressed as an additive function of individual glass units' hydration free energies (ΔG$_i$), as follows:

$$\Delta G = \sum_i x_i \Delta G_i \tag{5}$$

where x_i is the mole fraction of an individual glass unit (i). The hydration free energy was determined based on the assumption that the glass matrix is homogenous and the presence of crystalline phases, i.e., iron spinel, is of negligible effect on the hydration. This negligible effect is attributed to their isotropic nature that minimizes grain boundary dissolution [26]. All the metal oxides were converted into silicates except silicon, boron, aluminum, and iron and the individual hydration free energy at 90 °C was obtained as indicated by Perret et al. [35].

2.4. Leaching Mechanisms Evaluation

Glass leaching mechanisms were evaluated based on the analysis of the experimental data to a collective model that represents the cumulative leach fraction (CLF$_i$) of the structural elements, modifiers, and waste oxides as superimposed leaching processes that include a first-order reaction exchange between the leaching solution and bounded element on the matrix or the formed colloides, bulk diffusion of elements throughout the matrix, congruent dissolution, and instantaneous release of loosely bounded element from the surface [6,7,9–14,36]:

$$CLF_i = Q_{Oi}\left(1 - e^{-K_i t}\right) + \left(\frac{S}{V}\right)\left(2\sqrt{\frac{D_i t}{\pi}} + U_i t\right) + C, \tag{6}$$

where Q_{oi} is the initial exchangeable fraction of element on the surface of the waste form, K_i is the rate constant for the exchange reaction (h^{-1}), U_i is the glass network dissolution rate (m·h^{-1}), and D_i is the effective diffusion coefficient of the element (m^2·h^{-1}). This equation is used in conditions when saturation effects are not important, such as the initial stage of glass dissolution.

3. Results and Discussion

3.1. Leaching Behavior

Elemental releases (Ci) for all the studied elements show an increasing pattern with time characterized by an initial slow portion (within 7 days) followed by steep increase rates (Figures 1–3). The release of alkaline earth metals from MABS-Waste (20) and ABS-Waste (25) is characterized by very slow rates (Figure 1e,f) and their normalized release rates are in conformance with published data for different ABS-waste matrices [30]. Glass formers have higher releases than that of Al and Te (Figure 2), and increasing the metal oxide loading led to a reduction in the releases for metalloid, post-transition, and transition elements (Figure 2). Finally, for rare earth elements, releases are characterized by a slow increase as time passes (Figure 3). The normalized release rates of alkali metals (Tables 5–7) are in conformance with reported data for ABS-Reactor Bolshoy Moshchnosty Kanalny (RBMK), ABS-Water-Water Energetic Reactor (WWER), K-26, and composite glass [31,32]. Sample ABS-waste (17) has the highest normalized release rates for most of the studied elements, whereas ABS-waste (25) has the lowest normalized release rates for formers, alkaline earth elements, and transition elements. The low values of boron's normalized leach rates suggest the formation of smectite alteration phases in the three samples at extended leaching times [37]. From the abovementioned data, it can be concluded that the releases for all studied elements are monotonically increasing with time, and the changes in the slope of the release-time represent a possible change in the controlling leaching mechanisms [7,9,11–14,27].

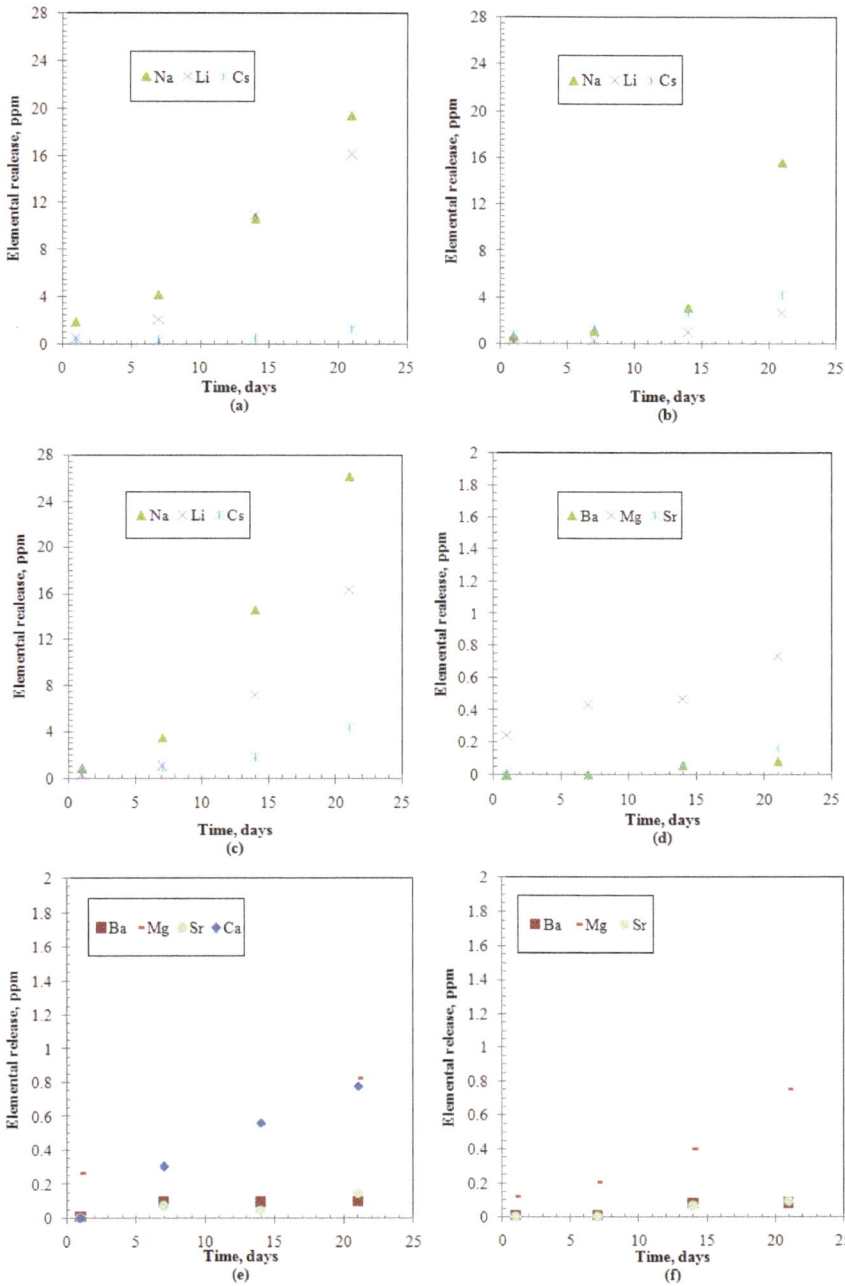

Figure 1. Elements releases from the studied samples: (**a**) Group I-Alkali-Borosilicate (ABS)-Waste (17); (**b**) Group I- Modified Alkali-Borosilicate (MABS)-Waste (20); (**c**) Group I-ABS-Waste (25); (**d**) Group II-ABS-Waste (17); (**e**) Group II-MABS- Waste (20); (**f**) Group II-ABS-Waste (25).

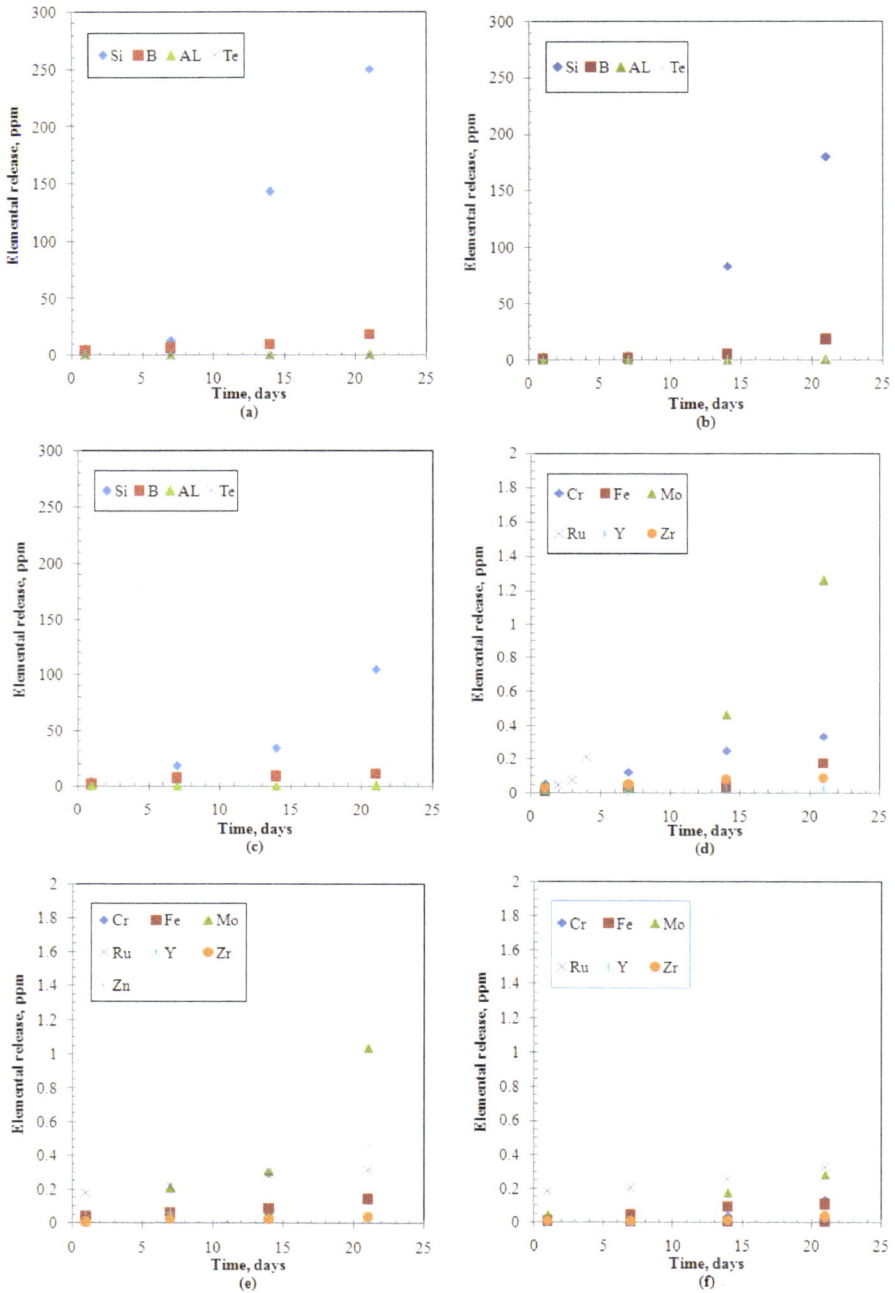

Figure 2. Elemental releases from the studied samples: (**a**) Metalloid and post-transition-ABS-Waste (17); (**b**) Metalloid and post-transition-MABS-Waste (20); (**c**) Metalloid and post-transition-ABS-Waste (25); (**d**) Transition-ABS-Waste (17); (**e**) Transition-MABS-Waste (20); (**f**) Transition-ABS-Waste (25).

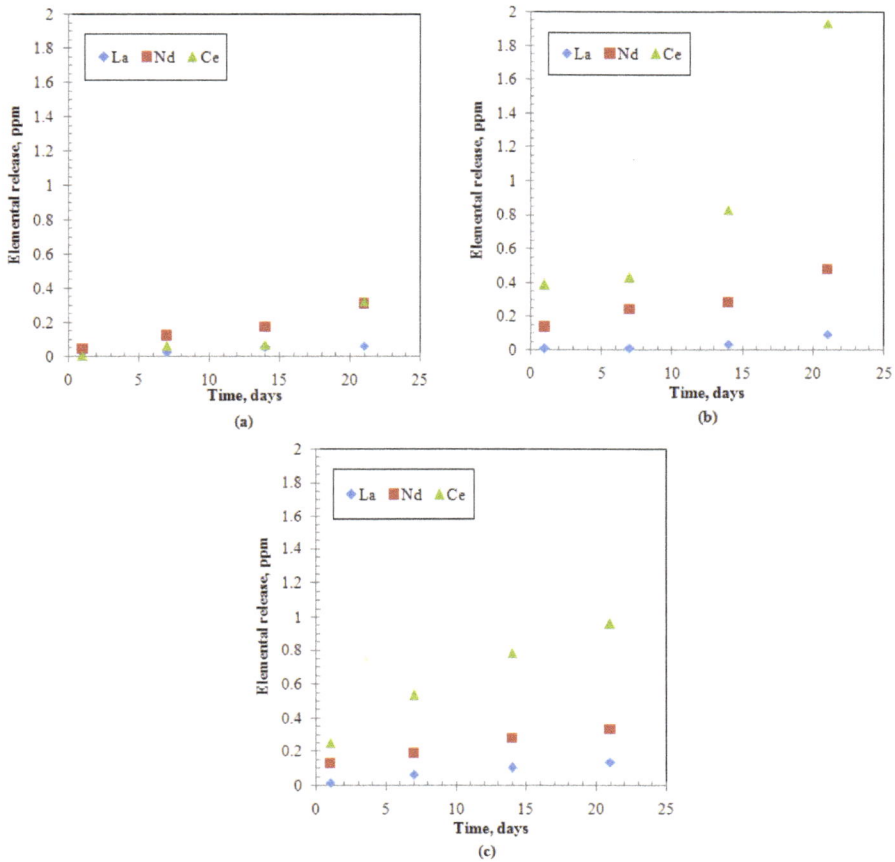

Figure 3. Rare earth elements releases from the studied samples: (**a**) ABS-Waste (17); (**b**) MABS-Waste (20); (**c**) ABS-Waste (25).

Table 5. The normalized release rate (mg·m^{-2}·d^{-1}) for structural elements and modifiers.

Compound	SiO$_2$	B$_2$O$_3$	Na$_2$O	Li$_2$O	CaO	ZnO
ABS-Waste (17)	50.814	36.024	28.301	37.940	-	-
MABS-Waste (20)	41.361	31.541	22.256	25.977	1.041	0.397
ABS-Waste (25)	22.918	20.121	40.363	32.122	-	-

Table 6. Normalized release rate (mg·m^{-2}·d^{-1}) for waste components: Alkali, alkaline, post-transitions, and metalloids.

Compound	Alkali	Alkaline Earth Metals			Post-Transitions and Metalloids	
	Cs$_2$O	BaO	MgO	SrO	Al$_2$O$_3$	TeO$_2$
ABS-Waste (17)	40.927	2.588	0.706	4.479	3.443	4.568
MABS-Waste (20)	47.539	1.288	1.587	3.447	2.038	1.493
ABS-Waste (25)	27.198	0.928	3.686	1.266	3.464	0.857

Table 7. Normalized release rate $(mg \cdot m^{-2} \cdot d^{-1})$ for waste components: transitions and rare earth elements.

Compound	Transition Metals					Rare Earth			
	Cr_2O_3	Fe_2O_3	MoO_3	RuO_2	ZrO_2	Y_2O_3	CeO_2	La_2O_3	Nd_2O_3
ABS-Waste (17)	15.648	1.766	12.841	6.458	0.670	2.818	3.781	6.639	8.578
MABS-Waste (20)	3.146	0.708	5.624	3.790	0.187	2.668	8.044	1.924	3.457
ABS-Waste (25)	3.590	0.696	0.811	3.726	0.071	0.702	3.852	2.017	1.670

NBO are formed in ABS-waste matrices due to the presence of alkali modifiers and the waste metal oxides (Equation (2)); a higher value of NBO fraction is indicative of glass matrix de-polymerization [15,34]. The silicon-to-boron (Si/B) ratio for all studied samples is greater than 2, which highlights the role of NBO in glass degradation and refers to the neglected effect of cluster detachment in this process [38]. ABS-waste (17) has the highest de-polymerization potential due to the presence of the largest fraction of higher field strength elements, i.e., Li and Mg represent 16.9%, that enhances BO_3 and NBO cluster formation [10,19,39,40]. The NBO are reduced during the progress of the leaching process due to modifiers and waste metal oxides releases; the overall NBO reduction is in the order ABS-Waste (25) > ABS-waste (17) > MABS-waste (20) (Figure 4a). It is noted that the MABS-waste (20) sample, which is the highest polymerized matrix, has a different NBO reduction pattern that is characterized by its slowest rate of NBO reduction within the first week. This behavior is accompanied by reduced silicon and boron releases (Figure 2b) and nearly unleached Zn (Figure 2e). This can be attributed to the nature of modifier incorporation in the matrix, where Ca incorporated in the vitreous structure of the matrix to compensate for the charge and Zn formed a spinel crystalline structure [10]. Although calcium has high field strength and is involved in the formation of NBO, the enhanced highest polymerization of this matrix might be related to the following [37,39–41]:

- The ratio between alkali and alkaline elements to boron is greater than 1, which led to enhanced calcium stabilization;
- Ca silicate has a lower hydration free energy compared to alkali elements silicates, which led to lower calcium hydration and subsequently a more stable sample.

Figure 4b quantifies the effect of glass former fraction evolution during the leaching process on glass matrix de-polymerization. A reducing linear pattern is noted, where the lowest NBO fractions (0.6–0.76) are noted for the unleached samples (higher glass former fraction content). As the leaching process continues, the glass former fraction is reduced and the NBO fraction increases. The linear dependency between the formers and NBO fractions indicates that both silicon and boron sites are linked to NBO [37]. ABS-waste (17) has the highest NBO fraction, which explains its higher normalized release rate, whereas the ABS-waste (25) has the lowest fraction. The linear regression coefficients are in the range (0.994–0.999), where the highest NBO fractions of fully degraded samples are in the range (1.7–1.8), and the degradation slope is in the order ABS-Waste (25) < MABS-waste (20) < ABS-waste (17).

Table 8 shows glass matrix degradation measures. It reveals that the fraction of the degraded glass increases with time and the highest degraded sample is ABS-waste (17), which is more stable than that of international simple glass [22]. The calculated alteration thickness for modified glass is similar to that of the experimentally deduced value of sample MABS-waste (20) [10]. The relations between the calculated ET values and the leaching time (t) and the Boron releases in terms of cumulative leach fraction of boron (CLF_B) were calculated via regression as illustrated in Table 8. The alteration thickness increases exponentially as the leaching period for ABS-waste (17) and MABS-waste (20) samples increases, whereas a linear dependence is noted with time for the ABS-Waste (25) sample. The linear dependence between the alteration thickness and the time was noted for some glass samples during a very short leaching experiment (t < 8 h) [19]. This indicates that the mechanism that controls that leaching process within the studied period is not diffusion [42,43]. It should be noted that the formed alteration layer is inhomogeneous, as it is formed under non-equilibrium conditions, and the main driving degradation force is the matrix chemical composition within the studied period [44]. The investigations of the relation between alteration layer thickness and boron cumulative leach fraction shows a linear dependency, where the formation of the alteration layer is the most sensitive in the case of ABS-waste (17).

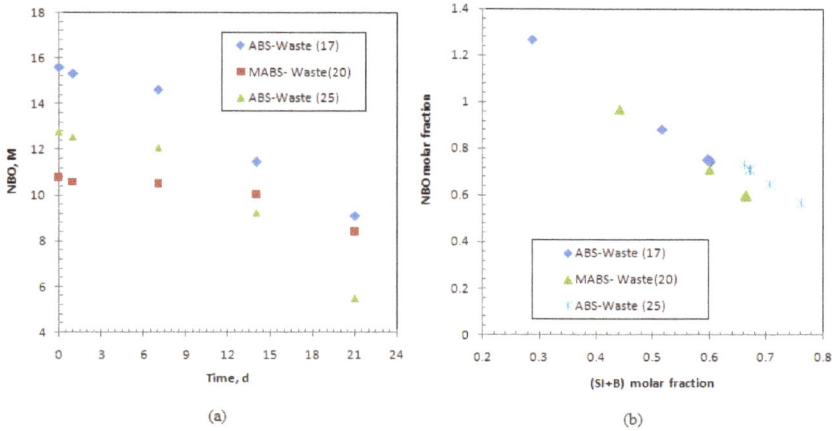

(a) (b)

Figure 4. The evolution of glass matrix composition during the leaching process: (**a**) temporal changes in non-bridging oxygen (NBO); (**b**) The NBO fraction as a function of former fraction.

Table 8. The evolution and dependency of glass degradation measures.

Glass Sample	AGF% $\times 10^{-4}$				ET (μm)			
	1 d	7 d	14 d	21 d	Time Dependency	R^2	Boron Release Dependency	R^2
ABS-Waste (17)	5.419	10.437	16.423	50.337	$ET = 0.466e^{0.106t}$	0.969	$ET = 9.032CLF_B - 0.842$	0.976
MABS-Waste (20)	0.780	1.832	6.889	42.672	$ET = 0.052e^{0.199t}$	0.983	$ET = 8.023CLF_B - 0.235$	0.989
ABS-Waste (25)	1.629	10.448	13.540	18.685	$ET = 0.080t + 0.244$	0.943	$ET = 5.395CLF_B - 0.072$	0.997

3.2. Hydration Free Energy of the Studied Matrices

The hydration free energies of the matrices were −6.7, −5.45, and −6.0 kcal/mol for ABS-waste (17), MABS-waste (20), and ABS-Waste (25), respectively. These values refer to the spontaneous nature of the hydration reaction that is reduced with increasing the metal oxide loading. The use of calcium and zinc additives has reduced this spontaneous nature of the reaction. The contribution of the glass constituents to the hydration free energy is shown in Figure 5. It is clear that the presence of the rare earth elements does not contribute to the hydration reaction, which is attributed to their low content and small hydration energy. These elements could be used to stabilize the hydration reaction. Alkali metals have the highest contribution to the hydration reaction and this contribution is reduced by increasing the metal oxide loading and additive presence. Transition metals have considerable effect on the hydration reaction, and this effect increases as the metal oxide loading increases. The contribution of alkaline metals, metalloids, and post-transition elements to the hydration reaction is slightly affected by the metal oxide loading or the additive presence. It should be noted that the contribution of Li and Mg to the overall hydration free energy of the sample ABS-waste (17) represents 46.73%, which is reduced to 18.92% and 27.22% for the samples MABS-waste (20), and ABS-Waste (25), respectively. So, it could be concluded that the presence of Li and Mg had led to the higher degradation of the sample ABS-waste (17), as their presence increases the thermodynamic instability of the sample by increasing the hydration free energy. Reported studies indicated that the presence of Na- and Mg-silicates have reduced the glass stability [35].

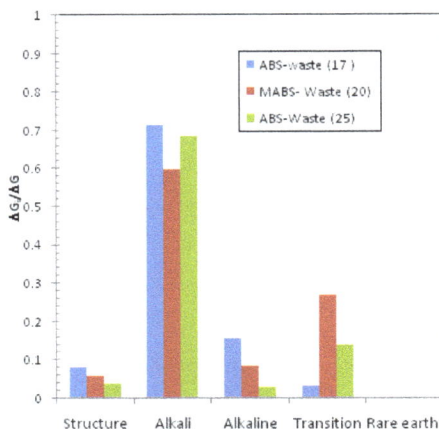

Figure 5. Contribution of the waste matrix constituents to the hydration free energy.

3.3. Leaching Mechanism of Structural Elments and Metaloids

The controlling leaching mechanism is preliminarily screened by plotting the release of structural glass elements (Si and B) as a function of square root time; linear plots indicate the diffusion-controlled process [7,9,11,24,26,43]. Visual examination of the experimental patterns for both silicon and boron show non-linear dependency between elemental release in the leachant and the square root of time for the sample that contains the lowest metal oxide loading. As the waste loading increases, a weak linear dependency starts to appear (Figure 6a–c). The mask of the linear dependency reflects that the dominant leaching process is congruent dissolution not diffusion [10,24,43]. This indicates that, as the metal oxide loading increases, the diffusion through the matrix or the ion-exchange mechanism plays an important role in determining the leaching characteristics. An earlier study on the characterization of sample MABS-waste (20) showed that ion-exchange contributed to the leaching mechanism after 7 days of the leaching experiment [10].

To identify the controlling mechanism and the effect of the metal oxide loading on the mechanisms, the experimental data were fitted to the collective leaching model. Tables 9–11 list the fitting parameters for metalloids and post-transition elements incorporated in the waste matrices; it is obvious that diffusion only contributes to the release of boron (i.e., the diffusion coefficient has a significant value) from the highest metal oxide waste. Silicon release takes place via dissolution and a first-order reaction independently on the mixed oxide incorporation percentage. This also applies to boron release, except for low metal oxide incorporation (sample ABS-waste (17%)), where some fraction of loosely bounded boron is released. The loosely bounded boron fraction is independent of time and could be related to the reduced polymerization due to the presence of Li and Mg [38]. The maximum dissolution rates for both elements are the highest for the ABS-waste (17) sample and decreased with increasing the metal oxide loading. Figure 7a shows that linear dissolution is the main leaching mechanism that causes the release of both structural elements from ABS-waste matrices (17 and 25%). This finding is in conformance with the interfacial dissolution-reprecipitation theory that proposes dissolution of structural elements as the controlling process in the initial stage of glass degradation [39,40,45]. For the MABS-waste (20) matrix, the main leaching process is a first-order reaction, which could be attributed to the absence of a large ring of silica tetrahedrons that limit the water diffusion into the matrix as a result of matrix modification [40,46]. It is clear that, as the waste loading increases, the contribution of the dissolution process to the overall release of silicon and boron decreases by 43.44 and 5.05%, respectively, and the presence of modifiers reduces this contribution by 56.19 and 65.60% for silicon and boron, respectively.

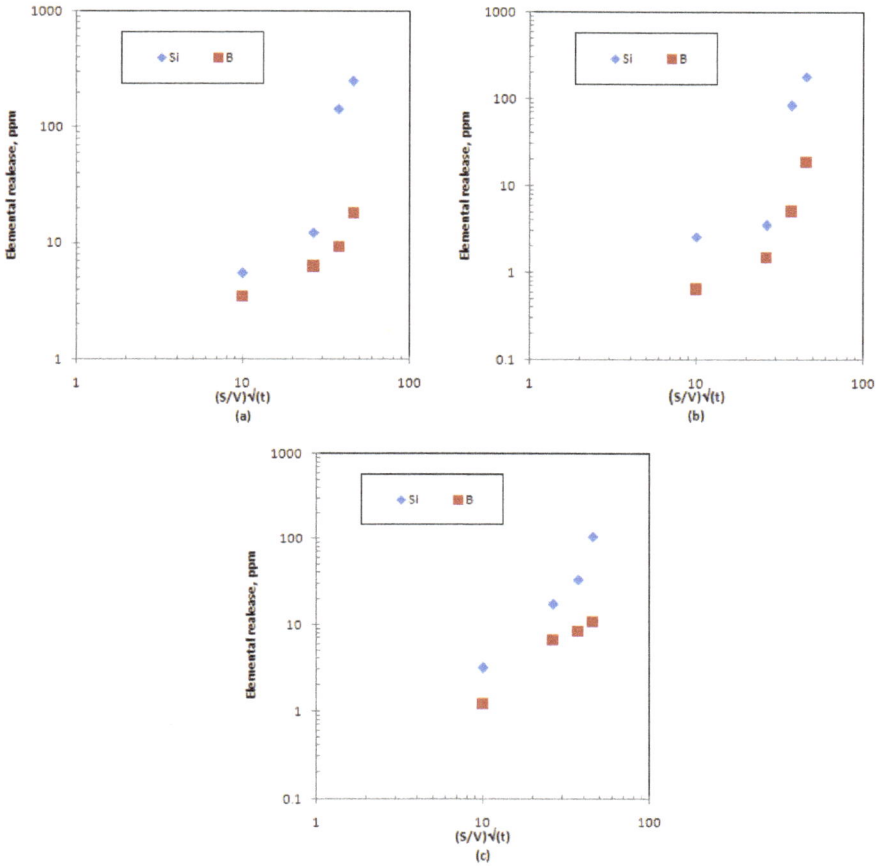

Figure 6. Preliminary investigation of structural element leaching mechanisms for samples: (**a**) ABS-waste (17); (**b**) MABS-waste (20); (**c**) ABS-waste (25).

Table 9. Nonlinear curve fitting parameters of the cumulative leach fraction of metalloid and post-transition elements: ABS-Waste (17).

Element	D (m²·h⁻¹), × 10⁻¹³	U (m·h⁻¹) × 10⁻⁷	Q$_o$,	K (h⁻¹) × 10⁻⁸	C × 10⁻⁴	R²
Si	0	1.583	0.004	938.143	0	0.912
B	0	1.013	0.001	619.759	62.200	0.940
Te	0	0.127	0.105	49.556	0	0.830
Al	0	0.074	0.058	0.845	22.400	0.879

Table 10. Nonlinear curve fitting parameters of the cumulative leach fraction of metalloid and post-transition elements: MABS-Waste (20).

Element	D (m²·h⁻¹), × 10⁻¹³	U (m·h⁻¹) × 10⁻⁷	Q$_o$,	K (h⁻¹) × 10⁻⁸	C × 10⁻⁴	R²
Si	0	1.211	0.094	10.397	0	0.934
B	0	0.851	0.152	4.011	0	0.899
Te	0	0.033	0.084	0.122	7.264	0.806
Al	0	0.065	0.009	14.107	1.576	0.917

Table 11. Nonlinear curve fitting parameters of the cumulative leach fraction of metalloid and post-transition elements: ABS-Waste (25).

Element	$D\ (m^2 \cdot h^{-1}),\ \times 10^{-13}$	$U\ (m \cdot h^{-1}) \times 10^{-7}$	Q_o	$K\ (h^{-1}) \times 10^{-8}$	$C \times 10^{-4}$	R^2
Si	0	0.648	0.003	1.630	0	0.868
B	0.678	0.303	3.20×10^{-4}	0.149	0	0.967
Te	N*	0.028	0.002	1.592	1.879	0.902
Al	0	0.102	8.05×10^{-4}	0.747	0	0.886

N* neglected value.

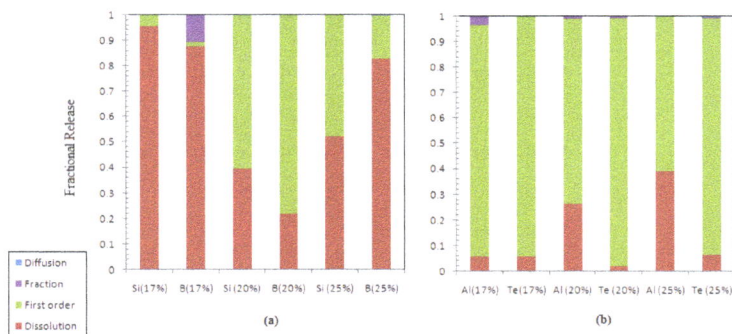

Figure 7. Contribution of different leaching processes to the fractional release: (**a**) structural element; (**b**) post-transition and other metalloid elements.

On the other hand, the fractional releases of Al as a post-transition element and Te as a metalloid waste component are mainly controlled by the first-order reaction (Figure 7b). A small fraction of Al release could be attributed to the instantaneous leaching of loosely bound Al in the sample (ABS-Waste (17)). This fraction was not noted for the other samples; this might be due to the effect of the modifier and the decreased Al loading, where a higher Al loading can create an Al cluster and large silicon rings [38]. The contribution of the dissolution process to Te release is fairly constant independently of its loading, except for the modified sample that has a lower contribution to the dissolution.

3.4. Leaching Characteristics of Alkali and Alkaline Earth Metals

Tables 12–14 list the estimated leaching parameters, revealing that the diffusion of alkali and alkaline earth metals does not play any role in controlling their leaching behavior at any waste loading. To quantify the role of each mechanism in the overall cumulative leaching fraction, the contribution of each mechanism was plotted and is shown in Figure 8. It is clear that congruent dissolution of Li and Cs is the major mechanism for ABS-waste (17) and MABS-waste (20). As the metal oxide loading increases, the first-order exchange reaction becomes a dominant leaching process. The increase in the waste loading from 17 to 25% reduced the contribution of the dissolution mechanism to the release by 38.07, 52.99, and 31.25% for Na, Li, and Cs, respectively. Alkaline metal leaching is controlled by a first-order exchange reaction. This notable change in the controlling leaching process for alkali and alkaline metals could be attributed to the higher field strength of the alkaline metals that leads to glass stabilization [46].

Table 12. Nonlinear curve fitting parameters of the cumulative leach fraction of alkali and alkaline earth metals: ABS-Waste (17).

Group	Element	D (m²·h⁻¹), × 10⁻¹³	U (m·h⁻¹), × 10⁻⁷	Q_o,	K (h⁻¹), × 10⁻⁸	C × 10⁻⁴	R^2
	Na	0	0.904	0.055	3.444	3.192	0.966
Alkali metals	Li	0	1.266	0.038	0.0001	0	0.952
	Cs	0	1.119	0.028	1.935	49.301	0.870
	Ba	0	0.075	0.034	0.112	0	0.907
Alkaline earth metals	Mg	1.570	0.009	0.223	2.407	0.975	0.914
	Sr	0	0.126	0.054	0.124	0	0.838

Table 13. Nonlinear curve fitting parameters of the cumulative leach fraction of alkali and alkaline earth metals: MABS-Waste (20).

Group	Element	D (m²·h⁻¹), × 10⁻¹³	U (m·h⁻¹), × 10⁻⁷	Q_o,	K (h⁻¹), × 10⁻⁸	C × 10⁻⁴	R^2
	Na	0	0.576	0.009	0.287	0	0.867
Alkali metals	Li	0	0.752	0.006	0.558	0	0.876
	Cs	0	1.399	0.002	0.193	0.009	0.974
	Ca	0	0.133	0.088	39.593	0.039	0.989
Alkaline earth metals	Ba	9.159	0.002	8.2*10⁻⁴	0.196	0	0.805
	Mg	0	0.041	0.084	0.289	5.844	0.968
	Sr	0	0.093	0.041	5.769	0	0.902

Table 14. Nonlinear curve fitting parameters of the cumulative leach fraction of alkali and alkaline earth metals: ABS-Waste (25).

Group	Element	D (m²·h⁻¹), × 10⁻¹³	U (m·h⁻¹), × 10⁻⁷	Q_o,	K (h⁻¹), × 10⁻⁸	C × 10⁻⁴	R^2
	Na	0	1.273	0.084	2.157	0	0.949
Alkali metals	Li	0	1.595	0.075	37.659	0	0.890
	Cs	0	0.743	0.008	3.074	0.003	0.867
	Ba	0	0.029	0.758	0.629	0.333	0.869
Alkaline earth metals	Mg	0	0.109	0.006	0.139	4.169	0.952
	Sr	0	0.042	0.001	359.254	0	0.907

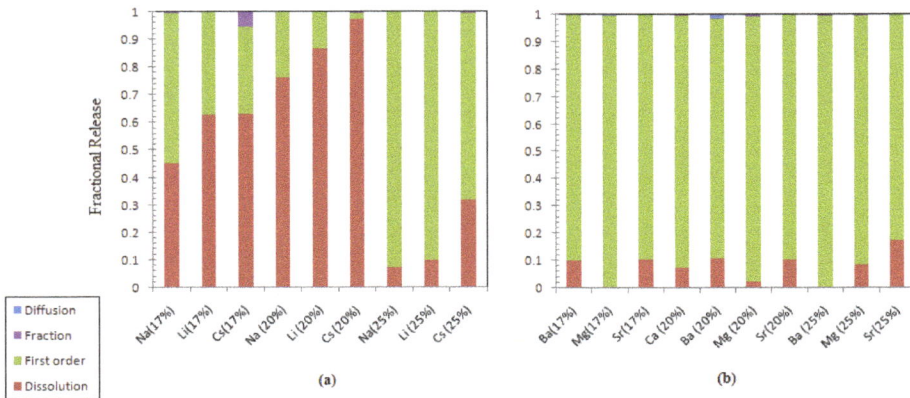

Figure 8. Contribution of different leaching processes to the fractional release: (**a**) alkali elements; (**b**) alkaline earth metals.

3.5. Leaching Characteristics of the Transition and Rare Earth Elements

The leaching parameters as estimated from the nonlinear regression of the experimental data to the collective model for transition and rare earth elements are given in Tables 15–17, and the contribution of each leaching process to the overall release fraction is presented in Figures 9 and 10. Ru and Mo release from ABS-Waste (17) sample is only controlled by the dissolution, and the rest of the releases are controlled by the first-order model. Increasing the metal oxide loading can lead to the formation of spinels that are used to immobilize transition metal ions [10].

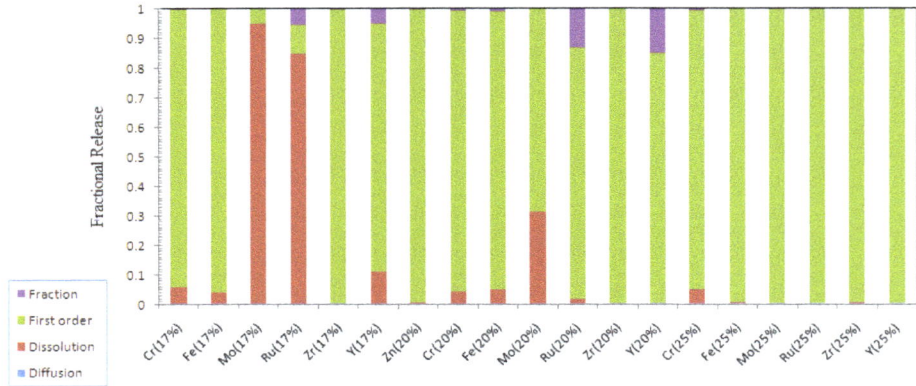

Figure 9. Contribution of different leaching processes to the fractional release of transition metals.

Table 15. Nonlinear curve fitting parameters of the cumulative leach fraction of transition and rare earth elements: ABS-Waste (17).

Group	Element	D (m²·h⁻¹), × 10⁻¹³	U (m·h⁻¹), × 10⁻⁷	Q_o,	K (h⁻¹), × 10⁻⁸	C × 10⁻⁴	R²
	Cr	0	0.516	0.415	2.883	16.600	0.994
	Fe	0	0.005	0.006	0.239	0	0.876
Transition	Mo	0	0.359	0.001	1.302	0	0.917
elements	Ru	0	0.172	0.001	0.747	5.654	0.898
	Zr	2.047	N*	0.022	0.609	0.600	0.961
	Y	0	0.063	0.0247	0.913	15.000	0.892
Rare earth	Ce	0	0.100	0.070	0.583	0	0.885
Elements	La	0	0.027	0.241	0.003	2.326	0.926
	Nd	0	0.026	0.110	0.352	1.033	0.971

N* neglected value.

Table 16. Nonlinear curve fitting parameters of the cumulative leach fraction of transition and rare earth elements: MABS-Waste (20).

Group	Element	D (m²·h⁻¹), × 10⁻¹³	U (m·h⁻¹), × 10⁻⁷	Q_o,	K (h⁻¹), × 10⁻⁸	C × 10⁻⁴	R²
	Zn	0	0.003	0.033	0.341	0.103	0.968
	Cr	0	0.078	0.087	2.343	8.234	0.857
Transition	Fe	0	0.018	0.017	0.105	2.321	0.914
elements	Mo	1.294	0.060	0.007	0.157	0	0.962
	Ru	27.347	0.005	0.014	42.141	22.100	0.837
	Zr	0.118	N*	0.034	0.137	0	0.946
	Y	0	0	0.005	358×10^3	8.762	0.953
Rare earth	Ce	0	0.212	0.048	0.005	24.600	0.929
Elements	La	0	0.005	0.020	350.372	0	0.83
	Nd	0	0.089	0.144	7.292	11.900	0.936

N* neglected value.

Table 17. Nonlinear curve fitting parameters of the cumulative leach fraction of transition and rare earth elements: ABS-Waste (25).

Group	Element	D (m²·h⁻¹), × 10⁻¹³	U (m·h⁻¹), × 10⁻⁷	Q_o	K (h⁻¹), × 10⁻⁸	C × 10⁻⁴	R^2
	Cr	0	0.085	0.008	742.683	6.376	0.721
	Fe	N*	0.019	0.021	21.050	0	0.974
Transition	Mo	0	0.025	0.072	9.901	0.313	0.871
elements	Ru	0	0.071	0.583	0.135	28.400	0.823
	Zr	0	0.023	0.018	2.098	0	0.919
	Y	0	0.002	0.015	4.655	0.128	0.815
Rare earth	Ce	64.407	0.003	0.062	393.147	0.207	0.983
Elements	La	2.602	0.045	0.010	0.135	0	0.993
	Nd	3.063	0.019	0.038	0.279	5.912	0.944

N* neglected value.

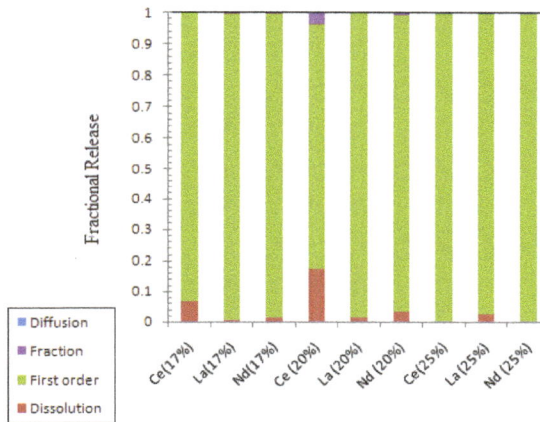

Figure 10. Contribution of different leaching processes to the fractional release of rare earth elements.

4. Conclusions

Leaching characteristics of different structural elements, modifiers, and waste components were investigated for three alkali-borosilicate-mixed oxide glasses that represent different waste loadings. The main concluding remarks from this work are as follows:

1. The normalized release rates of the studied elements are in conformance with data reported in the literature for borosilicate waste glass matrices.
2. Elemental releases monotonically increase with time; the changes in the slope of the Release-Time represent a possible change in the controlling leaching mechanism.
3. The high incorporation of Li and Mg in the ABS-waste (17) glass led to a high de-polymerization of glass and contributed to the thermodynamic instability of the matrix.
4. The MABS-waste (20) glass has the slowest rate of NBO reduction due to the incorporation of Ca as matrix modifier of low hydration free energy which increased the thermodynamic stability against a hydration reaction.
5. Rare earth elements could be used to stabilize the glass hydration reactions.
6. The alteration thickness increases exponentially with increasing the leaching period for the ABS-waste (17) and MABS-waste (20) samples, whereas a linear dependence is noted with time for the ABS-Waste (25) sample.
7. The alteration layer thickness is linearly dependent on boron's cumulative leach fraction and the formation of the alteration layer is the most sensitive in the case of ABS-waste (17) glass.

8. As the waste loading increases, the contribution of the dissolution process to the overall fractional release of structural elements decreases and the presence of modifiers reduces this contribution for all the studied metalloids.
9. The use of Zn and Ca modifiers could reduce the instantaneous release of Al.
10. The initial fractional release of alkaline earth metals and transition and rare earth elements is mostly controlled by the first-order reaction process, with notable exceptions for Mo and Ru.

Author Contributions: O.M.F. designed and performed the experiments; R.O.A.R. conceptualized the theoretical modeling and data analysis; O.M.F., M.I.O., A.M. and R.O.A.R. contributed equally to the writing of the original draft; M.I.O. and R.O.A.R. reviewed and edited the final draft.

Funding: This research was funded by the Center of Nuclear Engineering (CNE), Department of Materials, Imperial College London (EP/I036400/1).

Conflicts of Interest: The authors declare no conflict of interest.

References

1. Abdel Rahman, R.O.; Kozak, M.W.; Hung, Y.-T. Radioactive pollution and control. In *Handbook of Environment and Waste Management*, 1st ed.; Hung, Y.T., Wang, L.K., Shammas, N.K., Eds.; World Scientific Publishing Co.: Singapore, 2014; pp. 949–1027.
2. Abdel Rahman, R.O.; El Kamash, A.M.; Zaki, A.A.; El Sourougy, M.R. Disposal: A Last Step Towards an Integrated Waste Management System in Egypt. In Proceedings of the International Conference on the Safety of Radioactive Waste Disposal, Tokyo, Japan, 3–7 October 2005; IAEA-CN-135/81, pp. 317–324.
3. Abdel Rahman, R.O.; Saleh, H.M. Introductory chapter: Safety aspects in nuclear engineering. In *Principles and Applications in Nuclear Engineering: Radiation Effects, Thermal Hydraulics, Radionuclide Migration in the Environment*, 1st ed.; Abdel Rahman, R.O., Saleh, H.M., Eds.; IntechOpen: London, UK, 2018; pp. 1–14.
4. Abdel Rahman, R.O.; Rakhimov, R.Z.; Rakhimova, N.R.; Ojovan, M.I. *Cementitious Materials for Nuclear Waste Immobilisation*; Wiley: New York, NY, USA, 2014; pp. 1–26.
5. Abdel Rahman, R.O.; Ibrahim, H.A.; Hanafy, M.; Abdel Monem, N.M. Assessment of Synthetic Zeolite NaA-X as Sorbing Barrier for Strontium in a Radioactive Disposal Facility. *Chem. Eng. J.* **2010**, *157*, 100–112. [CrossRef]
6. Abdel Rahman, R.O.; Zein, D.H.; Abo Shadi, H. Cesium binding and leaching from single and binary contaminant cement-bentonite matrices. *Chem. Eng. J.* **2014**, *245*, 276–287. [CrossRef]
7. Abdel Rahman, R.O.; Zaki, A.A. Comparative study of leaching conceptual models: Cs leaching from different ILW cement based matrices. *Chem. Eng. J.* **2011**, *173*, 722–736. [CrossRef]
8. Abdel Rahman, R.O.; El-Kamash, A.M.; Zaki, A.A. Modeling the Long Term Leaching Behavior of 137Cs, 60Co, and 152,154Eu Radionuclides from Cement-Clay Matrices. *Hazard. Mater.* **2007**, *145*, 372–380. [CrossRef]
9. Abdel Rahman, R.O.; Zein, D.H.; Abo Shadi, H. Assessment of strontium immobilization in cement–bentonite matrices. *Chem. Eng. J.* **2013**, *228*, 772–780.
10. Farid, O.M.; Abdel Rahman, R.O. Preliminary Assessment of Modified Borosilicate Glasses for Chromium and Ruthenium Immobilization. *Mater. Chem. Phys.* **2017**, *168*, 462–469. [CrossRef]
11. Ojovan, M.I.; Lee, W.E. *An Introduction to Nuclear Waste Immobilisation*, 2nd ed.; Elsevier: Amsterdam, The Netherlands, 2014; p. 362.
12. Lee, W.E.; Ojovan, M.I.; Jantzen, C.M. *Radioactive Waste Management and Contaminated Site Clean-Up: Processes, Technologies and International Experience*; Woodhead: Cambridge, UK, 2013; p. 924.
13. Ojovan, M.I. *Handbook of Advanced Radioactive Waste Conditioning Technologies*; Woodhead: Cambridge, UK, 2011; p. 512.
14. The National Academies Press. *Waste Forms Technology and Performance: Final Report. Committee on Waste Forms Technology and Performance*; National Research Council: Washington, DC, USA, 2011; p. 340, ISBN 0-309-18734-6.
15. *Engineered Barrier Systems and the Safety of Deep Geological Repository: State of the Art Report*; Nuclear Energy Agency: Paris, France, 2003; ISBN 92-64-18498-8.

16. Ma, T.; Jivkov, A.P.; Li, W.; Liang, W.; Wang, Y.; Xu, H.; Han, X. A mechanistic model for long-term nuclear waste glass dissolution integrating chemical affinity and interfacial diffusion barrier. *J. Nucl. Mater.* **2017**, *486*, 70–85. [CrossRef]

17. Jantzen, C.M.; Trivelpiece, C.L.; Crawford, C.L.; Pareizs, J.M.; Pickett, J.B. Accelerated Leach Testing of GLASS (ALTGLASS): I. Informatics approach to high level waste glass gel formation and aging. *Int. J. Appl. Glass Sci.* **2017**, *8*, 69–83. [CrossRef]

18. Neeway, J.J.; Rieke, P.C.; Parruzot, B.P.; Ryan, J.V.; Asmussen, R.M. The dissolution behavior of borosilicate glasses in far-from equilibrium conditions. *Geochim. Cosmochim. Acta* **2018**, *226*, 132–148. [CrossRef]

19. Guoa, R.; Brigdena, C.T.; Ginb, S.; Swantonc, S.W.; Farnana, I. The effect of magnesium on the local structure and initial dissolution rate of simplified UK Magnox waste glasses. *J. Non-Cryst. Solids* **2018**, *497*, 82–92. [CrossRef]

20. Geisler, T.; Nagel, T.; Kilburn, M.R.; Janssen, A.; Icenhower, J.P.; Fonseca, R.O.C.; Grange, M.; Nemchin, A.A. The mechanism of borosilicate glass corrosion revisited. *Geochim. Cosmochim Acta* **2015**, *158*, 112–129. [CrossRef]

21. Vienna, J.D.; Neeway, J.J.; Ryan, J.V.; Kerisit, S.N. Impacts of glass composition, pH, and temperature on glass forward dissolution rate. *NPJ Mater. Degrad.* **2018**, *2*, 22. [CrossRef]

22. Gin, S.; Jollivet, P.; Fournier, M.; Berthon, C.; Wang, Z.; Mitroshkov, A.; Zhu, Z.; Ryan, J.V. The fate of silicon during glass corrosion under alkaline conditions: A mechanistic and kinetic study with the International Simple Glass. *Geochim. Cosmochim. Acta* **2015**, *151*, 68–85. [CrossRef]

23. Inagaki, Y.; Kikunaga, T.; Idemitsu, K.; Arima, T. Initial Dissolution Rate of the International Simple Glass as a Function of pH and Temperature Measured Using Microchannel Flow-Through Test Method. *Int. J. Appl. Glass Sci.* **2013**, *4*, 317–327. [CrossRef]

24. Van Iseghem, P.; Aertsens, M.; Gin, S.; Deneele, D.; Grambow, B.; McGrail, P.; Strachan, D.; Wicks, G. *A Critical Evaluation of the Dissolution Mechanisms of High-level Waste Glasses in Conditions of Relevance for Geological Disposal (GLAMOR)*; EUR 23097; European Commission: Paris, France, 2007.

25. Grambow, B.; Muller, R. First order dissolution rate law and the role of surface layers in glass performance assessment. *J. Nucl. Mater.* **2001**, *298*, 112–124. [CrossRef]

26. Jantzen, C.M.; Brown, K.G.; Pickett, J.B. Durable Glass for Thousands of Years. *Int. J. Appl. Glass Sci.* **2010**, *1*, 38–62. [CrossRef]

27. Manaktala, H.K. *An Assessment of Borosilicate Glass as a High-Level Waste Form*; CNWRA 92-017; Nuclear regulatory commission-USA: San Antonio, TX, USA, 1992.

28. Padmaja, G.; Kistaiah, P. Optical absorption and EPR spectroscopic studies of $(30−x)Li_2O-xK_2O-10CdO-59B_2O_3-1Fe_2O_3$: An evidence for mixed alkali effect. *Solid State Sci.* **2010**, *12*, 2015–2019. [CrossRef]

29. Heath, P.G.; Corkhill, C.L.; Stennett, M.C.; Hand, R.J.; Whales, K.M.; Hyatt, N.C. Immobilisation of Prototype Fast Reactor raffinate in a barium borosilicate glass matrix. *J. Nucl. Mater.* **2018**, *508*, 203–211. [CrossRef]

30. Lifanov, F.A.; Ojovan, M.I.; Stefanovsky, S.V.; Burc, R. Cold crucible vitrification of NPP operational waste. *Mater. Res. Soc. Symp. Proc.* **2003**, *757*, 1–6. [CrossRef]

31. Ojovan, N.V.; Startceva, I.V.; Barinov, A.S.; Ojovan, M.I.; Bacon, D.H.; McGrail, B.P.; Vienna, J.D. Product consistency test of fully radioactive high-sodium content borosilicate glass K-26. *Mater. Res. Soc. Symp. Proc.* **2004**, *824*, 1–6. [CrossRef]

32. Aloy, A.S.; Iskhakova, O.A.; Trofimenko, A.V.; Shakhmatkin, B.A.; Jardine, L.J. Chemical durability of borosilicate and phosphate glasses with high content of plutonium. UCRL-JC-135066. In Proceedings of the European Nuclear Society '99, Radioactive Waste Management: Commitment to the Future Environment Antwerp, Brussels, Belgium, 10–14 October 1999.

33. Kim, C.W.; Lee, B.G. Feasibility Study on Vitrification for Rare Earth Wastes of PyroGreen Process. *J. Korean Radioact. Waste Soc.* **2013**, *11*, 1–9. [CrossRef]

34. Geldart, R.W.; Kindle, C.H. *The Effects of Composition on Glass Dissolution Rates: The Application of Four Models to a Data Base*; PNL, 6333; Pacific Northwest Lab.: Richland, WA, USA, 1988.

35. Perret, D.; Crovisier, J.L.; Stille, P.; Shields, G.; Advocat, U.M.T.; Schenk, K.; Chardonnens, M. Thermodynamic stability of waste glasses compared to leaching behavior. *Appl. Geochem.* **2003**, *18*, 1165–1184. [CrossRef]

36. Drace, Z.; Mele, I.; Ojovan, M.I.; Abdel Rahman, R.O. An overview of research activities on cementitious materials for radioactive waste management. *Mater. Res. Soc. Symp. Proc.* **2012**, *1475*, 253–264. [CrossRef]

37. Inagaki, Y.; Shinkai, A.; Idemistu, K.; Arima, T.; Yoshikawa, H.; Yui, M. Aqueous alteration of Japanese simulated waste glass P0798: Effects of alteration-phase formation on alteration rate and cesium retention. *J. Nucl. Mater.* **2006**, *354*, 171–184. [CrossRef]

38. Pierce, E.M.; Windisch, C.F.; Burton, S.D.; Bacon, D.H.; Cantrell, K.J.; Serne, R.J.; Kerisit, S.N.; Valenta, M.M.; Mattigod, S.V. *Integrated Disposal Facility FY2010 Glass Testing Summary Report*; PNNL-19736; U.S. Department of Energy: Washington, DC, USA, 2010.

39. Wu, J.; Stebbins, J.F. Temperature and modifier cation field strength effects on aluminoborosilicate glass network structure. *J. Non-Cryst. Solids* **2013**, *362*, 73–81. [CrossRef]

40. Bunker, B.C. Molecular mechanisms for corrosion of silica and silicate glasses. *J. Non-Cryst. Solids* **1994**, *179*, 300–308. [CrossRef]

41. Wu, J.; Stebbins, J.F. Cation Field Strength Effects on Boron Coordination in Binary Borate Glasses. *J. Am. Ceram. Soc.* **2014**, *97*, 2794–2801. [CrossRef]

42. Bouakkaz, R.; Abdelouas, A.; Grambow, B. Kinetic study and structural evolution of SON68 nuclear waste glass altered from 35 to 125 °C under unsaturated H_2O and D_2O_{18} vapour conditions. *Corros. Sci.* **2018**, *134*, 1–16. [CrossRef]

43. Ojovan, M.I.; Hand, R.J.; Ojovan, N.V.; Lee, W.E. Corrosion of alkali-borosilicate waste glass K-26 in non-saturated conditions. *J. Nucl. Mater.* **2005**, *340*, 12–24. [CrossRef]

44. Frankel, G.S.; Vienna, J.D.; Lian, J.; Scully, J.R.; Gin, S.; Ryan, J.V.; Wang, J.; Kim, S.H.; Wind, W.; Du, J. A comparative review of the aqueous corrosion of glasses, crystalline ceramics, and metals. *NPJ Mater. Degrad.* **2018**, *2*, 15. [CrossRef]

45. Hellmann, R.; Cotte, S.; Cadel, E.; Malladi, S.; Karlsson, L.S.; Lozano-Perez, S.; Cabié, M.; Seyeux, A. Nanometre-scale evidence for interfacial dissolution–reprecipitation control of silicate glass corrosion. *Nat. Mater.* **2015**, *14*, 307–311. [CrossRef] [PubMed]

46. Gin, S.; Neill, L.; Fournier, M.; Frugier, P.; Ducasse, T.; Tribet, M.; Abdelouas, A.; Parruzot, B.; Neeway, J.; Wall, N. The controversial role of inter-diffusion in glass alteration. *Chem. Geol.* **2016**, *440*, 115–123. [CrossRef]

materials

MDPI

Article

The Effect of Heavy Ion Irradiation on the Forward Dissolution Rate of Borosilicate Glasses Studied In Situ and Real Time by Fluid-Cell Raman Spectroscopy

Mara Iris Lönartz [1,*], Lars Dohmen [2], Christoph Lenting [1], Christina Trautmann [3], Maik Lang [4] and Thorsten Geisler [1]

[1]   Institut für Geowissenschaften und Meteorologie, Universität Bonn, Poppelsdorfer Schloss, Meckenheimer Allee 169, 53115 Bonn, Germany; cl@uni-bonn.de (C.L.); tgeisler@uni-bonn.de (T.G.)
[2]   SCHOTT AG, Hattenbergstr. 10, 55122 Mainz, Germany; lars.dohmen@schott.com
[3]   GSI Helmholtzzentrum, 64291 Darmstadt and Technische Universität Darmstadt, 64287 Darmstadt, Germany; c.trautmann@gsi.de
[4]   Department of Nuclear Engineering, University of Tennessee, Knoxville, TN 37996, USA; mlang2@utk.edu
*   Correspondence: mara.loenartz@uni-bonn.de; Tel.: +49-0228-73-2761

Received: 27 March 2019; Accepted: 1 May 2019; Published: 7 May 2019

Abstract: Borosilicate glasses are the favored material for immobilization of high-level nuclear waste (HLW) from the reprocessing of spent fuel used in nuclear power plants. To assess the long-term stability of nuclear waste glasses, it is crucial to understand how self-irradiation affects the structural state of the glass and influences its dissolution behavior. In this study, we focus on the effect of heavy ion irradiation on the forward dissolution rate of a non-radioactive ternary borosilicate glass. To create extended radiation defects, the glass was subjected to heavy ion irradiation using ^{197}Au ions that penetrated ~50 μm deep into the glass. The structural damage was characterized by Raman spectroscopy, revealing a significant depolymerization of the silicate and borate network in the irradiated glass and a reduction of the average boron coordination number. Real time, in situ fluid-cell Raman spectroscopic corrosion experiments were performed with the irradiated glass in a silica-undersaturated, 0.5 M $NaHCO_3$ solution at temperatures between 80 and 85 °C (initial pH = 7.1). The time- and space-resolved in situ Raman data revealed a 3.7 ± 0.5 times increased forward dissolution rate for the irradiated glass compared to the non-irradiated glass, demonstrating a significant impact of irradiation-induced structural damage on the dissolution kinetics.

Keywords: borosilicate glass corrosion; heavy ion irradiation; in situ fluid-cell Raman spectroscopy; forward dissolution rate

1. Introduction

Borosilicate glass is the favored material for the geological disposal of high-level nuclear waste (HLW) from the reprocessing of spent fuel used in nuclear power plants [1–3]. Vitrification of HLW is currently the treatment of choice for immobilization of radionuclides for the following reasons: (a) high capability of glass to reliably incorporate a wide spectrum of isotopes with different ionic charges and sizes, (b) simple and economic production technology adapted from glass manufactures, (c) small volume of the resulting waste form, (d) high chemical durability of waste form glasses in contact with natural waters, and (e) high tolerance of these glasses to self-irradiation damage. Vitrification is also attracting great interest for other types of wastes, such as operational radioactive wastes from nuclear power plants as well as radioactive and toxic legacy waste from medicine [4]. In a deep geological disposal, the vitrified nuclear waste can come into contact with infiltrating ground waters once the protective metallic containers are broken or corroded. Numerous laboratory experiments were conducted to identify the key mechanisms that lead to glass breakdown in aqueous solutions and to determine the reaction kinetics [1,2,5–12]. There is still an intensive debate on the reaction and transport mechanisms controlling glass alteration and the formation of silica-based surface alteration layers (SALs) over geological time scales, reflected by the formulation of several different mechanisms, including the fundamentally different and highly debated leaching and interface-coupled dissolution-precipitation (ICDP) models [5–13]. While the leaching model is based on the fact that the SAL forms by diffusion-controlled chemical leaching of the glass, the ICDP model is based on the notion that the glass initially dissolves congruently until a surface solution boundary layer is supersaturated with amorphous silica, which eventually precipitates at the surface of the glass. Both, glass dissolution and silica precipitation are coupled in space and time, resulting in the formation of an inwardly migrating dissolution-precipitation front. In a recent study, however, Lenting et al. [14] presented a unifying model based on an ICPD mechanism as the SAL forming process, but also includes an interdiffusion zone that develops ahead of the ICDP front inside the glass once the ICDP reaction slowed down due to chemical transport limitations.

Regardless of the mechanistic details, several kinetic stages have been identified during nuclear glass corrosion as schematically shown in Figure 1 [2,15,16]. During the first kinetic stage, the corrosion rate r_0, defined as the amount of glass lost per unit time, is assumed to be constant until it linearly drops to a residual rate r_r [3,16,17]. However, Ojovan et al. [4] proposed an exponential decrease of the elemental release rate already from the beginning of reaction. A late-stage resumption of the corrosion rate can occur under yet unclear circumstances, which are in focus of current research [18–20].

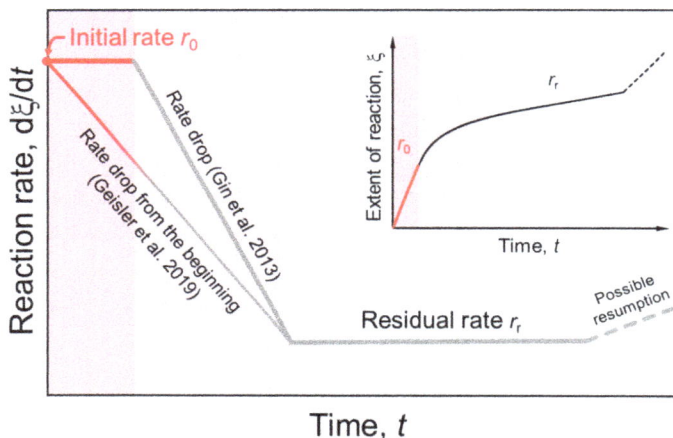

Figure 1. Kinetic regimes of glass corrosion that have been identified to date [2], illustrated by a plot of the reaction rate and the extent of reaction, ξ, (inset diagram), as a function of time. Here, ξ represents the amount of glass dissolved into solution or transformed into a silica-based surface layer (SAL). In the present work, we are mainly concerned with the initial reaction rate, r_0, that represents the forward dissolution rate of the glass. Gin and coworkers [2] proposed a constant r_0 for a certain time until an SAL is formed. In contrast, Geisler et al. [21] recently observed a linearly decreasing r_0 over time already from the beginning of the reaction without the formation of an SAL.

Using a unique experimental setup that allows the in situ observation of the corrosion process by fluid-cell confocal Raman spectroscopy, Geisler et al. [21] found a linear decrease of the dissolution rate with time already from the beginning of the reaction (Figure 1). The in situ experiments with a ternary Na borosilicate glass revealed an initially continuously retreating glass-water interface without any evidence for the formation of an SAL, suggesting that the glass initially dissolved congruently and that affinity effects control the initial rate decrease until a several micrometer-thick SAL had formed and the rate dramatically dropped to zero. The initial dissolution rate r_0 was determined from the intercept of a linear fit with the rate axis and interpreted to be the forward dissolution rate of the investigated borosilicate glass far away from equilibrium under the given physicochemical conditions (Figure 1).

To reliably assess the chemical durability of nuclear waste glasses over geological time scales, it is critical to understand the impact of self-irradiation damage on the glass properties and particularly its effect on the aqueous dissolution kinetics. Self-irradiation damage from alpha-decay falls into two categories: (i) the transfer of the energy from the damaging energetic alpha particle to the electrons of the glass (ionization and electronic excitations); and (ii) the transfer of energy to the atomic nuclei, primarily by ballistic processes involving elastic collisions. The more massive, but low energy (70–100 keV) α-recoil particles account for most of the total number of displacements produced by ballistic processes in HLW glasses. The α-recoils lose nearly their entire energy in elastic collisions over a very short range, producing highly localized damage (displacement cascades) [22–25]. For the simulation of the absorbed dose from both types of electronic or nuclear collisions, the irradiation of non-radioactive glass with heavy or light ions is a scientifically valuable method [13]. In particular, ion accelerator experiments allow an assessment of the effect of a single process on the material structure that is produced during the radioactive decay under well-controlled conditions. The ability to isolate defect mechanisms is crucial to the study of complex irradiation-damage problems, where both competing and synergistic contributions from the irradiation environment of temperature and stress exist [26].

In this study, Au ions of ~1 GeV energy were chosen to evaluate the effect of purely electronic collisions on the glass network and ultimately on the initial forward dissolution rate r_0. The use of swift heavy ions has the advantage that a relatively thick layer of irradiated material can be produced (Figure 2). For this, we performed two in situ and real time, fluid-cell Raman spectroscopic corrosion experiments with an ion-irradiated Na borosilicate glass in a 0.5 mol $NaHCO_3$ solution, following the method described in detail in Geisler et al. [21]. These authors already carried out corrosion experiments with non-irradiated glass samples of the same composition and under very similar physicochemical conditions, which serve as reference. In addition, we repeated one experiment with a non-irradiated sample to test the reproducibility of the method.

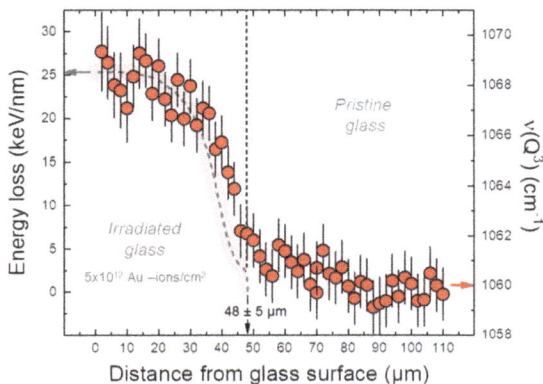

Figure 2. The energy loss of 4.8 MeV/u Au ions calculated with the SRIM code [27] (dashed gray line) and the frequency of the Q^3 Si-O stretching mode (red circles) as a function of depth. Note the good agreement between the SRIM calculation, indicating a maximum Au penetration depth of 48 ± 5 μm, and structural changes reflected by the shift of the frequency of the Q^3 band.

2. Materials and Methods

2.1. Glass Synthesis

The irradiation experiment was carried out with a polished, homogenous, ternary Na borosilicate glass (TBG) coupon ($13 \times 5 \times 3$ mm^3) that was prepared from a second production batch of the glass used by Geisler et al. [18]. The nominal composition of the glass was 57.9 wt.% SiO_2, 19.7 wt.% Na_2O, and 22.3 wt.% B_2O_3 [21]. It was synthesized from SiO_2, B_2O_3, and Na_2CO_3 precursors in a platinum crucible at 1400 °C. The glass was molten a second time after crushing and milling it in a ball mill to ensure chemical homogeneity. After quenching the melt in a pre-heated stainless steel mold, it was tempered at 560 °C for 6 h and then letting it cool down over night by switching off the oven.

2.2. Heavy Ion Bombardment

One side of the polished sample was irradiated at room temperature and at normal incidence with ^{197}Au ions at the beamline M3 of the UNILAC at the GSI Helmholtz Center for Heavy Ion Research in Darmstadt, Germany. The generated ^{197}Au ions had an energy of 4.8 MeV/u (with u being the mass unit) and an effective energy of 4.5 MeV/u at the sample surface, i.e., a kinetic energy of 945.6 MeV. The samples were irradiated to a fluence of 5×10^{12} ions/cm^2 with an uncertainty in the order of 10–20%. The penetration depth of the ^{197}Au ions in the Na borosilicate glass (2.51 g/cm^3) was 48 ± 5 μm as calculated with the SRIM 2013.00 code [27] (Figure 2). For the two corrosion experiments, the irradiated glass coupon was cut in two smaller coupons with a size of about $6 \times 5 \times 3$ mm^3 and $7 \times 5 \times 3$ mm^3, respectively.

2.3. Raman Spectroscopy

All Raman measurements were conducted with a high resolution Horiba Scientific HR800 confocal Raman system at the Institute for Geosciences and Meteorology of the University in Bonn, Germany. A 2 W solid state Nd:YAG laser (532.09 nm) with about 600 mW power at the sample surface was used. The scattered light was detected with an electron multiplier charge-coupled device detector after having passed a 1000 μm confocal aperture and a 100 μm spectrometer entrance slit, and being dispersed by a grating of 600 grooves/mm. A 100× long working distance (LWD) microscope objective with a numerical aperture of 0.8 was used for all measurements, yielding an empirically determined lateral resolution of about 6 μm at the depth of the in situ Raman measurements [21]. The Raman signal was measured in the wavenumber ranges from 200 to 1800 and from 2800 to 4000 cm^{-1}. The first frequency range includes a Ne line at 1706.06 cm^{-1} that was recorded as an internal wavenumber standard to correct each spectrum for any spectrometer shift during long term measurements of up to 75 h (resulting from ±0.5 °C temperature variations in the laboratory) by placing a Ne lamp in the beam path of the scattered light.

To determine the relative fractions of different structural glass units in the non-irradiated and irradiated glass from Raman intensities, the measured Raman intensities $I(\nu)$ were corrected for (1) the wavelength-dependent instrumental sensitivity (white light correction); (2) the thermal population of the excited states by the Bose–Einstein temperature factor $B = [1 - \exp(-h\nu c/kT)]$ with h, k, c, and T being the Planck and Boltzmann constant, the speed of light, and the absolute temperature, respectively; (3) the scattering factor $q_s = (\nu_e - \nu)^{-3}$ with ν_e and ν being the excitation wavenumber and Raman shift, respectively; (4) the frequency factor ν; and, finally, (5) for background signals (stray light, fluorescence) that were fitted with a third order polynomial function. The corrected spectrum, the so-called reduced or $R(\nu)$ spectrum, is directly proportional to the relative scattering activity in terms of mass-weighted normal coordinates and thus most closely matches the vibrational density of states.

2.4. Experimental Set-Up and Determination of Glass Retreat/Dissolution Rate

Three in situ experiments were performed with two irradiated and one non-irradiated reference glass coupon under similar physiochemical conditions using a home-made fluid-cell (Figure 3a) [21].

Before each experiment, the fluid-cell was disassembled, cleaned with 35% HCl solution and rinsed with MilliQ© water. The glass coupon was placed in a sample holder that is integrated in the sealing cap. In the closed fluid-cell the accurate positioning of the glass coupon, perpendicular to the fused silica window, is of great importance for the reliable determination of the glass retreat, i.e., the glass dissolution rate. Furthermore, the distance between the window and the glass sample surface was kept as small as possible in order to minimize corrosive processes in this region and to obtain the best possible spatial resolution. In the first experimental set up the distance between the fused silica window and the sample surface was ~60 μm. In the second experiment the distance was ~20 μm. The fluid-cell was then filled with 13.0 ± 0.2 mL of 0.5 M NaHCO$_3$ solution (initial pH of 7.1 at 85 °C; [21]) and heated to nominal temperature of 90 °C. Due to a slight temperature gradient in the cell, the actual temperature at the position of the Raman measurements was between about 80 and 85 °C (Table 1), which was accurately determined from the temperature-related frequency shifts of the $\nu_5(HCO_3)$ band as described in detail in [21].

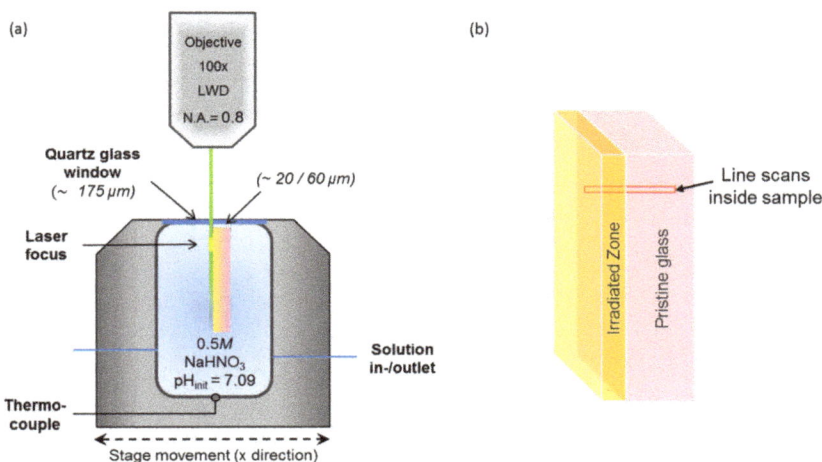

Figure 3. (**a**) Schematic drawing of the fluid cell (not to scale) used in this study, and (**b**) the location of the Raman spectroscopic line scans cross the heavy ion bombarded (5×10^{12} Au ions/cm^2) surface layer into the non-irradiated glass.

Point-by-point parallel Raman line scans were performed every ~2 h about 100 μm below the surface of the glass monolith with a 2-μm step size by automatically moving the stage in y–x direction (Figure 3a). The total counting time for each point was about 25 s, involving the two spectral windows that were each measured for 2 × 6 s. For the determination of the glass retreat as a function of time, the integrated total intensity of the Q^n species between 1000 and 1250 cm^{-1} was determined from white light- and background-corrected $I(\nu)$ spectra. The resulting intensity profiles across the glass-water interface were than fitted with an exponential function from which the glass retreat was determined with an error better than about ± 1 μm. For a more detailed description of the determination of the glass retreat we refer to Geisler et al. [21].

3. Results

3.1. Structural State of the Irradiated Glass Samples

The Au irradiation of the pristine glass with a kinetic energy of ~950 MeV (at the surface) and a fluence of 5×10^{12} Au ions/cm^2 did not result in any macroscopically visible changes of the sample. However, an effect of heavy ion irradiation on the structure of the ternary Na borosilicate glass (TBG) was detected by Raman spectroscopy. Figure 4a shows a representative, average $R(\nu)$ Raman spectrum from both the irradiated and non-irradiated TBG, which were extracted from the first Raman point-by-point line scan from the irradiated glass surface into the non-irradiated glass that was measured at the beginning of each experiment (Figure 3b). To quantify spectral changes after irradiation, which are already evident by visible inspection of the Raman spectra, we fitted the background-corrected $R(\nu)$ spectra shown in Figure 4a with 15 Gauss-Lorentz functions in the wavenumber range between 300 and 1600 cm^{-1}. It is noted that the least-squares fitting procedure of silicate glass vibrational bands is still highly debated [28]. For instance, some researchers favor the use of Gauss functions to fit Raman bands of silicate glasses (e.g., [29,30]), whereas others use a convolution of a Gauss and Lorentz function (e.g., [28,31]), which takes into account phonon dumping on the density of states which is particularly relevant in disordered materials where phonon life times are short [28]. Furthermore, the broad vibrational bands from silicate glasses make it often difficult to decide how many vibrational modes contribute to a broad band profile. In this study, we applied the deconvolution procedure proposed by Manara et al. [31], who used 14 Gauss-Lorentz functions to fit the Raman spectra of a series of ternary Na borosilicate glasses in the wavenumber range between 300 and 1600 cm^{-1}. The 15th band observed in this study was located near 1555 cm^{-1} and reflects the stretching mode of molecular oxygen. The assignment of the 14 vibrational modes of typical structural units in Na borosilicate glasses, which is given in Figure 4a, was also adapted from Manara et al. [31], but their assignment also bases on earlier work (e.g., [32–34]).

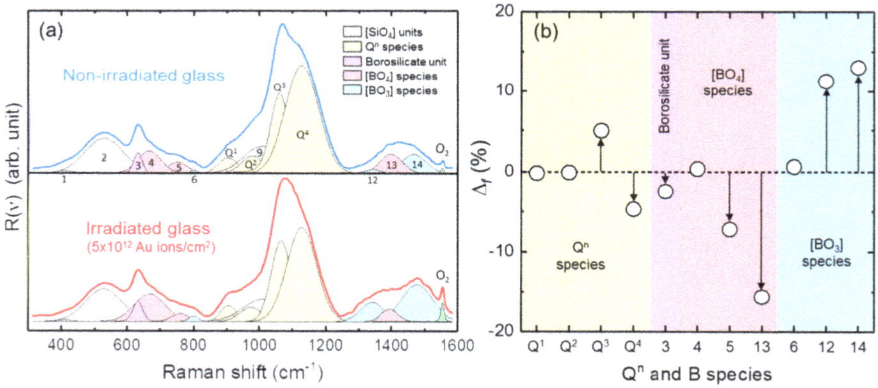

Figure 4. (**a**) Representative background-corrected $R(\nu)$ Raman spectra (average of 10 individual spectra) from the non-irradiated glass (upper spectrum) and the irradiated surface layer (lower spectrum) of the ternary Na borosilicate glass (TBG) used in this study for dissolution experiments along with the result of a least-squares fit with 15 Gauss–Lorentz functions. The assignment of the Raman bands to a vibrational mode of a structural glass unit and to O$_2$ is indicated by colored Gauss-Lorentz profiles. The different glass bands are labeled by numbers and by the Q^n notation in the cases of the stretching motions of [SiO$_4$] groups with n = 1, 2, 3, or 4 bridging oxygen atoms. Note the significantly higher intensity of the O$_2$ band near 1555 cm^{-1} in the spectrum from the irradiated sample. (**b**) Diagram showing the difference, $\Delta_f = f_{irr} - f_{non-irr}$, between the intensity fraction of different silicate- and borate-related bands of the irradiated, f_{irr}, and non-irradiated, $f_{non-irr}$, TBG. The fractions were determined from the integrated intensities and the intensity sum of the respective Raman bands that are shown and labeled in Figure 4a.

The band near 495 cm^{-1} in the spectrum of the non-irradiated glass—also called R band (band 2 in Figure 4a)—was assigned to bending and rocking modes of Si–O–Si bonds. A wavenumber upshift by 10 cm^{-1} was observed for this band in the spectra from the irradiated part of the sample, which is in perfect agreement with findings of Mir et al. [35] on borosilicate glass irradiated with ^{129}Xe ions, and indicates a decreased mean Si-O-Si angle due to irradiation (Figure 4a). Vibrational stretching motions of the silicate network are represented by bands near 905, 978, 1061, and 1128 cm^{-1} in the non-irradiated TBG, which have been assigned to Q^1, Q^2, Q^3, and Q^4 species in the silicate network, respectively (the superscript refers to the number of bridging oxygen atoms). Another Raman band contributes to the overall profile in this wavenumber region, namely a band near 1008 cm^{-1} (band 9 in Figure 4a) that was considered to represent Si–O^0 bridging oxygen stretching motions [34] or alternatively vibrations of structural units associated with Na$^+$ ([32,33]). In general, the structural units with a high degree of polymerization contribute to the high-frequency side of the Q^n region, whereas the units with a low degree of polymerization contribute to the low frequency side [35]. Here, we observed a 5.0 ± 0.5% increase of the Q^3 fraction in the irradiated part of the TBG (Figure 4b), and a simultaneous −4.7 ± 0.5% decrease of the Q^4 fraction, clearly indicating a depolymerization of the SiO$_4$ glass network. On the other hand, the Q^1 and Q^2 fractions did not significantly change. The Q^3 band showed the largest frequency shift (~8 cm^{-1}; Figure 2) among the Q^n bands in response to Au ion irradiation. The variation of the frequency of the Q^3 band as a function of depth perfectly mimics the energy loss curve for Au ions in the TBG calculated by the SRIM code (Figure 2).

We consider the Raman bands assigned to the different borate species in borosilicate glasses. The pronounced band near 632 cm^{-1} (band 3 in Figure 4a) has been associated with bending motions of danburite-like borosilicate ring units [31]. The intensity fraction of this band on the total intensity of all Raman bands assigned to boron species (Figure 4a) was 2.5 ± 0.4% lower in the spectra from the irradiated surface layer (Figure 4b). However, the intensity fraction of the bands near 1342 and 1466 cm^{-1} (bands 12 and 14 in Figure 4a, respectively), that were assigned to the BO$_3$ units in boroxol rings, was significantly higher in the spectra from the irradiated surface layer (Figure 4b). This change was complemented by significantly lower intensity fractions of BO$_4$-related bands (Figure 4b), i.e., bands 4, 5, and 14 (Figure 4a). A detailed inspection of the spectra furthermore revealed that the broad band profile between 1250 and 1650 cm^{-1} in the spectra from the irradiated surface layer were characterized by a number of new inflection points. This suggests that apart from an average reduction of the boron coordination number, the B–O bond length ordering has also changed due to irradiation. These changes are clear evidence for significant structural alterations of the borate sub-network as a result of heavy ion irradiation. In this respect, we also noted a significantly higher intensity of the O$_2$ band in the spectrum from the irradiated compared to the non-irradiated part of the samples (Figure 4a). Whereas the O$_2$ band in the spectra from the non-irradiated glass areas solely stems from air located between the glass surface and the objective, the higher intensity of the O$_2$ band in the spectra from the irradiated part of the TBG indicates the occurrence of additional molecular oxygen within the irradiated glass structure. This is consistent with Raman spectroscopic results of Mir et al. [32], who also observed the O$_2$ band after irradiation of a ternary Na borosilicate glass with ^{129}Xe ions.

3.2. Glass Retreat and forward Dissolution Rate

Two in situ fluid-cell Raman spectroscopic corrosion experiments were conducted with the irradiated TBG samples for ~45 and ~75 h (Figure 5a,b). Another short-term experiment was performed with a non-irradiated TBG coupon for 7 hours to test the reproducibility of the method by comparing results with those from Geisler et al. [21]. The results are summarized in Table 1. Both experiments with irradiated samples started with a remarkably higher initial forward dissolution rate than experiments with the non-irradiated glass (Figure 5), clearly demonstrating a significant effect of radiation damage and associated structural modifications on the glass dissolution kinetics. In agreement with results from Geisler et al. [21], an approximately linear decrease of the glass retreat rate is observed from the beginning of both experiments until about 12 to 15 h, when a sharp rate drop occurred. Two rate-time regimes can thus be distinguished that were each individually fitted with a linear function (red and blue lines in Figure 5). Since the drop of the glass retreat rates occurred just about when the dissolution of the glass reached the maximum penetration depth of the Au ions (Figure 5 insets), the drop can reliably be linked to the time when the irradiated surface layer is dissolved completely so that the non-irradiated glass is exposed to the solution. In this respect, it is also noteworthy that first silica signals were observed only in the second, longer experiment after about 60 h, which is fully in line with findings of Geisler et al. [21]. Thus, extrapolation of the linear fits to the initial rate data of both experiments (red lines in Figure 5) to $t = 0$ delivered the initial forward dissolution rate of the irradiated glass, i.e., $r_0{}^{irr}$ = 5.1 ± 0.4 and 5.5 ± 0.6 µm/h, respectively (Figure 5; Table 1). The relative large uncertainty of the dissolution rates at the beginning reflects the fast dissolution kinetics with respect to the scanning time and thickness of the irradiated surface layer. The intercept of the best fit line of both kinetic regimes should correspond to the initial dissolution rate of the non-irradiated TBG. It also gives the time when the rate drop occurred, i.e., after 15 ± 3 and 12 ± 2 h, respectively (Figure 5). Indeed, we obtain r_0 values of 1.8 ± 0.2 and 1.5 ± 0.1 µm/h that are comparable with the forward dissolution rates obtained from the non-irradiated TBG (Table 1). Figure 6 shows a compilation of all data in an Arrhenius diagram to consider the slightly different temperatures of the experiments (Table 1).

Table 1. Summary of forward dissolution rates of non-irradiated, r_0, and irradiated TBG, $r_0{}^{irr}$, obtained by in situ fluid-cell Raman spectroscopy. The uncertainties were obtained from the least-squares fit (Figure 5) and are given at the 1-sigma level.

Sample	Non-Irradiated TBG			Irradiated TBG	
T (°C)	85.2 ± 0.2	81.7 ± 0.1	85.1 ± 0.7	80.0 ± 0.2	82.6 ± 0.3
$r_0{}^{irr}$ (µm/h)	-	-	-	5.1 ± 0.4	5.5 ± 0.6
r_0 (µm/h)	1.61 ± 0.04	1.00 ± 0.03	2.10 ± 0.03	1.8 ± 0.2	1.5 ± 0.1
Reference	[21]			This study	

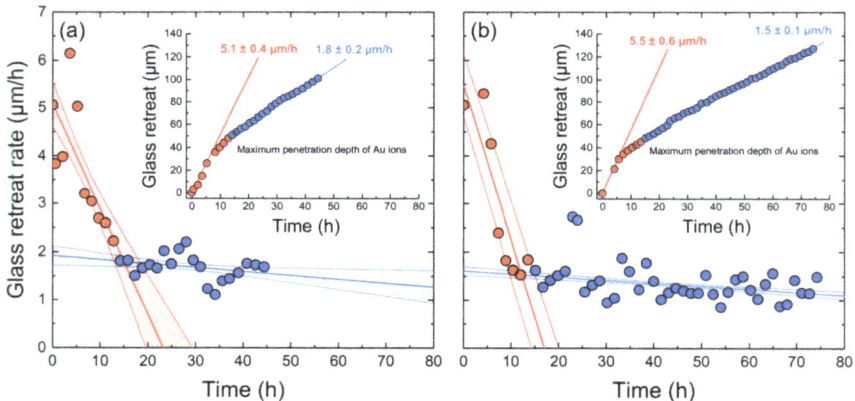

Figure 5. The dissolution rate and glass retreat (inset) as a function of time for (**a**) the first and (**b**) the second experiment with the irradiated TBG. The rate data clearly define two distinct trends that were individually fitted with a linear function. Light blue data points in Figure 5b were identified as outliers and excluded from the fit. The two trends can be related to the congruent (stoichiometric) dissolution of the irradiated surface layer (red symbols) and the underlying non-irradiated glass (blue symbols). The intercept of the red line with the y axis defines the forward dissolution rate, r_0, of the irradiated glass, whereas the intercept of the blue line with the red line gives the forward dissolution rate of the non-irradiated glass and defines the time when glass dissolution reached the non-irradiated part of the glass.

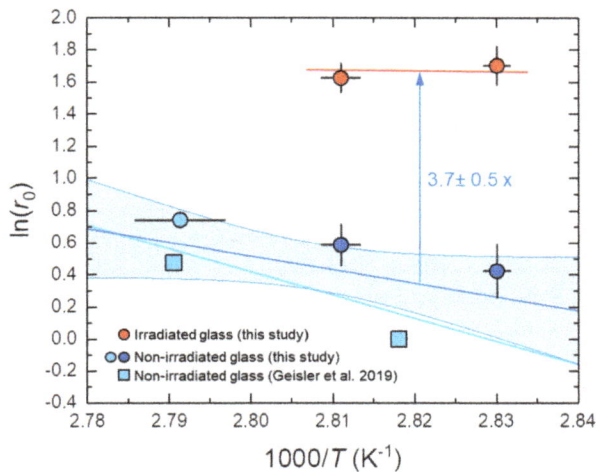

Figure 6. Arrhenius diagram with the initial or forward dissolution rates r_0 obtained for the non-irradiated (blue symbols) and irradiated (red symbols) TBG. Data from Geisler et al. [21] are also plotted. The dark and light blue lines represent an unweighted and an error-weighted linear fit to the data, respectively. For the unweighted fit the 1-sigma confidence interval is also shown as we use this fit to estimate r_0^{irr}/r_0 and its error.

The significantly higher forward dissolution rate of the irradiated glass compared to the non-irradiated counterpart is immediately apparent, but the rate data from the non-irradiated TBG themselves are highly scattered. This excess scatter may reflect (i) the slightly different chemistry of the glass coupons used in this study compared with those used by Geisler et al. [21], and/or (ii) the observation that the dissolution rate may locally vary as reflected by etch pits often observed in glass dissolution studies [12]. An unweighted linear least squares fit to the data yielded a slope that corresponds to an activation energy of 71 ± 66 kJ/mol, whereas the slope of an error-weighted linear fit yielded an activation energy of 116 ± 83 kJ/mol (Figure 6). Both values are statistically not significantly different from zero, but are nevertheless in the typical range of activation energies for hydrolysis reactions at borosilicate glass surfaces that were calculated from first principles density functional theory (DFT) simulations and measured in neutral solutions ([36], and references therein). However, it is not expected that the dissolution rates obtained for the non-irradiated glass from the experiment with the surface-irradiated TBG perfectly fit to those obtained with non-irradiated glass coupons, since at the time when the irradiated surface layer is completely dissolved, the solution chemistry is different from the initial chemistry. At that time the solution was no longer free of silica (but still undersaturated with respect to amorphous silica) and the pH near the surface changed, so the chemical affinity or driving force for the dissolution reactions also changed [34]. In any case, to obtain a reliable activation energy for the dissolution of this TBG under close to neutral conditions, more experiments under a wider temperature range are necessary. The unweighted fit yielded a ratio between r_0^{irr} and r_0 of 3.7 ± 0.5 (Figure 6). Thus, an almost four-fold increase of the forward dissolution rate was observed for the irradiated TBG.

4. Discussion and Conclusions

In this study, the irradiated glass structure was found to dissolve 3.7 ± 0.5 times faster than the corresponding non-irradiated glass, verifying previous studies that also reported an increased dissolution kinetics of radiation-damaged silicate glasses [37–39]. In contrast, other studies claim that neither the alpha activity nor the alpha decay dose has a significant impact on the initial dissolution rate [40,41]. The reasons for such a disagreement are not evident and should be subject of future investigations, particularly given the importance of this knowledge to assess the chemical durability of nuclear glasses in a natural nuclear repository. Raman spectroscopic measurements of the Au-irradiated TBG have revealed (1) a significant modification of the short-range order around the main network formers, (2) a decrease of the average boron coordination number, (3) an accumulation of molecular oxygen, and (4) an increase of the Q^3 fraction at the expense of the Q^4 species, indicating depolymerization of the silicate network due to radiation damage. Such structural and chemical modifications are in line with findings of Mir et al. [35] and known to be accompanied by changes in the physical properties, including hardness and Young's modulus, density, and fracture toughness [23,37,42–47]. Our study reveals that such structural modifications due to irradiation have a profound effect on the glass dissolution property.

The increased dissolution kinetics can principally be explained by the less interlinked, irradiated glass network that offers a larger free volume for the hydrolysis of Si–O and B–O bonds in the glass by hydrogen species from solution. During heavy ion bombardment, several micrometer-long cylindrical damage trails, so-called ion tracks [48], are created, which have been made visible by transmission electron microscopy [24] and atomic force microscopy [49]. In particular, the Si–O and B–O bonds inside the ion tracks are likely high energy sites that can preferentially be hydrolyzed. In our samples, these ion tracks must be parallel oriented and elongated in the direction of the dissolution front movement due to the direction of bombardment. Considering the long-range ballistic character of heavy ions in the material, it would be interesting to test whether the dissolution kinetics depends on track orientation? This question directly relates to the general debate about the scientific value of external heavy ion irradiation experiments for the simulation of α-recoil damage [37]. Charpentier et al. [50], for example, observed a higher damage level in silicate glasses irradiated with Au ions than in short-lived actinide-doped glasses that accumulated an equivalent dose of recoil events. A possible explanation for this observation could be partial self-healing processes which arise from alpha particles involved in alpha decay events [49]. External irradiation with energetic heavy ions may also induce the removal of alkali ions even from depths that correspond to the ion penetration depth [51]. At high fluences, the entire near surface region may be depleted of alkali ions to depths even exceeding the ion range [51]. However, for our samples electron microprobe measurements across the irradiated surface layer did not reveal any Na loss from areas deeper than ~1 µm (lateral resolution of the measurements) below the surface. Moreover, a very critical point is that in real case scenarios, radiolysis reactions in the surface boundary solution may significantly affect the glass dissolution kinetics as also observed to affect the dissolution of UO_2, respectively spent fuel [39,52,53].

All these observations challenge the assignability of results from heavy ion irradiated samples to the real world of self-irradiation in nuclear glasses. However, the irradiation with swift heavy ions requires less experimental effort and represents an effective approach to simulate radiation damage over micrometer length scale and thus allows studying the effects of radiation damage on the long-term durability of borosilicate glasses. On the other hand, experiments with borosilicate glasses doped with short- and long-lived actinide and/or fission products appear to be most important for the assessment of the long-term durability of a nuclear glass in a natural repository and, not to forget, to compare the results with those obtained from externally irradiated samples. Here, fluid-cell Raman spectroscopy can potentially open up new possibilities to study the corrosion of nuclear glass in situ without disturbing or interrupting the reaction. As shown by Geisler et al. [21], and partly in this work, fluid-cell Raman spectroscopy thereby provides a wealth of chemical and structural information with a spatial resolution at the micrometer-scale. Moreover, light stable isotopes, such as ^{2}H and ^{18}O, can be used to in situ trace distinct reaction and transport processes, as demonstrated by Geisler et al. [21], who quantified the transport of molecular water through a silica-based surface alteration layer by using a ^{2}H-labelled solution. A single experiment thus delivers the information of numerous quench experiments with different samples so that complicated sample syntheses and preparation that require a radiation-protected environment, as well as costs and new nuclear waste, are minimized. Such experiments can potentially also be performed over longer durations to be able to also investigate the corrosion process of nuclear waste glasses as they are kinetically more stable against aqueous corrosion than the glass used in the present study.

Author Contributions: T.G. initiated and planned the study, and obtained funding; M.I.L., L.D., and C.L. carried out the experiments; M.I.L., L.D., C.L., and T.G. refined and analyzed the Raman spectroscopic data; M.L. performed the SRIM calculations; C.T. carried out the irradiation experiments at the Helmholtz Centre in Darmstadt; M.I.L. and T.G. wrote the first draft of the manuscript, but all authors contributed to the final version.

Funding: This research has financially been supported by Schott AG Mainz, Germany, and the German Research Foundation (grant no. GE1094/21-1).

Acknowledgments: We thank G. Paulus (Schott AG) for synthesizing and characterizing the borosilicate glass, and D. Lülsdorf and H. Blanchard (University of Bonn) as well as W. Bauer (Schott AG) for helping with the design and construction of the fluid cell. The results presented here are based on a UMAT experiment performed at the M-branch of the UNILAC at the GSI Helmholtzzentrum für Schwerionenforschung, Darmstadt (Germany) in the frame of FAIR Phase-0. We acknowledge support by M. Bender and D. Severin during the ion irradiation. Part of this research is being performed using funding received from the DOE Office of Nuclear Energy's Nuclear Energy University Program under US-DOE, contract DE-NE0008694.

Conflicts of Interest: The authors declare no conflict of interest.

References

1. Grambow, B. Nuclear waste glasses—How durable? *Elements* **2006**, *2*, 357–364. [CrossRef]
2. Gin, S.; Abdelouas, A.; Criscenti, L.J.; Ebert, W.L.; Ferrand, K.; Geisler, T.; Harrison, M.T.; Inagaki, Y.; Mitsui, S.; Mueller, K.T.; et al. An international initiative on long-term behavior of high-level nuclear waste glass. *Mater. Today* **2013**, *16*, 243–248. [CrossRef]
3. Icenhower, J.P.; Steefel, C.I. Experimentally determined dissolution kinetics of SON68 glass at 90 °C over a silica saturation interval: Evidence against a linear rate law. *J. Nucl. Mater.* **2013**, *439*, 137–147. [CrossRef]
4. Ojovan, M.I.; Lee, W.E. Glassy wasteforms for nuclear waste immobilization. *Metall. Mater. Trans. A* **2011**, *42*, 837–851. [CrossRef]
5. Grambow, B. A General rate equation for nuclear waste glass corrosion. *MRS Proc.* **1984**, *44*, 15–27. [CrossRef]
6. Hellmann, R.; Cotte, S.; Cadel, E.; Malladi, S.; Karlsson, L.S.; Lozano-Perez, S.; Cabié, M.; Seyeux, A. Nanometre-scale evidence for interfacial dissolution–reprecipitation control of silicate glass corrosion. *Nat. Mater.* **2015**, *14*, 307–311. [CrossRef]
7. Grambow, B.; Müller, R. First-order dissolution rate law and the role of surface layers in glass performance assessment. *J. Nucl. Mater.* **2001**, *298*, 112–124. [CrossRef]
8. Lanford, W.A.; Davis, K.; Lamarche, P.; Laursen, T.; Groleau, R.; Doremus, R.H. Hydration of soda-lime glass. *J. Non Cryst. Solids* **1979**, *33*, 249–266. [CrossRef]
9. Geisler, T.; Janssen, A.; Scheiter, D.; Stephan, T.; Berndt, J.; Putnis, A. Aqueous corrosion of borosilicate glass under acidic conditions: A new corrosion mechanism. *J. Non Cryst. Solids* **2010**, *356*, 1458–1465. [CrossRef]
10. Doremus, R.H. Interdiffusion of hydrogen and alkali ions in a glass surface. *J. Non Cryst. Solids* **1975**, *19*, 137–144. [CrossRef]
11. Abraitis, P.K.; Livens, F.R.; Monteith, J.E.; Small, J.S.; Trivedi, D.P.; Vaughan, D.J.; Wogelius, R.A. The kinetics and mechanisms of simulated british magnox waste glass dissolution as a function of pH, silicic acid activity and time in low temperature aqueous systems. *Appl. Geochem.* **2000**, *15*, 1399–1416. [CrossRef]
12. Geisler, T.; Nagel, T.; Kilburn, M.R.; Janssen, A.; Icenhower, J.P.; Fonseca, R.O.; Grange, M.; Nemchin, A.A. The mechanism of borosilicate glass corrosion revisited. *Geochim. Cosmochim. Acta* **2015**, *158*, 112–129. [CrossRef]
13. Gin, S.; Patrick, J.; Magaly, T.; Sylvain, P.; Schuller, S. Radionuclides containment in nuclear glasses: An overview. *Radiochim. Acta* **2017**, *105*, 927–959. [CrossRef]
14. Lenting, C.; Plümper, O.; Kilburn, M.; Guagliardo, P.; Klinkenberg, M.; Geisler, T. Towards a unifying mechanistic model for silicate glass corrosion. *Npj. Mater. Degrad.* **2018**, *2*, 28. [CrossRef]
15. Poinssot, C.; Gin, S. Long-term behavior science: The cornerstone approach for reliably assessing the long-term performance of nuclear waste. *J. Nucl. Mater.* **2012**, *420*, 182–192. [CrossRef]
16. Vernaz, E.; Gin, S.; Jégou, C.; Ribet, I. Present understanding of R7t7 glass alteration kinetics and their impact on long-term behavior modeling. *J. Nucl. Mater.* **2001**, *298*, 27–36. [CrossRef]

17. Frugier, P.; Gin, S.; Minet, Y.; Chave, T.; Bonin, B.; Godon, N.; Lartigue, J.-E.; Jollivet, P.; Ayral, A.; De Windt, L.; et al. SON68 Nuclear glass dissolution kinetics: Current state of knowledge and basis of the new GRAAL model. *J. Nucl. Mater.* **2008**, *380*, 8–21. [CrossRef]

18. Fournier, M.; Gin, S.; Frugier, P. Resumption of nuclear glass alteration: State of the art. *J. Nucl. Mater.* **2014**, *448*, 348–363. [CrossRef]

19. Ribet, S.; Gin, S. Role of neoformed phases on the mechanisms controlling the resumption of SON68 glass alteration in alkaline media. *J. Nucl. Mater.* **2004**, *324*, 152–164. [CrossRef]

20. Frugier, P.; Fournier, M.; Gin, S. Modeling resumption of glass alteration due to zeolites precipitation. *Procedia Earth Planet. Sci.* **2017**, *17*, 340–343. [CrossRef]

21. Geisler, T.; Dohmen, L.; Lenting, C.; Fritzsche, M.B.K. Real-time in situ observations of reaction and transport phenomena during silicate glass corrosion by fluid-cell Raman spectroscopy. *Nat. Mater.* **2019**, *18*, 342–348. [CrossRef]

22. Advocat, T.; Jollivet, P.; Crovisier, J.L.; Del Nero, M. Long-term alteration mechanisms in water for SON68 radioactive borosilicate glass. *J. Nucl. Mater.* **2001**, *298*, 55–62. [CrossRef]

23. Weber, W.J.; Ewing, R.C.; Angell, C.A.; Arnold, G.W.; Cormack, A.N.; Delaye, J.M.; Griscom, D.L.; Hobbs, L.W.; Navrotsky, A.; Price, D.L.; et al. Radiation effects in glasses used for immobilization of high-level waste and plutonium disposition. *J. Mater. Res.* **1997**, *12*, 1948–1978. [CrossRef]

24. Hobbs, L.W.; Pascucci, M.R. Radiolysis and defect structure in electron-irradiated α-quartz. *J. Phys. Colloq.* **1980**, *41*, 237–242. [CrossRef]

25. Hobbs, L.W. Topology and geometry in the irradiation-induced amorphization of insulators. *Nucl. Instrum. Methods Phys. Res. Sect. B Beam Interact. Mater. At.* **1994**, *91*, 30–42. [CrossRef]

26. Lang, M.; Tracy, C.L.; Palomares, R.; Zhang, F.; Severin, D.; Bender, M.; Trautmann, C.; Park, C.; Prakapenka, V.B.; Skuratov, V.A.; et al. Characterization of ion-induced radiation effects in nuclear materials using synchrotron x-ray techniques. *J. Mater. Res.* **2015**, *30*, 1366–1379. [CrossRef]

27. Ziegler, J.F.; Ziegler, M.D.; Biersack, J.P. SRIM—The stopping and range of ions in matter (2010). *Nucl. Instrum. Methods Phys. Res. Sect. B Beam Interact. Mater. At.* **2010**, *268*, 1818–1823. [CrossRef]

28. Efimov, A. Vibrational spectra, related properties, and structure of inorganic glasses. *J. Non Cryst. Solids* **1999**, *253*, 95–118. [CrossRef]

29. Mysen, B.O.; Finger, L.W.; Virgo, D.; Seifert, F.A. Curve-fitting of Raman spectra of silicate glasses. *Am. Mineral.* **1982**, *67*, 686–695.

30. Mckeown, D.; Muller, I.; Buechele, A.; Pegg, I.L.; Kendziora, C. Structural characterization of high-zirconia borosilicate glasses using Raman spectroscopy. *J. Non Cryst. Solids* **2000**, *262*, 126–134. [CrossRef]

31. Manara, D.; Grandjean, A.; Neuville, D.R. Advances in understanding the structure of borosilicate glasses: A Raman spectroscopy study. *Am. Mineral.* **2009**, *94*, 777–784. [CrossRef]

32. Fukumi, K.; Hayakawa, J.; Komiyama, T. Intensity of Raman band in silicate glasses. *J. Non Cryst. Solids* **1990**, *119*, 297–302. [CrossRef]

33. Neuville, D. Viscosity, structure and mixing in (Ca, Na) silicate melts. *Chem. Geol.* **2006**, *229*, 28–41. [CrossRef]

34. Frantza, J.D.; Mysen, B.O. Raman spectra and structure of BaO-SiO₂ SrO-SiO₂ and CaO-SiO₂ melts to 1600 °C. *Chem. Geol.* **1995**, *121*, 155–176. [CrossRef]

35. Mir, A.H.; Monnet, I.; Boizot, B.; Jégou, C.; Peuget, S. Electron and electron-ion sequential irradiation of borosilicate glasses: Impact of the pre-existing defects. *J. Nucl. Mater.* **2017**, *489*, 91–98. [CrossRef]

36. Zapol, P.; He, H.; Kwon, K.; Criscenti, L. First-Principles Study of Hydrolysis Reaction Barriers in a Sodium Borosilicate Glass. *Int. J. Appl. Glass Sci.* **2013**, *4*, 395–407. [CrossRef]

37. Peuget, S.; Tribet, M.; Mougnaud, S.; Miro, S.; Jégou, C. Radiations effects in ISG glass: from structural changes to long-term aqueous behavior. *Npj. Mater. Degrad.* **2018**, *2*, 23. [CrossRef]

38. Burns, W.G.; Hughes, A.E.; Marples, J.A.C.; Nelson, R.S.; Stoneham, A.M. Effects of radiation on the leach rates of vitrified radioactive waste. *J. Nucl. Mater.* **1982**, *107*, 245–270. [CrossRef]

39. Weber, W.J.; McVay, G.L.; Wald, J.W. Effects of alpha-radiolysis on leaching of a nuclear waste glass. *J. Am. Ceram. Soc.* **1985**, *68*, 253–255. [CrossRef]

40. Peuget, S.; Broudic, V.; Jégou, C.; Frugier, P.; Roudil, D.; Deschanels, X.; Rabiller, H.; Y. Noel, P. Effect of alpha radiation on the leaching behaviour of nuclear glass. *J. Nucl. Mater.* **2007**, *362*, 474–479. [CrossRef]

41. Wellman, D.; Icenhower, J.; Weber, W. Elemental Dissolution Study of Pu-Bearing Borosilicate Glasses. *J. Nucl. Mater.* **2005**, *340*, 149–162. [CrossRef]

42. Marples, J.A.C. Dose rate effects in radiation damage to vitrified radioactive waste. *Nucl. Instrum. Methods Phys. Res. Sect. B Beam Interact. Mater. At.* **1988**, *32*, 480–486. [CrossRef]

43. Delaye, J.-M.; Kerrache, A.; Gin, S. Topography of borosilicate glass reacting interface under aqueous corrosion. *Chem. Phys. Lett.* **2013**, *588*, 180–183. [CrossRef]

44. Dewan, L.; Hobbs, L.W.; Delaye, J.-M. Topological Analysis of the Structure of Self-irradiated Sodium Borosilicate Glass. *J. Non Cryst. Solids* **2012**, *358*, 3427–3432. [CrossRef]

45. Peuget, S.; Maugeri, E.A.; Charpentier, T.; Mendoza, C.; Moskura, M.; Fares, T.; Bouty, O.; Jégou, C. Comparison of radiation and quenching rate effects on the structure of a sodium borosilicate glass. *J. Non Cryst. Solids* **2013**, *378*, 201–212. [CrossRef]

46. Karakurt, G. Effect of alpha radiation on the physical and chemical properties of silicate glasses. Ph.D. Thesis, Ecole des Mines de Nantes, Nantes, France, 2014.

47. Mir, A.H.; Peuget, S.; Toulemonde, M.; Bulot, P.; Jégou, C.; Miro, S.; Bouffard, S. Defect recovery and damage reduction in borosilicate glasses under double ion beam irradiation. *EPL Europhys. Lett.* **2015**, *112*, 36002. [CrossRef]

48. Kluth, P.; Schnohr, C.S.; Pakarinen, O.H.; Djurabekova, F.; Sprouster, D.J.; Giulian, R.; Ridgway, M.C.; Byrne, A.P.; Trautmann, C.; Cookson, D.J.; et al. Fine structure in swift heavy ion tracks in amorphous SiO_2. *Phys. Rev. Lett.* **2008**, *101*, 175503. [CrossRef]

49. Mir, A.H.; Monnet, I.; Toulemonde, M.; Bouffard, S.; Jégou, C.; Peuget, S. Mono and sequential ion irradiation induced damage formation and damage recovery in oxide glasses: Stopping power dependence of the mechanical properties. *J. Nucl. Mater.* **2015**, *469*, 244–250. [CrossRef]

50. Charpentier, T.; Martel, L.; Mir, A.H.; Somers, J.; Jégou, C.; Peuget, S. Self-healing capacity of nuclear glass observed by NMR spectroscopy. *Sci. Rep.* **2016**, *6*, 25499. [CrossRef] [PubMed]

51. Arnold, G.W. Alkali depletion and ion beam mixing in glasses. *Nucl. Instrum. Methods Phys. Res. Sect. B Beam Interact. Mater. At.* **1984**, *1*, 516–520. [CrossRef]

52. Stroes-Gascoyne, S.; King, F.; Betteridge, J.S.; Garisto, F. The effects of alpha-radiolysis on UO_2 dissolution determined from electrochemical experiments with 238Pu-doped UO_2. *Radiochim. Acta* **2002**, *90*, 9–11. [CrossRef]

53. McVay, G.L.; Weber, W.J.; Pederson, L.R. Effects of radiation on the leaching behavior of nuclear waste forms. *Nucl. Chem. Waste Manag.* **1981**, *2*, 103–108. [CrossRef]

materials

MDPI

Review

Ceramic Mineral Waste-Forms for Nuclear Waste Immobilization

Albina I. Orlova [1] and Michael I. Ojovan [2,3,*]

1   Lobachevsky State University of Nizhny Novgorod, 23 Gagarina av.,
    603950 Nizhny Novgorod, Russian Federation
2   Department of Radiochemistry, Lomonosov Moscow State University, Moscow 119991, Russia
3   Imperial College London, South Kensington Campus, Exhibition Road, London SW7 2AZ, UK
*   Correspondence: m.ojovan@imperial.ac.uk

Received: 31 May 2019; Accepted: 12 August 2019; Published: 19 August 2019

Abstract: Crystalline ceramics are intensively investigated as effective materials in various nuclear energy applications, such as inert matrix and accident tolerant fuels and nuclear waste immobilization. This paper presents an analysis of the current status of work in this field of material sciences. We have considered inorganic materials characterized by different structures, including simple oxides with fluorite structure, complex oxides (pyrochlore, murataite, zirconolite, perovskite, hollandite, garnet, crichtonite, freudenbergite, and P-pollucite), simple silicates (zircon/thorite/coffinite, titanite (sphen), britholite), framework silicates (zeolite, pollucite, nepheline /leucite, sodalite, cancrinite, micas structures), phosphates (monazite, xenotime, apatite, kosnarite (NZP), langbeinite, thorium phosphate diphosphate, struvite, meta-ankoleite), and aluminates with a magnetoplumbite structure. These materials can contain in their composition various cations in different combinations and ratios: Li–Cs, Tl, Ag, Be–Ba, Pb, Mn, Co, Ni, Cu, Cd, B, Al, Fe, Ga, Sc, Cr, V, Sb, Nb, Ta, La, Ce, rare-earth elements (REEs), Si, Ti, Zr, Hf, Sn, Bi, Nb, Th, U, Np, Pu, Am and Cm. They can be prepared in the form of powders, including nano-powders, as well as in form of monolith (bulk) ceramics. To produce ceramics, cold pressing and sintering (frittage), hot pressing, hot isostatic pressing and spark plasma sintering (SPS) can be used. The SPS method is now considered as one of most promising in applications with actual radioactive substances, enabling a densification of up to 98–99.9% to be achieved in a few minutes. Characteristics of the structures obtained (e.g., syngony, unit cell parameters, drawings) are described based upon an analysis of 462 publications.

Keywords: crystalline ceramics; nuclear waste; immobilization; sintering; spark plasma sintering

1. Introduction

Crystalline ceramics, aiming to immobilize high-level radioactive waste (HLW), are important for the current stage of development of modern nuclear technology in the world.

The crystal-chemical principle is used to design multicomponent ceramics with needed structures. The approach to designing mineral-like crystalline materials is based upon the structural features of materials and isomorphism concept. The choice of the structural forms of compounds for discussion here was based upon the following criteria:

(1)   The ability of the structure to include various cations in different combinations and ratios.
(2)   Known high stability of structure to the action of the destructive factors of the environment during their prolonged exposure ("mineral-like" compounds preferred while "the nature suggests") such as high temperatures, thermal "stresses", radiation levels, the corrosive action of water and other chemical solutions. Criteria for the resistance of materials to such effects are justified by the features of the crystal structure of materials including small interatomic distances, and the

possibility of their controlled variation in the desired direction when cations and/or anions of given sizes are included in the crystallographic positions. Most of the crystalline matrices discussed in the present work meet these criteria in full or in part. Herewith the classification criteria for crystalline ceramics were based on considering first simple and then more complex structures, e.g., starting with oxides (from simple oxides to complex oxides) and moving to salt compositions (from simple salt to complex ones).

The concept of immobilizing the radioactive elements of nuclear waste in an assemblage of mineral phases was originally introduced by Hatch [1] at Brookhaven National Laboratory in 1953. The feasibility of making a ceramic of natural, mineralogically-stable phases was demonstrated by McCarthy [2,3] and Roy [4] at the Pennsylvania State University between 1973 and 1976. Since that time, a number of other mineralogic-ceramic assemblages have been developed [5]. Among these are the Sandia titanate-based ceramic [6], the Australian ceramic "SYNROC" [7–10], the silicate-phosphate supercalcine ceramics [11], the alumina-based tailored ceramics [12,13] and the Pu pyrochlores [14,15]. The structural types of monazite, kosnarite (NZP), langbeinite and other ones were considered as matrices for the incorporation of simulated wastes containing f-elements and that also contain uni-, bi-, and trivalent elements involved in radiochemical processes [16–27]. Cold pressing and sintering, as well as hot isostatic pressing often result in ceramics containing an intergranular glassy phase with radionuclides preferentially migrating to the glassy phase [28–36]. The radionuclides that are incorporated in the intergranular glassy phase(s) will then have leaching rates at about the same order as those from a glassy waste-form.

Crystalline waste-forms synthesized at moderate temperatures such as within 700 to 750 °C have not been investigated as intensely as those formed at high temperatures [11], although crystalline waste-forms made from clay have been studied almost continuously since the 1953 work of Hatch [1,11]. Supercalcine ceramics synthesized at high temperatures often contained sodalite-cancrinite mineral assemblages. Roy [37] proposed in 1981 a low solubility phase assemblage as a waste-form [37] using a low temperature hydrothermal process. The assemblage consisted of micas, apatite, pollucite, sodalite-cancrinite and nepheline, many of which could be produced using various clay minerals such as kaolin, bentonite and illite mixed with radioactive waste. However there were no continuous commercial technologies available at that time that could process the waste/clay mixtures in a hydrothermal environment, and clay-based crystalline waste-forms were not pursued. The situation changed in 1999 when Studsvik had built in Erwin a commercial facility to continuously process radioactive wastes by pyrolysis at moderate temperatures in a hydrothermal steam environment [38,39]. This facility utilizes Fluidized Bed Steam Reforming (FBSR) technology to pyrolyze ^{137}Cs- and ^{60}Co-contaning spent organic ion exchange resins produced by commercial nuclear facilities. FBSR technology can also process a wide variety of solid and liquid radioactive wastes, including spent organic ion exchange resins, charcoal, graphite, sludge, oils, solvents and cleaning solutions with contaminations up to radiation levels reaching 4 Sv/h (400 R/hr). The waste organics are destroyed, creating steam and CO_2. The clay serves in the FBSR process as a mineralizing agent, and feldspathoid minerals (sodalite, nosean and nepheline) are formed by the nanoscale reaction of waste components with clay. The phases formed act as hosts for waste contaminants such as Cl, I, F, ^{99}Tc from SO_4 alkali (Na, K, Cs) bearing wastes [40–44]. The mineralization occurs at moderate temperatures used within the range when most clays become amorphous at the nanoscale level, e.g., kaolin, bentonite (montmorillonite), and illite. The octahedral Al^{3+} cation in the clay structure is destabilized, and clays become amorphous as confirmed by X-ray diffraction (XRD) analysis, losing their hydroxyl (OH–) groups. The alkalis from waste act as activators of unstable Al^{3+} cations, and form new mineral phases catalyzing the mineralization. In the absence of steam many of these mineral phases can only be formed if temperatures are above 1200 °C.

Many of the compounds under consideration have structures similar to those of natural minerals (the so-called mineral-like compounds). Others of the discussed ones are not structural analogs of any known minerals (that its, of what is known today, as there are examples of compounds being developed for the radioactive waste immobilization that were obtained synthetically, and many years later a mineral was discovered, whose structural analog they became. For example, the mineral kosnarite $KrZr_2(PO_4)_3$ was discovered in 1991, and then kosnarite-like compounds (for example, NZP and NASICON) were synthesized and investigated many years before the discovery of this mineral).

Ceramic waste-forms can range from single phase, i.e., UO_2 and single phase solid solutions, i.e., (U, Th, Pu)O_2, to multiphase ceramics formulated in a such way that each waste radionuclide can substitute on a given host lattice in the various phases used.

2. Theoretical Aspects of Substitution

The crystal-chemical substitutions in crystalline waste-forms must be electrically balanced [45,46] which is important when relying on the long range order (LRO) of crystals accounting for the size and coordination of the crystallographic site, which will act as host to a given radionuclide, or its decay product upon transmutation (see [15] for natural analogs). Moreover, if a monovalent cation transmutes to a divalent one, the substitutions must be coupled to retain the electrical balance of the host phase without destroying the integrity of the phase. It means that the lattice site must be of suitable size and have a bond coordination able to accept the cation resulting from transmutation. The bond system of a crystalline ceramic can only maintain its charge balance if:

(1) Sufficient lattice vacancies exist in the structure or,
(2) A variable valence cation such as Fe or Ti is present in a neighboring lattice site balancing the charge.

Both above ways assume that the variable valence cations do not change lattice sites, and that the charge balancing cations are in the nearby lattice sites of the same host phase. The lattice site must be of close size flexible enough to accommodate the transmuting cation. Better flexibility is characteristic to host phases with lattice sites having irregular coordination or are distorted, as shown in some examples below. The flexibility (solubility) of waste-form mineral phase(s) as hosts for a different valence substituted cation can be analyzed by performing coupled substitutions. When the number of cations changes during the substitution, a vacancy is either created or consumed, however the substitution must maintain electrical neutrality. These types of substitution are characteristic for polymorphic changes such as [47], where □ denotes a vacancy:

$$\square + Ba^{2+} \rightarrow 2K^+, \text{ or } \square + Ca^{2+} \rightarrow 2Na^+, \text{ or } \square + Na^+ + 2Ca^{2+} \rightarrow 3Na^+ + Ca^{2+}$$

In these coupled substitutions it is implicit that the exchanging cations occupy the same lattice sites, have the same coordination, and thus the crystallographic symmetry is maintained. These substitutions are typically written using Roman numerals that designate the number of oxygen atoms that coordinate around a given cation, e.g., ^{VIII}Ca designates the octahedral VIII-fold coordination for the Ca^{2+} lattice site in oxyapatites:

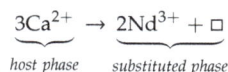

$$\underbrace{3Ca^{2+}}_{host\ phase} \rightarrow \underbrace{2Nd^{3+} + \square}_{substituted\ phase}$$

Calcium-neodymium-coupled substitutions were proven successful in the apatite $(Ca_6[SiO_4]_3)$ structure, resulting in a completely substituted $Nd_4\square_2[SiO_4]_3$, where 2/3 of the lattice sites have Nd^{3+} and 1/3 are vacant [45–47]. Ca^{2+} is normally in VIII-fold coordination in the apatite and has a 1.12 Å atomic radius [47–50]. The Nd^{3+} cation in VIII-fold coordination also has an atomic radius of 1.11 Å [50], which is very close to the Ca^{2+} atomic radius in VIII-fold coordination. It has been shown that the rare earth elements from La^{3+} through Lu^{3+} can substitute for Ca^{2+} and form oxyapatites,

$RE_{4.67}\square_{0.33}[SiO_4]_3O$ [51]. It was also shown [3] that even more complex but coupled substitutions were possible in the oxyapatite structure, such as:

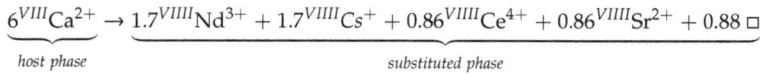

$$\underbrace{6^{VIII}Ca^{2+}}_{\text{host phase}} \rightarrow \underbrace{1.7^{VIIII}Nd^{3+} + 1.7^{VIIII}Cs^+ + 0.86^{VIIII}Ce^{4+} + 0.86^{VIIII}Sr^{2+} + 0.88\,\square}_{\text{substituted phase}}$$

where the atomic radius, r, of Cs^+ in VIII-fold coordination is 1.74 Å, Ce^{4+} in VIII-fold coordination is 0.97 Å, and Sr^{2+} in VIII-fold coordination is 1.26 Å. In this case small radii cations e.g., Ce^{4+} are mixed with larger radii cations such as Cs^+, so that individual lattice sites can distort without perturbing the entire crystal structure of the host mineral. It should be noted that the exchanging cations are always in the same lattice site of the same host phase [3,45,46,51].

The substitutions such as those given above for the oxyapatites were also demonstrated to be possible in many other Ca-bearing mineral phases such as larnite (Ca_2SiO_4 or b-C_2S), alite (calcium trisilicate or Ca_3SiO_5 or C_3S), C_3A ($Ca_3Al_2O_6$) and C_4AF ($Ca_4Al_2Fe_2O_{10}$), characteristic for cements [45,46]. This allowed Jantzen, et. al. [52,53] to make substitutions for Ca^{2+} in each phase (up to ~15 mole%) and prove possible the following additional substitutions:

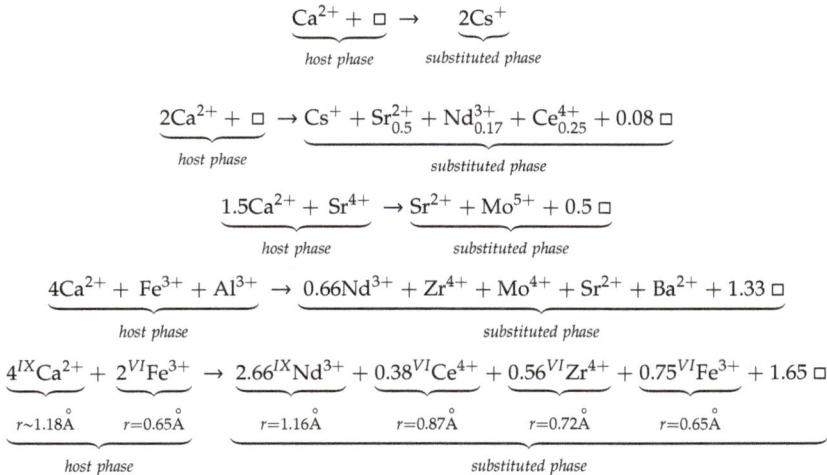

$$\underbrace{Ca^{2+} + \square}_{\text{host phase}} \rightarrow \underbrace{2Cs^+}_{\text{substituted phase}}$$

$$\underbrace{2Ca^{2+} + \square}_{\text{host phase}} \rightarrow \underbrace{Cs^+ + Sr^{2+}_{0.5} + Nd^{3+}_{0.17} + Ce^{4+}_{0.25} + 0.08\,\square}_{\text{substituted phase}}$$

$$\underbrace{1.5Ca^{2+} + Sr^{4+}}_{\text{host phase}} \rightarrow \underbrace{Sr^{2+} + Mo^{5+} + 0.5\,\square}_{\text{substituted phase}}$$

$$\underbrace{4Ca^{2+} + Fe^{3+} + Al^{3+}}_{\text{host phase}} \rightarrow \underbrace{0.66Nd^{3+} + Zr^{4+} + Mo^{4+} + Sr^{2+} + Ba^{2+} + 1.33\,\square}_{\text{substituted phase}}$$

$$\underbrace{\underbrace{4^{IX}Ca^{2+}}_{r\sim1.18\text{Å}} + \underbrace{2^{VI}Fe^{3+}}_{r=0.65\text{Å}}}_{\text{host phase}} \rightarrow \underbrace{\underbrace{2.66^{IX}Nd^{3+}}_{r=1.16\text{Å}} + \underbrace{0.38^{VI}Ce^{4+}}_{r=0.87\text{Å}} + \underbrace{0.56^{VI}Zr^{4+}}_{r=0.72\text{Å}} + \underbrace{0.75^{VI}Fe^{3+}}_{r=0.65\text{Å}} + 1.65\,\square}_{\text{substituted phase}}$$

It should be noted that the number of lattice sites have to be equivalent on the left-hand side and right hand site of the above equations.

These types of crystal-chemical substitutions have been studied in several waste-forms including SYNROC (SYNthetic ROCk) titanate phases containing zirconolite ($CaZrTi_2O_7$), perovskite ($CaTiO_3$), and hollandites (nominally $Ba(Al,Ti)_2Ti_6O_{16}$) [54], and in high alumina-tailored ceramic phases such as magnetoplumbites. Notable that magnetoplumbites were also found as a minor component of SYNROC, which immobilizes waste with high contents of Al [55].

Hollandite is the Cs^+ host phase in the SYNROC phase assemblages. Its structure can be written as $Ba_xCs_y(Al,Fe)_{2x+y}Ti_{8-2xy}O_{16}$ where x + y must be <2 [56]. It has two types of octahedral sites, one of which accommodates trivalent cations like Al^{3+}, Ti^{3+} and Fe^{3+}, while the other accommodates Ti^{4+}. The Cs^+ is accommodated in tunnels that normally accommodate the Ba^{2+} cation, and Cs-Ba lattice sites are VIII-fold coordinated [54,56]. On synthesis the substitution orders and incommensurate superstructures result when Cs^+ substitutes for Ba^{2+} [55]. Cs has been experimentally substituted for Ba when Fe^{3+} is substituted for Ti^{4+} in the VI-fold sites of hollandite. The species

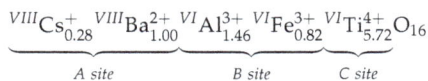

$$\underbrace{^{VIII}Cs^+_{0.28}\ ^{VIII}Ba^{2+}_{1.00}}_{A\ site}\ \underbrace{^{VI}Al^{3+}_{1.46}\ ^{VI}Fe^{3+}_{0.82}}_{B\ site}\ \underbrace{^{VI}Ti^{4+}_{5.72}}_{C\ site}O_{16}$$

has been synthesized by the sintering (frittage) of precursors in air at 1320 °C [56]. Ba–Al hollandite $(Ba_{1.16}Al_{2.32}Ti_{5.68}O_{16})$ was irradiated with 1–2.5 MeV electrons and β-irradiated up to summary doses of 4×10^8 to 7×10^9 Gy, after which it was found to contain Ti^{3+} centers and O_2- superoxide ions that confirmed the mechanism of charge balance during transmutation [56]. Theoretically, the limiting value of Cs in hollandite is y = 0.81, which corresponds to a 9.54 wt% waste loading of Cs_2O [57].

3. Synthesis of Ceramic Waste-Forms

Research and development of ceramic materials based upon compounds on the base of the oxides and salt compositions were carried out for the immobilization of high-level wastes and the transmutation of minor actinides. Structures of such materials provide the incorporation of various cations and anions, either individually, or in various combinations and ratios. Structural forms in which can be implemented a wide isomorphism of cations and anions (including in different crystallographic positions) deserve special attention.

Among such structures the type $NaZr_2(PO_4)_3$ (NZP) (analog—Mineral kosnarite) is regarded. NZP solid solutions may include more than half of the elements of Periodic Table of Elements in various combinations and ratios. The SYNROC developer calls them "near-universal solvent" [23], wherein this form of the consolidation of waste components is mono-phase in contrast to the multiphase SYNROC.

Ceramic materials are synthesized using the following methods: Pressing and sintering (frittage), hot isostatic or hot uniaxial pressing and other variants. Method Spark Plasma Sintering is the perspective for this aim. It provides a formation of virtually no porous ceramics having a relative density close to 99–100% for short time intervals (from 3 to 15 min). Reducing the porosity reduces the free surface, and therefore reduces the reaction surface and reactivity in heterogeneous systems with the participation of such materials. This in turn increases the heat, radiation and chemical stability of the ceramic.

Ceramic forms characteristics are presented here with their structures.

4. Crystalline Ceramic Phase:

4.1. Simple Oxides

1. Silica, SiO_2 [58–75], Figure 1.

Silicon dioxide, commonly known as silica (and/or quartz), is a prevalent element in the Earth's crust, a mineral of most igneous and metamorphic rocks. The formula "SiO_2" is commonly known as silicon dioxide. Silicon dioxide has a wide range of purposes, the main one being glass manufact-uring. In nature, silicon dioxide is commonly found as sand and quartz. Silica has polymorphism. It is stable under normal conditions of polymorphic modification—α-quartz (low temperature). Accordingly, β-quartz is called a high-temperature modification. Silica (α-quartz) possesses the rhombohedral structure, sp. gr. R3. Various elements with various oxidation states may attend in quartz: Li, Na, K. Mg, Ca, Mn, Cu, Ni, Pb B, Al, Fe, Cr, Ti, Zr and Te. Materials based on silicon oxide SiO_2, Silica (quartz) were prepared in ceramic form by using methods: Hot isostatic pressing, laser sintering, cold pressing and sintering at 1500 °C, cold pressing and ultra-low temperature sintering at T = 554–600 °C (30 min) and Spark Plasma Sintering.

Materials on the base of Silica can serve as a matrix for the immobilization of radioactive Iodine I-129 (half-life $T_{1/2} = 15.7 \times 10^3$ years).

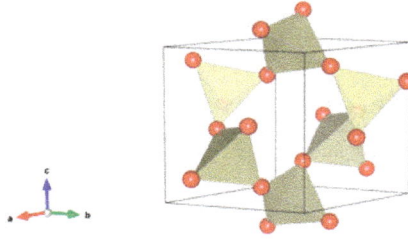

Figure 1. Silica, SiO_2. α-quartz (low temperature modification), structure rhombohedra, Sp. gr. R3. β-quartz (high-temperature modification, it forms from α-quartz at 846 K, stable up to 1140 K). Structure hexagonal, Sp. gr. $P6_222$. Cations can be Li, Na, K. Mg, Ca, Mn, Cu, Ni, Pb B, Al, Fe, Cr, Ti, Zr and Te.

2. Oxides Fluorite, XO_2 [76–93], Figure 2.

ZrO_2, UO_2, ThO_2, HfO_2, PuO_2, α-U_2O_3 and Np_2O_3 have the simple fluorite cubic structure, sp. gr. Fm3m. Fluorite has physical properties that allow it to be used for a wide variety of chemical, metallurgical and ceramic processes. The waste ceramics with high zirconia and alumina contents, and Y_2O_3-stabilized zirconia with fluorite structure, are the main host phases for actinide, rare earth elements, as well as Cs, Sr in high-level radioactive waste (HLW). Ceramics were made by HIP, HUP, press and sinter, melting and crystallization and by Spark Plasma Sintering with high relative density (up to 97–99%).

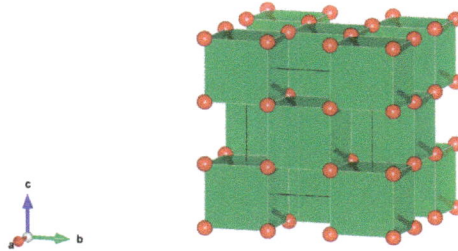

Figure 2. Fluorite, ZrO_2. Structure cubic, Sp. gr. Fm3m. Cations can be Zr, Hf, Th, U, Np and Pu.

4.2. Complex Oxides

3. Pyrochlore [86,94–117], Figure 3.

Many compounds with $A_2B_2O_7$ stoichiometry adopt the pyrochlore structure. A derivative of the fluorite structure type, $A_2B_2O_7$, where the A-site contains large cations (Na, Ca, U, Th, Y and lanthanides) and the B-site contains smaller, higher valence cations (Nb, Ta, Ti, Zr and Fe^{3+}). Structure: Cubic, Sp. gr. $Fd\bar{3}m$, z = 8. Ceramics were prepared by cold pressing and sintering.

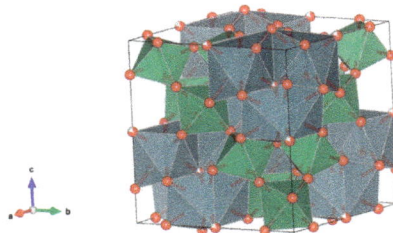

Figure 3. Pyrochlore. $A_2B_2O_7$. Structure cubic, Sp. gr. Fd3m. A-site-cations can be Na, Ca, Y, lanthanides, Th and U, while on the B-site—cations can be Fe^{3+}, Ti, Zr, Nb and Ta.

4. Murataite [104,106,108,118–131], Figure 4.

Murataite is a derivative of the isometric fluorite structure $A_6B_{12}C_5TX_{40-x}$, with multiple units of the fluorite unit cell; hosts U, Np, Pu, Am, Cm and REE, including Gd, a neutron absorber. It forms in solid solution with pyrochlore. Structure: Cubic, Sp. gr. $F4\bar{3}m$, z = 4. Ceramics were prepared by cold pressing and sintering.

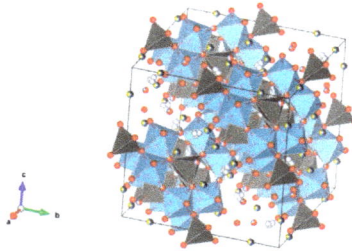

Figure 4. Murataite. $A_6B_{12}C_5TX_{40-x}$. Structure: Cubic, Sp. gr. F4m. Cations can be U, Np, Pu, Am, Cm and REE, including Gd (a neutron absorber).

5. Zirconolite [112,113,132–150], Figure 5.

Monoclinic $CaZrTi_2O_7$, has a fluorite-derived structure closely related to pyrochlore, where Gd, Hf, Ce, Th, U, Pu and Nb may be accommodated on the Ca/Zr-sites, as in the case of $Ca(Zr,Pu)Ti_2O_7$. Structure: Trigon., Pr. gr. C2/c. Ceramics were prepared by cold pressing and sintering.

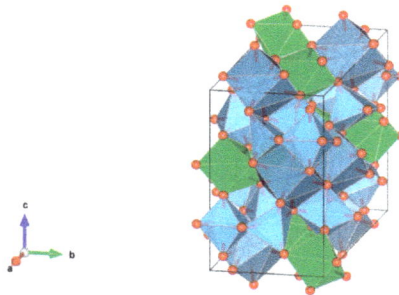

Figure 5. Zirconolite. $CaZrTi_2O_7$, Structure monoclinic, Sp. gr. C2/c. Cations can be Gd, Hf, Ce, Th, U, Pu and Nb.

6. Perovskite [110,134,140,151–159], Figure 6.

$CaTiO_3$ has a wide range of compositions as stable solid-solutions; orthorhombic; consists of a 3-dimensional network of corner-sharing TiO_6 octahedra, with Ca occupying the large void spaces between the octahedra (the corner-sharing octahedra are located on the eight corners of a slightly distorted cube). Plutonium, other actinides and rare-earth elements can occupy the Ca site in the structure, as in $(Ca,Pu)TiO_3$. The octahedra can also tilt to accommodate larger cations in the Ca site. Structure: Cubic, sp. gr. Pm3m; rombohedral, Sp. gr. Pnma; may include: Ca, Y, REE, Ti, Zr, U and Pu. Ceramics were prepared by cold pressing and sintering, and by hot pressing enabling densities up to 90–98% of theoretical.

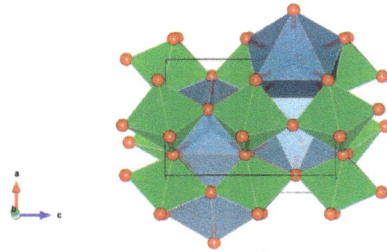

Figure 6. Perovskite. $CaTiO_3$, Structure cubic, Sp. gr. Pm3m. Cations can be Ca, Y, REE, Ti, Zr, U and Pu.

7.   Hollandite [160–169], Figure 7.

$Ba_{1.2}(Al,Ti)_8O_{16}$ tunnels between TiO_6 octahedra to accommodate ^{133}Ba, ^{137}Cs and ^{90}Sr. Structure: Tetragon, Sp. gr. I4/m, Z = 4 and monocl., Sp. gr. I2/m, z = 1; may include: Na, K, Cs, Mg, Ca, Ba, Al, Fe, Mn^{3+}, Si, Ti and Mn^{4+}. Ceramics were prepared by cold pressing and sintering.

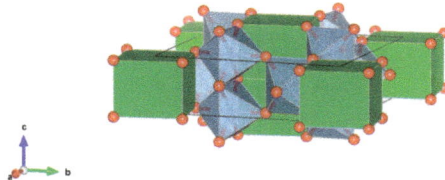

Figure 7. Hollandite. $Ba_{1.2}(Al,Ti)_8O_{16}$. Structure tetragon, Sp. gr. I4/m, monocl, Sp. gr. I2/m. Cations can be Na, K, Cs, Mg, Ca, Sr, Ba, Al, Fe, Mn^{3+}, Si, Ti and Mn^{4+}.

8.   Garnet [87,89,104,105,170–194], Figure 8.

(1) $^{[8]}A_3{}^{[6]}B_2[TiO_4]_3$, e.g., $^{[8]}$(Ca,Gd, actinides)$^{[6]}Fe_2{}^{[4]}Fe_3O_{12}$.

(2) $A_3B_2(XO_4)_3$; distorted cubic structure; BO_6 octahedra and XO_4 tetrahedra establish a framework structure alternately sharing corners; A and B sites can host actinides, REs, Y, Mg, Ca, Fe^{2+}, Mn^{2+} and X = Cr^{3+}, Fe^{3+}, Al^{3+}, Ga^{3+}, Si^{4+}, Ge^{4+} and V^{5+}. Structure: Cubic, Sp. gr. Ia3d, z = 8. Ceramics were prepared by cold pressing and sintering and using Spark Plasma Sintering with high relative density up to 98–99% of theoretical.

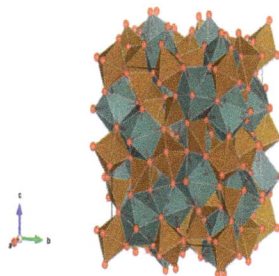

Figure 8. Garnet, $Ca_3Al_2Si_3O_{12}$. Structure cubic, Sp. gr. Ia3d. Cations can be Mg, Ca, Mn, Co, Cd, Al, Sc, Fe, Ga, Y, In, La, REE, Ti, Zr, Ru, Sn, N, P, V and As.

9. Crichtonite [131,195–202], Figure 9.

(Sr,Pb,La,Ce,Y)(Ti,Fe^{3+},Mn,Mg,Zn,Cr,Al,Zr,Hf,U,V,Nb,Sn,Cu,Ni)$_{21}$O$_{38}$. Sr, La, Ce, Y positions are indicated by the solid circles. Other cations are in the octahedral positions. Structure: Rombohedral, Sp. gr. R3. Ceramics were prepared by hot pressing.

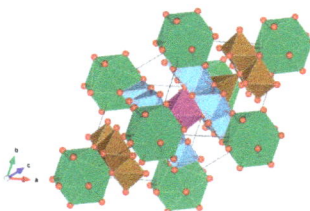

Figure 9. Crichtonite. Sr(Mn,Y,U)Fe$_2$(Ti,Fe,Cr,V)$_{18}$(O,OH)$_{38}$. Structure rombohedral, Sp. gr. R3. Cations can be Mg, Mn, Ni, Cu, Mn, Sr, Pb, Cr, Fe, Y, La, Ce, Ti, Zr, Hf, U, V and Nb.

10. Freudenbergite [153,155,203,204], Figure 10.

Na$_2$Al$_2$(Ti,Fe)$_6$O$_{16}$ a spinel-based phase suitable for incorporating Al-rich wastes from Al fuel cladding/decladding. The A site can accommodate Na and K while the different octahedral sites can accommodate Mg, Co, Ni, Zn, Al, Ti^{3+}, Cr, Fe, Ga, Si and Nb. Structure: Monocl., Sp. gr. C12/m1. Ceramics were prepared by cold pressing and sintering, $\rho = 90\%$.

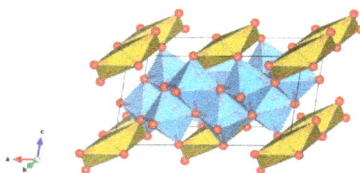

Figure 10. Freudenbergite (spinel). Na$_2$Al$_2$(Ti,Fe)$_6$O$_{16}$;. Structure monocl. Sp. gr. C12/m1. Cations can be Na, K, Mg, Co, Ni, Zn, Al, Ti^{3+}, Cr, Fe, Ga, Si and Nb.

11. P-Pollucite [205–215], Figure 11.

The ability of the pollucite structure to include large 1-, 2- and 3-valent cations allows flexibility to select the desired model composition. When replacing the cations it will be becomes possible to use cheap components; the introduction of small cations increases the concentration of cesium in the composition of the mono-phase product. Structure: Cubic, sp. gr. I4$_1$32, z = 16; may include: Li, Na, K, Rb, Cs, Tl, Be, Mg, Sr, Ba, Cd, Mn, Co, Ni, Cu, Zn, B, Al, Fe, Si, Ti, P, V, Nb and Ta. Compounds are hydrolytically and radiation-wise stable. Ceramics were prepared by cold pressing and sintering and Spark Plasma Sintering with high relative density (at last those up to 98–99%).

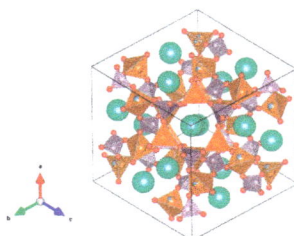

Figure 11. P-Pollucite. (Na,K,Rb,Cs)MgAl$_{0.5}$P$_{1.5}$O$_6$; Structure cubic, Sp. gr. I4$_1$32. Cations can be Li, Na, K, Rb, Cs, Tl, Be, Mg, Mn, Co, Ni, Cd, Sr, Ba, Sr. Ba, B, Al, Fe, Si, Ti, P, V, Nb and Ta.

12. Magnetoplumbites (aluminates) [13,55,216–224], Figure 12.

Nominally $X(Al,Fe)_{12}O_{19}$, where X = Sr, Ba, $(Cs_{0.5} + La_{0.5})$ and $(Na_{0.5} + La_{0.5})$. The X site is XII-fold coordinated and both Cs^+/Ba^{2+}-Fe^{3+}/Fe^{2+} or Cs^+/Ba^{2+}-Ti^{4+}/Ti^{3+} type substitutions can occur. Accommodating structures because they are composed of spinel blocks with both IV-fold and VI-fold coordinated sites for multivalent cations, and interspinel layers which have unusual V-fold sites for small cations. The interspinel layers also accommodate large cations of 1.15–1.84 Å, replacing oxygen in XII-fold sites in the anion close packed structure. The large ions may be monovalent, divalent, or trivalent with balancing charge substitutions either in the interspinel layer $(Na_{0.5} + La_{0.5})$ or between the interspinel layer and the spinel blocks $(Cs^+/Ba^{2+}$–Fe^{3+}/Fe^{2+} or Cs^+/Ba^{2+}–$Ti^{4+}/Ti^{3+})$. Structure: Hexagon., Sp. gr. $P6_3/mmc$, z = 2; may include: Na, Cs, Mg, Sr, Ba, Pb, Mn, Co, Cu, Al, Fe, Sc, Y, La, Ce, Sm, Gd, Yb, Lu, actinides, Si, Ti and Sn. Ceramics were prepared by cold pressing and sintering and by hot pressing.

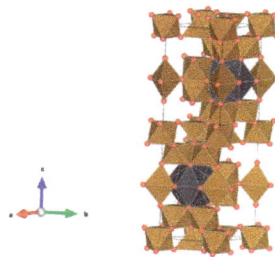

Figure 12. Magnetoplumbite. $(Sr,Ba, ((Na,Cs)_{0.5}+La_{0.5}))(Al,Fe)_{12}O_{19}$. Structure hexagon., Sp. gr. $P6_3/mmc$. Cations can be Na, Cs, Mg, Sr, Ba, Pb, Mn, Co, Cu, Al, Fe, Sc, Y, La, Ce, Sm, Gd, Yb, Lu, An, Si, Ti and Sn.

13. Zircon/Thorite/Coffinite [83,110,140,225–235], Figure 13.

$ZrSiO_4/ThSiO_4/USiO_4$; zircon is an extremely durable mineral that is commonly used for U/Pb age-dating, as high uranium concentrations (up to 20,000 ppm) may be present; the $PuSiO_4$ end member is known, and Ce, Hf and Gd have been found to substitute for Zr. Structure: Tetragon. Sp. gr. I41/amd, z = 4; may include: REE, Th, U, Pu; Na, Mg, Ca, Mn, Co, Fe, Ti, P, V, Se and Mo. Ceramics were prepared by hot pressing, ρ = 99.1% and by Spark Plasma Sintering, ρ = 99%

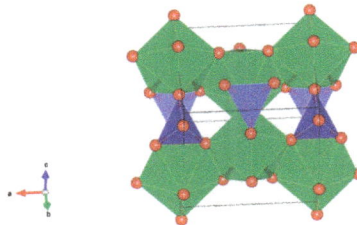

Figure 13. Zircon/Thorite/Coffinite. $ZrSiO_4/ThSiO_4/USiO_4$. Structure tetragon., Sp. gr. I41/amd. Cations can be Na, Tl, Mg, Ca, Mn, Co, Fe, Ti, REE, Ti, Th, U, Pi, P, V, Mo and Se.

14. Titanite (sphene) [104,110,236–238], Figure 14.

$CaTiSiO_5$ $[CaTiO(SiO_4)]$. Structure: Monocl. Sp. gr. $P2_1/a$, Z = 4; may include: Mg, Ca, Sr, Ba, Mn, Al, Fe, Cr, Ce, Y, Zr, Th and F. Ceramics are known as a matrix for actinide immobilization, and were prepared by cold pressing and sintering.

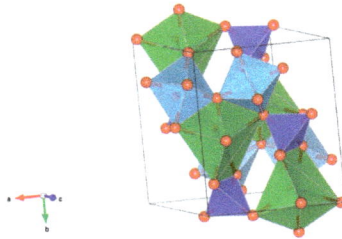

Figure 14. Titanite (sphene). CaTiSiO$_5$ [CaTiO(SiO$_4$)]. Structure monocl., Sp. gr. P$_2$I/a. Cations can be Mg, Ca, Sr, Ba, Mn, Al, Fe, Cr, Ce, Y, Zr, Th and F.

15. Britholite (silicate apatite; also known as oxy-apatite in the literature) [3,46,51,239–249], Figure 15.

(REE,Ca)$_5$(SiO$_4$,PO$_4$)$_3$(OH,F); i.e., Ca$_2$Nd$_8$(SiO$_4$)$_6$O$_2$, Ca$_2$La$_8$(SiO$_4$)$_6$O$_2$; based on ionic radii of Nd^{3+}, La^{3+} and Pu^{3+}, an extensive range of solubility for Pu^{3+} substitution for the Nd or La, particularly on the *6h* site, is expected. Since there is an extensive range in the Ca/RE ratio in these silicate apatites, a fair amount of Pu^{4+} substitution may be possible; La^{3+} through Lu^{3+} can substitute for Ca^{2+} and form oxyapatites, RE$_{4.67}$□$_{0.33}$[SiO$_4$]$_3$O; can also accommodate Cs, Sr, B, Th, U and Np. Structure: Monocl., Sp. gr. P2$_1$ and hexagon. Sp. gr. P6$_3$/m. Ceramics were prepared by cold pressing and sintering, ρ = 95%.

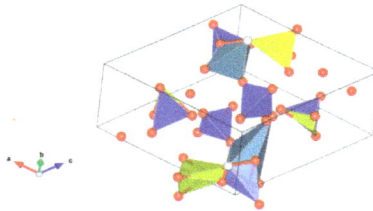

Figure 15. Britholite (silicate apatite, oxy-apatite). (REE,Ca)$_5$(SiO$_4$,PO$_4$)$_3$(OH,F)-Structure monoclin. Sp. gr. P2$_1$/hexagonal, Sp. gr. P6$_3$/m. Cations can be Cs, Sr, B, REE, Th, U, Np and Pu.

4.3. Framework Silicates

16. Zeolites [75,250–266], Figure 16.

(X$_{x/n}$[(AlO$_2$)$_x$(SiO$_2$)$_y$] where X is the charge balancing counter-ion, n is the charge of the counter-ion, x is the number of charge-deficient alumina sites, and y is the number of charge-neutral silica sites. Zeolites are characterized by internal voids, channels, pores, and/or cavities of well-defined size in the nanometer range, ≈4–13 Å. The channels and/or cavities may be occupied by charge compensating ions and water molecules. Zeolites like Ag-Mordenite selectively sorbs I$_2$ (^{129}I); certain zeolites can be converted to condensed oxide ceramics by heating. This process is particularly attractive for waste-form synthesis because contaminants capture and immobilization is performed with minimal steps. Structure of Zeolite-A showing alternate Al and Si atom ordering but omitting the tetrahedral oxygens around each Al and Si may include Na, K, NH$_4$$^+$, Cs, Mg, Ca, Sr, Co, Fe, Ga, REE and Ti. 45 natural zeolites and 100 artificial ones are known. Ceramics were prepared by hot pressing.

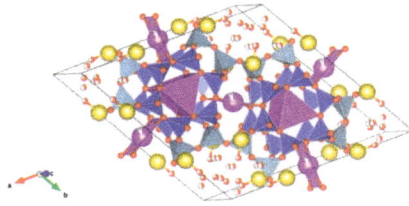

Figure 16. Zeolites. $X_{x/n}[(AlO_2)_x(SiO_2)_y]$ (where Xn+ is the charge balancing counter-ion). Structure depends on chemical composition. Cations can be Na, K, NH^{4+}, Cs, Mg, Ca, Sr, Co, Fe, Ga, REE and Ti.

17.   Pollucite [37,87,212,214,215,259,267–293], Figure 17.

$(Ca,Na)_2Al_2Si_4O_{12}\cdot 2H_2O$; host for fission products such as ^{137}Cs. Structure: Cubic, Sp. gr. Ia3d, z = 16; may include: Li, Na, K, Rb, Cs, Tl, Be, Mg, Sr, Ba, Cd, Mn, Co, Ni, Cu, Zn, B, Al, Fe, Si, Ti, P, V and Nb. Ceramics were prepared by Spark Plasma Sintering with high relative density (up to 96%).

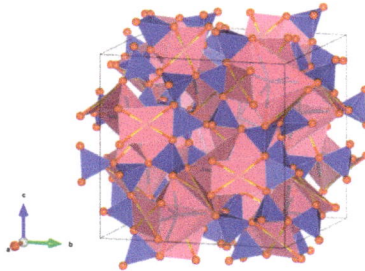

Figure 17. Pollucite. $(Ca,Na)_2Al_2Si_4O_{12}\cdot 2H_2O$. Structure cubic, Sp. gr. Ia3d. Cations can be Li, Na, K, Rb, Cs, Tl, Be, Mg, Sr, Ba, Cd, Mn, Co, Ni, Cu, Zn, B, Al, Fe, Si, Ti, P, V and Nb.

18.   Nepheline/Leucite [37,58,61,73,155,294–297], Figure 18.

$NaAlSiO_4$ silica "stuffed derivative" ring type structure; some polymorphs have large nine-fold cation cage sites, while others have 12-fold cage-like voids that can hold large cations (Cs, K, Ca). Natural nepheline structure accommodates Fe, Ti and Mg. Two-dimensional representation of the structure of nepheline showing the smaller 8 oxygen sites that are occupied by Na and the larger 9 oxygen sites that are occupied by K and larger ions, such as Cs and Ca. Structure may include: Li, Na, K, Rb, Cs, Be, Mg, Ca, Ba, Pb, Mn, Co, Ni, Al, Fe, Cr, Si, Ti and V. Structure: Hexagon. Sp. gr. P6$_3$, z = 2. Leucite. Structure: Tetragon. Sp. gr. I4$_1$/a and I4$_1$/acd; cubic, Sp. gr. Ia3d, z = 16.

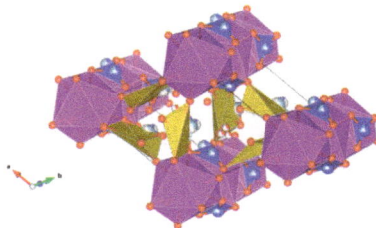

Figure 18. Nepheline/Leucite. $(Na, K)AlSiO_4/K[AlSi_2O_6]$. Structure hexagon., Sp. gr. P6$_3$/tetragonal, Sp. gr. I4$_1$/a and I4$_1$/acd or cubic, Sp. gr. Ia3d. Cations can be Li, Na, K, Rb, Cs, Be, Mg, Ca, Ba, Pb, Mn, Co, Ni Al, Fe, Cr, Si, Ti and V.

19. Sodalite Group (name of mineral changes with anions sequestered in cage structure) [37,264,295,298–313], Figure 19.

(1) Sodalite $Na_8Cl_2Al_6Si_6O_{24}$, also written as $(Na,K)_6[Al_6Si_6O_{24}]\cdot(2NaCl)$ to demonstrate that 2Cl and associated Na atoms are in a cage structure defined by the aluminosilicate tetrahedra of six adjoining $NaAlSiO_4$, is a naturally occurring feldspathoid mineral. It incorporates alkali, alkaline earths, rare earth elements, halide fission products and trace quantities of U and Pu. Sodalite was and it is being investigated as a durable host for the waste generated from electro-refining operations deployed for the reprocessing of metal fuel. Supercalcines which are high temperature, silicate-based "natural mineral" assemblages proposed for HLW waste stabilization in the United States in 1973–1985, contained sodalites as minor phases retaining Cs, Sr and Mo, e.g., $Na_6[Al_6Si_6O_{24}](NaMoO_4)_2$. Sodalite structures are known to retain B, Ge, I, Br and Re in the cage-like structures. Structure of Sodalite showing (a) two-dimensional projection of the (b) three-dimensional structure and (c) the four fold ionic coordination of the Na site to the Cl-ion and three framework oxygen bonds. Structure: Cubic, Sp. gr. $P\bar{4}3n$, z = 1; may include: Na, K, Mg, Ca, Mn, Fe, Al, Si, Ti, Cl, SO_4 and CO_3. Ceramics were prepared by cold pressing and sintering; by HIP.

(2) Nosean, $(Na,K)_6[Al_6Si_6O_{24}](Na_2SO_4)$), silica "stuffed derivative" sodalite cage type structure host mineral for sulfate or sulfide species.

(3) Hauyne, $(Na)_6[Al_6Si_6O_{24}]((Ca,Na)SO_4)_{1-2}$ sodalite family; can accommodate either Na_2SO_4 or $CaSO_4$.

(4) Helvite $(Mn_4[Be_3Si_3O_{12}]S$: Be (beryllium) can be substituted in place of Al and S_2 in the cage structure along with Fe, Mn and Zn.

(5) Danalite $(Fe_4[Be_3Si_3O_{12}]S)$.

(6) Genthelvite $(Zn_4[Be_3Si_3O_{12}]S)$.

(7) Lazurite, $(Ca,Na)_6[Al_6Si_6O_{24}]((Ca,Na)S,SO_4,Cl)_x$; can accommodate either SO_4 or S_2, Ca or Na and Cl.

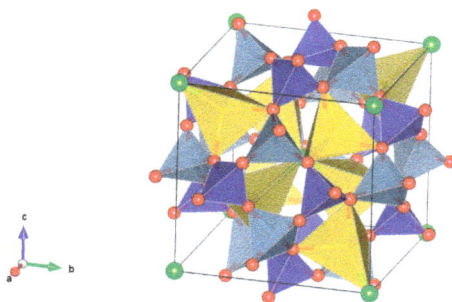

Figure 19. Sodalite.group minerals. Sodalite/Nosean/Hauyne/Helvite/Danalite/Genthelvite/Lazurite. $(Na,K)_6[Al_6Si_6O_{24}]\cdot(2NaCl)/(Na,K)_6[Al_6Si_6O_{24}](Na_2SO_4)/(Na)_6[Al_6Si_6O_{24}]((Ca,Na)SO_4)_{1-2}/$ $(Mn_4[Be_3Si_3O_{12}]S/(Fe_4[Be_3Si_3O_{12}]S)/(Zn_4[Be_3Si_3O_{12}]S)/(Ca,Na)_6[Al_6Si_6O_{24}]((Ca,Na),S,SO_4,Cl)_x$; Structure cubic, Sp. gr. P3n Cations and anions can be Na, K, Be, Mg, Ca, Mn, Fe, Al, Si, Ti, Cl, SO_4 and CO_3.

20. Cancrinite [37,314–319], Figure 20.

Cancrinite is a complex carbonate and silicate of sodium, calcium and aluminum with the formula $(Na,Ca,K)_6[Al_6Si_6O_{24}]((Na,Ca,K)_2CO_3)_{1.6}\cdot2.1H_2O$. It is classed as a member of the feldspathoid group of minerals. Cancrinite is unusual in that it is one of the few silicate minerals to have a carbonate ion

$(CO_3{}^{2-})$ present in its structure. Mineral cancrinite will also contain some percentages of sulfate ions $(SO_4{}^{2-})$ and a chlorine ion (Cl^-). Structure: Hexagonal, Sp. gr. $P6_3$.

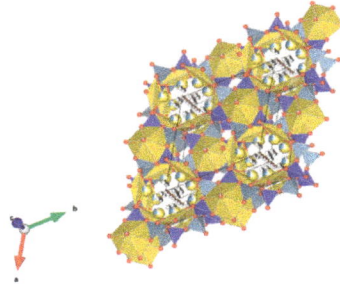

Figure 20. Cancrinite. $(Na,Ca,K)_6[Al_6Si_6O_{24}]((Na,Ca,K)_2CO_3)_{1.6} \cdot 2.1H_2O$. Structure hexagonal, Sp. gr. $P6_3$. Cations and anions can be Na, K, Ca. Al, Si, SO_4 and Cl.

21. Crystalline SilicoTitanate (CST) [73,110,273–275,277,320–324], Figure 21.

$[(Ca,N2a,K,Ba)AlSiO_4$ incorporates Na, K, Cs, Ca, Sr, Ba, Pb, Al, REE, Bi, Ti, Zr, Nb and Ta. Crystal structure of Cs exchanged Nb–titanium silicate. Structure: Cubic, sp. gr. Pm3m up to 105 °C, after tetragon. Sp. gr. I4/mcm or $P4_2$/mcm. Ceramics were prepared by hot isostatic pressing.

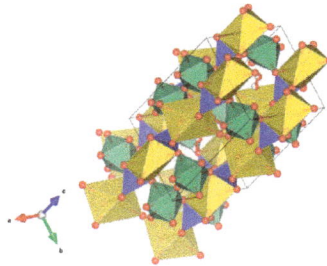

Figure 21. SilicoTitanate (CST). $SiTiO_4$. Structure cubic, Sp. gr. Pm3m up to 105 °C, after-tetragonal Sp. gr. I4/mcm or $P4_2$/mcm. Cations can be Na, K, Cs, Ca, Sr, Ba, Pb, Al, REE, Bi, Ti, Zr, Nb and Ta.

22. Micas (Dehydroxylated) [37,325–330], Figure 22.

The following dehydroxylated micas have been synthesized phase pure: $LiAl_3Si_3O_{11}$, $NaAl_3Si_3O_{11}$, $KAl_3Si_3O_{11}$, $RbAl_3Si_3O_{11}$, $CsAl_3Si_3O_{11}$, $TlAl_3Si_3O_{11}$, $Ca_{0.5}\square_{0.5}Al_3Si_3O_{11}$, $Sr_{0.5}\square_{0.5}Al_3Si_3O_{11}$, $Ba_{0.5}\square_{0.5}Al_3Si_3O_{11}$ and $La_{0.33}\square_{0.66}Al_3Si_3O_{11}$. In the Cs mica up to 30 wt% Cs_2O can be accommodated, in the Rb-mica up to 22 wt% Rb_2O can be accommodated, and in the Ba-mica up to 19 wt% BaO can be accommodated. Mg, Fe^{2+}, Fe^{3+}, Mn, Li, Cr, Ti and V can substitute for VI-fold coordinated Al^{3+}. Structure: Monoclinic. Sp. gr. C2/c.

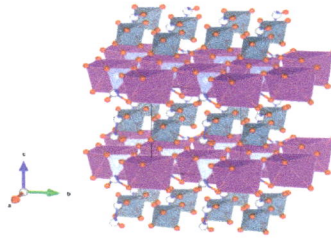

Figure 22. Micas (Dehydroxylated). $XY_{2-3}Z_4O_{10}(OH, F)_2$ with X = K, Na, Ba, Ca, Cs, (H_3O) and (NH_4); Y = Al, Mg, Fe^{2+}, Li, Cr, Mn, V and Zn; and Z = Si, Al, Fe^{3+}, Be and Ti. Structure monoclinic, Sp. gr. C2/c.

4.4. Phosphates

23.　Monazite [12,16–18,87,89,140,141,231,235,244,293,331–359], Figure 23.

$CePO_4$ or $LaPO_4$ are corrosion-resistant materials and can incorporate a large range of radionuclides including actinides and toxic metals into its structure. Monazite was proposed as a potential host phase for excess weapons plutonium and radionuclides, and toxic metals in glass ceramic waste-forms for low-level and hazardous wastes. Monazite structure (monazite mineral $CePO_4$) has wide capacity isomorphous through which the cerium and phosphorus can be substituted for other elements, e.g.,: Ce → Li, Na, K, Rb, Mg, Ca, Sr, Ba, Cd, Pb, Bi, Y, La, Pr, Nd, Sm, Eu, Gd, Tb, Yb, Am, Cm, Cf, Es, Ge, Zr, Th, Np, U and Pu; P → Cr, Si, Se, V, As and S. Alternating chains of PO_4 tetrahedra and REO_9 polyhedra. Structure: Monoclinic. Sp. gr. $P2_1/n$. Ceramics were prepared by cold pressing and sintering (ρ = 90–95%), hot pressing (ρ = 97%) and Spark Plasma Sintering with high relative density (up to 98–99%).

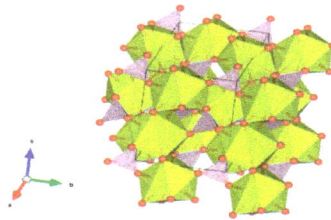

Figure 23. Monazite. $(Ce,La,Nd,Th)(PO_4,SiO_4)$. Structure monoclinic, Sp. gr. $P2_1/n$. Cations can be Li, Na, K, Rb, Mg, Ca, Sr, Ba, Cd, Pb, Bi, Y, La, Pr, Nd, Sm, Eu, Gd, Tb, Yb, Am, Cm, Cf, Es, Ge, Zr, Th, U, Np, Pu, Cm; Si, Se, V, As and S.

24.　Xenotime [231,334,344,360–363], Figure 24.

YPO_4. Structure: Tetragonal. Sp.gr. $I4_1/amd$, z = 4, C.N.Y-O_n, n = 8. Isomorph including: Be, Ca, Al, Sc, La, Ce, Er, Dy–Lu, Zr, Th and U. Ceramics were prepared by cold pressing and sintering.

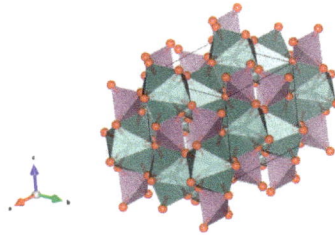

Figure 24. Xenotime (YPO$_4$). Ce,La,Nd,Th)(PO$_4$,SiO$_4$). Structure tetragonal, Sp. gr. I41/amd. Cations can be Be, Ca, Al, Sc, La, Ce, Er, Dy–Lu, Zr, Th and U.

25. Apatite [3,37,87,240,241,332,364–378], Figure 25.

Ca$_{4-x}$RE$_{6+x}$(SiO$_4$)$_{6-y}$(PO$_4$)$_y$(O,F)$_2$ can be actinide-host phases in HLW glass, glass-ceramic waste-forms, ceramic waste-forms and cements. The actinides can readily substitute in apatite for rare-earth elements as in Ca$_2$(Nd,Cm,Pu)$_8$(SiO$_4$)$_6$O$_2$, and fission products are also readily incorporated. However, the solubility for tetravalent Pu may be limited without other charge compensating substitutions.

Apatite has been proposed as a potential host phase for Pu and high-level actinide wastes. Structure: Hexagonal, Sp. gr. P6$_3$/m or monoclinic, Sp. gr. P2$_1$/b; may include: Na, K, Cs, Mg, Ca, Sr, Ba, Mn, Ni, Cd, Hg, Pb, Cr, Y, REE, Th, U, Si, P, V, As, S, F, Cl, OH and CO$_3$. Ceramics were prepared by cold pressing and sintering, ρ = 95%; by HIP.

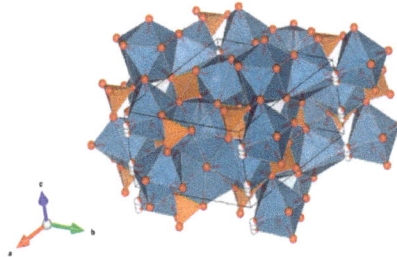

Figure 25. Apatite. Ca$_5$(PO$_4$)$_3$(OH,F,Cl). Apatite. Structure hexagonal, Sp. gr. P63/m, monoclinic, Sp. gr. P21/b. Cations and anions can be Na, K, Cs, Mg, Ca, Mn, Ni, Sr, Ba, Cd, Hg, Pb, Cr, Y, REE, Cm, Si, Th, U, P, V, As, S, F, Cl, OH and CO$_3$.

26. Sodium zirconium phosphate (NZP) [17–24,87,89,155,209,211,293,379–416], Figure 26.

The first studies of materials with such a structure were carried out by the authors [379–383] in 1976–1987. They substantiated the crystal-chemical approach when choosing the composition of substances and their structural modifications with ion-transforming properties (Li+, Na+, etc.): NASICON, Langbeinite. Such materials have a frame structure: Na$_{1+x}$Zr$_2$Si$_x$P$_{3-x}$O$_{12}$, Na$_3$M$_2$ (PO$_4$)$_3$ (M = Sc, Cr, Fe), Na$_5$Zr(PO$_4$)$_3$, Li$_x$Fe$_2$(WO$_4$)$_3$, Li$_x$Fe$_2$(MoO$_4$)$_3$. Elements in oxidation states 3–6 were introduced into the frame positions: Sc, Cr, Fe, Si, Zr, P, W and Mo. It was also the first time in 1987 that the rationale for the use of such structural analogs for the consolidation of HLW and transmutation of minoractinides [384] was presented. The development of such materials—Structural analogues of NASICON, NZP, Langbeinite—and their research, was continued in subsequent years.

NaZr$_2$(PO$_4$)$_3$. The NZP structure can incorporate a complex variety of cations, including plutonium; a three dimensional network of corner-sharing ZrO$_6$ octahedra and PO$_4$ tetrahedra in which plutonium can substitute for Zr, as in Na(Zr,Pu)$_2$(PO$_4$)$_3$. Complete substitution of Pu^{4+} for Zr has been demonstrated in NZP. Cs and Sr can substitute for Na, while fission products and actinides substitute for Zr in octahedral positions. P is tetrahedral. Phosphates with the mineral kosnarite structure (NaZr$_2$(PO$_4$)$_3$ type, NZP) form a wide family. They can contain various cations in the oxidation state from 1+ to 5+. The structure consists of several positions and so many various cations can occupy it. These are MI = Li, Na, K, Rb, Cs; H, Cu(I) and Ag; MII = Mg, Ca, Sr, Ba, Mn, Co, Ni, Cu, Zn, Cd and Hg; MIII = Al, Ga, In, Sc, Y, La, Ce-Lu, Am, Cm, V, Cr, Fe, Sb and Bi; MIV = Ge, Sn, Ti, Zr, Hf, Mo, Ce, Th, U, Np and Pu; MV = Sb, Nb and Ta. Structure: Rhombohedral, Sp. gr. $R\bar{3}c$, R3c, R3. This fact is extremely important, and can be useful for the synthesis of single-phase crystalline products of the solidification of radioactive waste whose cationic composition, as a rule, is extremely complicated. Ceramics were prepared by cold pressing and sintering (ρ = 80–98%), hot pressing (ρ = 96%) and Spark Plasma Sintering with high relative density (up to 98–99.9%).

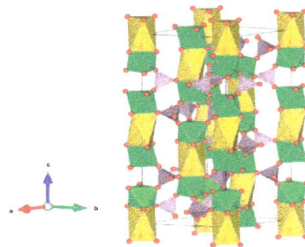

Figure 26. Sodium zirconium phosphate (NZP), NaZr$_2$(PO$_4$). Structure rhombohedral, Sp. gr. Rc, R3c, R3. Cations can be Li, Na, K, Rb, Cs, Cu and Ag; Mg, Ca, Mn, Zn, Sr and Ba; Mn, Co, Ni, Cu, Zn and Cd; Sc, Fe, Bi, Ce–Lu, Am and Cm; Zr, Hf, Th, U, Np and Pu; V, Nb, Sb and Ta; Ti, Ge, Zr, Hf, U, Np, Pu, Mo and Sn; Al, Sc, Cr, Fe, Ga, Y and In; Gd, Tb, Dy, Er and Yb; Mg; Na and K; Si, P, S, Mo and W.

27. Langbeinite [18,87,89,211,293,416–420], Figure 27.

Langbeinite is a potassium magnesium sulfate mineral with the formula: K$_2$Mg$_2$(SO$_4$)$_3$. It may include much of cesium and other large 1- and 2-valent elements. The structure is a framework type, also as for its kosnarite structure. Structure: Cubic, Sp. gr. P2$_1$3; may include: Na, K, Rb, Cs, Tl, NH$_4$, Mg, Sr, Ba, Pb, Mn, Co, Ni, Zn, Al, Fe Cr, Ti^{3+}, Ga, V^{3+}, Rh, In, REE, Bi, Sn, Ti, Zr, Hf, P, Nb, Ta and S. Ceramics were prepared by cold pressing and sintering, ρ = 88%.

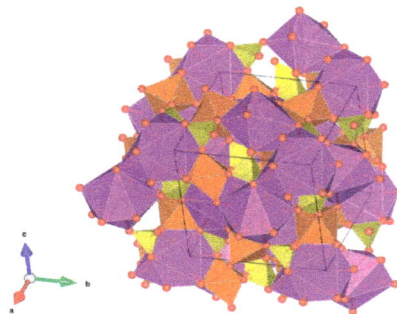

Figure 27. Langbeinite. K$_2$Mg$_2$(SO$_4$)$_3$. Structure cubic, Sp. gr. P2$_1$3. Cations can be Na, K, Rb, Cs, Tl, NH$_4$, Mg, Sr, Ba, Pb, Mn, Co, Ni, Zn, Al, Fe Cr, Ti^{3+}, Ga, V^{3+}, Rh, In, REE, Bi, Sn, Ti, Zr, Hf, P, Nb, Ta and S.

28. Whitlockite [87,89,421–432], Figure 28.

Phosphates with the whitlockite structure (analog β-Ca$_3$(PO$_4$)$_2$) were proposed as matrices for radioactive waste immobilization. Their origin is both biogenic and cosmogenic. Whitlockite samples from meteorites, rocks of the Moon, Mars and other cosmogenic bodies, preserve the crystalline form under the action of natural thermal "stress" and cosmic radiation. They contain small amounts of uranium and thorium, and it is presumed to contain plutonium. It is known to form isostructural compounds with H, Li, Na, K, Cu, Mg, Ca, Sr, Ba, Al, Sc, Cr, Fe, Ga, In, La, Ce, Sm, Eu, Gd, Lu, Th and Pu. Thermal stability is up to 1200 °C, thermal expansion up to 1×10^{-5} deg^{-1} (25–1000 °C) are close to Synroc and zirconolite; hydrothermal stable – leach rates at 90 °C up to 10^{-5} g·sm^{-2}·day^{-1}, radiation stable. Structure: Trigonal, Sp. gr. R3c. Ceramics were prepared by cold pressing and sintering (ρ = 92–97%) and Spark Plasma Sintering with high relative density (up to 95–98%).

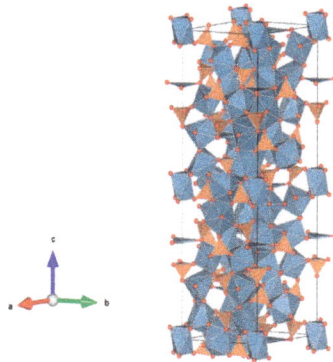

Figure 28. Whitlockite. $Ca_3(PO_4)_2$. Structure trigonal, Sp. gr. R3c.Cations can be H, Li, Na, K, Cu, Mg, Ca, Sr, Ba, Al, Sc, Cr, Fe, Ga, In, La, Ce, Sm, Eu, Gd, Lu, Th, U and Pu.

29.   Thorium phosphate/Diphosphate (TPD) [155,244,336,337,433–439], Figure 29.

$Th_4(PO_4)_4P_2O_7$; a unique compound for the immobilization of plutonium and uranium; partial substitution of Pu for Th has been demonstrated to up to 0.4 mole fraction, complete substitution is not possible. Structure: Orthorhombic, Sp. gr. Pbcm, Pcam, z = 2; may include: U, Np, Pu, Am and Cm. Ceramics were prepared by cold pressing and sintering (ρ = 87–93%).

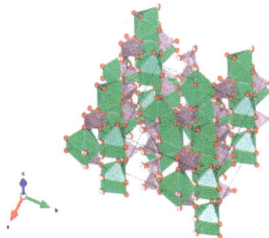

Figure 29. Thorium phosphate/Diphosphate (TPD). $Th_4(PO_4)_4P_2O_7$. Structure orthorhombic. Sp. gr. Pbcm and Pcam. Cations can be U, Np, Pu, Am and Cm.

4.5. Tungstate, Molybdates

30.   Scheelite [89,440–457], Figure 30.

Materials with the structure of the scheelite mineral (calcium tungstate $CaWO_4$) based on individual molybdates and tungstates and solid solutions may contain elements in oxidation degrees from 1+ to 7+: Li, Na, K, Rb, Cs and Tl; Ca, Sr, Ba, Mn and Cu; Fe, Ce, La–Lu and Y; Th, U, Np and Pu; Nb, Ta-in Ca-positions and Mo, W, Re, I, V and Ge in W-positions. The structural analog $CaWO_4$ crystallizes in the tetragonal structure, Sp. gr. I4/c. The structure is constructed of CaO_8 polyhedral and WO_4 tetrahedrals connected through common oxygen vertices. For some compounds ceramics were prepared by the Spark Plasma Sintering (SPS) method, with a relative density of 92%.

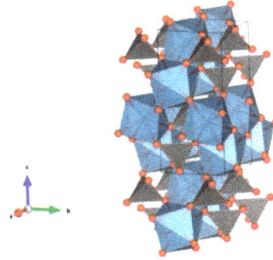

Figure 30. Scheelite. $CaWO_4$. Structure tetragonal, Sp. gr. I4/c. Cations can be Li, Na, K, Rb, Cs, Tl, Ca, Sr, Ba, Mn, Cu, Fe, Ce, La–Lu, Y, Ge, Th, U, Np, Pu, Nb, Ta, V, Mo, W, Re and I.

5. Summary of Crystalline Ceramic Waste-forms

Crystalline materials including oxides-simple and complex, salts-silicates, phosphates, tungstates with various compositions and different structural modifications (30 structure forms) intended for nuclear waste immobilization were developed using various approaches and accounting for criteria of enough high durability (see e.g., [15,238,458–460]) requested for nuclear wasteforms. These are presented in Table 1.

Table 1. Crystalline ceramic materials as potential forms for nuclear waste immobilization.

Type of Chemical Compound	Structure		Compound Cations
	Structural Type	Syngony, Sp. gr.	
Oxide Compounds			
Simple oxides			
SiO_2	Silica	rhombohedral, R3	Li, Na, K, Mg, Ca, Mn, Cu, Ni, Pb B, Al, Fe, Cr, Ti, Zr, Te
CeO_2	Fluorite	cubic, Fm3m	Cs, Sr, Ce, Y, Zr, U, Th, Hf, Pu, U, Np
$A_2B_2O_7$	Pyrochlore	cubic, Fd_3m	A: Na, Ca, U, Th, Y, Ln; B: Nb, Ta, Ti, Zr, Fe^{3+}
$A_6B_{12}C_5TX_{40-x}$	Murataite	cubic, $F4_3m$	U, Np, Pu, Am, Cm, REE
$CaZrTi_2O_7$	Zirconolite	trigonal, C2/c	Gd, Hf, Ce, Th, U, Pu, Nb
$CaTiO_3$	Perovskite	cubic, Pm3m; rhombohedral, Pnma	Ca, Y, REE, Ti, Zr, U, Pu
Complex oxides			
$Ba_{1.2}(Al,Ti)_8O_{16}$	Hollandite	tetragonal, I4/m	Na, K, Cs, Mg, Ca, Ba, Al, Fe, Mn^{3+}, Si, Ti, Mn^{4+}
[8]A_3[6]B_2[TiO_4]$_3$ [8](Ca,Gd, actinides)[6]Fe_2[4]Fe_3O_{12}	Garnet	cubic, Ia3d	A, B: REE, An, Y, Mg, Ca, Fe^{2+}, Mn^{2+}; X: Cr^{3+}, Fe^{3+}, Al^{3+}, Ga^{3+}, Si^{4+}, Ge^{4+}, V^{5+}
$(Sr,Pb,La,Ce,Y)(Ti,Fe^{3+},Mn,Mg,Zn,Cr,Al,Zr,Hf,U,V,Nb,Sn,Cu,Ni)_{21}O_{38}$	Crichtonite	rhombohedral, R3	Mg, Co, Ni, Zn, Al, Ti^{3+}, Cr, Fe, Ga, Si, Nb
$Na_2Al_2(Ti,Fe)_6O_{16}$	Freudenbergite spinel based phase	monoclinic, C12/m1	Li, Na, K, Rb, Cs, Tl, Be, Mg, Sr, Ba, Cd, Mn, Co, Ni, Cu, Zn, B, Al, Fe, Si, Ti, P, V, Nb, Ta
	P-Pollucite	cubic, $I4_132$	REE, Th, U, Pu: Na, Mg, Ca, Mn, Co, Fe, Ti, P, V, Se, Mo
Salt compounds			
$ZrSiO_4/ThSiO_4/USiO_4$	Zircon/Thorite/Coffinite	tetragonal, I41/amd	Mg, Ca, Sr, Ba, Mn, Al, Fe, Cr, Ce, Y, Zr, Th, F
$CaTiSiO_5$ [$CaTiO(SiO_4)$]	Titanite (sphene)	monoclinic, P21/a	Cs, Sr, B, Th, U, Np, Nd^{3+}, La^{3+}, Pu^{3+}
$(REE,Ca)_5(SiO_4,PO_4)_3(OH,F)$	Britholite (oxy-apatite)	monoclinic, sp. gr. P2$_1$, hexagonal, P6$_3$/m	
$(X_{x/n}[(AlO_2)_x(SiO_2)_y]$	Zeolites		Na, K, NH_4^+, Cs, Mg, Ca, Sr, Co, Fe, Ga, REE, Ti
Framework Silicates			
$(Ca,Na)_2Al_2Si_4O_{12}\cdot2H_2O$	Pollucite	cubic, Ia3d	Li, Na, K, Rb, Cs, Tl, Be, Mg, Sr, Ba, Cd, Mn, Co, Ni, Cu, Zn, B, Al, Fe, Si, Ti, P, V, Nb

Table 1. Crystalline ceramic materials as potential forms for nuclear waste immobilization.

Type of Chemical Compound	Structure		Compound Cations
	Structural Type	Syngony, Sp. gr.	
$NaAlSiO_4$	Nepheline/Leucite	Nepheline: hexagonal, $P6_3$; Leucite: tetragonal, $I4_1/a$, $I4_1/acd$; cubic, $Ia3d$	Li, Na, K, Rb, Cs, Be, Mg, Ca, Ba, Pb, Mn, Co, Ni, Al, Fe, Cr, Si, Ti, V
$Na_8Cl_2Al_6Si_6O_{24}$	Sodalite	cubic, $P\bar{4}3n$	Na, K, Mg, Ca, Mn, Fe, Al, Si, Ti, Cl, SO_4, CO_3 Cl^-, SO_4^{2-},
$(Na,Ca,K)_6[Al_6Si_6O_{24}]((Na,Ca,K)_2CO_3)_{1.6}\cdot 2.1H_2O$	Cancrinite	hexagonal, $P6_3$	Na, K, Cs, Ca, Sr, Ba, Pb, Al, REE, Bi, Tl, Zr, Nb, Ta
$[(Ca,Na,K,Ba)AlSiO_4]$	Crystalline SilicoTitanate (CST)	cubic, sp. gr. $Pm3m$ up to 105 °C, after tetragon. symm., sp. gr. $I4/mcm$ or $P4_2/mcm$	
$LiAl_3Si_3O_{11}$, $NaAl_3Si_3O_{11}$, $KAl_3Si_3O_{11}$, $RbAl_3Si_3O_{11}$, $CsAl_3Si_3O_{11}$, $TlAl_3Si_3O_{11}$, $Ca_{0.5}\square_{0.5}Al_3Si_3O_{11}$, $Sr_{0.5}\square_{0.5}Al_3Si_3O_{11}$, $Ba_{0.5}\square_{0.5}Al_3Si_3O_{11}$, $La_{0.33}\square_{0.66}Al_3Si_3O_{11}$	Micas (Dehydroxylated)	monoclinic, $C2/c$	Cs, Rb, Ba, Mg, Fe^{2+}, Fe^{3+}, Mn, Li, Cr, Ti, V
$CePO_4$	Monazite	monoclinic, $P2_1/n$	Ce: Li, Na, K, Rb, Mg, Ca, Sr, Ba, Cd, Pb, Bi, Y, La, Pr, Nd, Sm, Eu, Gd, Tb, Yb, Am, Cm, Cf, Es, Ge, Zr, Th, Np, U, Pu; P: Cr, Si, Se, V, As, S
YPO_4	Xenotime	tetragonal, $I4_1/amd$	Be, Ca, Al, Sc, La, Ce, Er, Dy–Lu, Zr, Th, U
$Ca_{4-x}RE_{6+x}(SiO_4)_{6-y}(PO_4)_y(O,F)_2$	Apatite	hexagonal, $P6_3/m$; monoclinic, $P2_1/b$	Na, K, Cs, Mg, Ca, Sr, Ba, Mn, Ni, Cd, Hg, Pb, Cr, Y, REE, Th, U, Si, P, V, As, S, F, Cl, OH, CO_3
$NaZr_2(PO_4)_3$	Sodium zirconium phosphate (NZP)	rhombohedral, $R\bar{3}c$, $R3c$, $R3$	Li, Na, K, Rb, Cs; H, Cu(I), Ag, Mg, Ca, Sr, Ba, Mn, Co, Ni, Cu, Zn, Cd, Hg, Al, Ga, In, Sc, Y, La, Ce-Lu, Am, Cm, V, Cr, Fe, Sb, Bi, Ge, Sn, Ti, Zr, Hf, Mo, Ce, Th, U, Np, Pu, Sb, Nb, Ta
$K_2Mg_2(SO_4)_3$	Langbeinite	cubic, $P2_13$	Na, K, Rb, Cs, Tl, NH4, Mg, Sr, Ba, Pb, Mn, Co, Ni, Zn, Al, Fe Cr, Ti^{3+}, Ga, V^{3+}, Rh, In, REE, Bi, Sn, Ti, Zr, Hf, P, Nb, Ta, S
$\beta\text{-}Ca_3(PO_4)_2$	Whitlockite	trigonal, $R3c$	H, Li, Na, K, Cu, Mg, Ca, Sr, Ba, Al, Sc, Cr, Fe, Ga, In, La, Ce, Sm, Eu, Gd, Lu, Th, Pu

Phosphates

Materials **2019**, *12*, 2638

Table 1. Crystalline ceramic materials as potential forms for nuclear waste immobilization.

Type of Chemical Compound	Structure		Compound Cations
	Structural Type	Syngony, Sp. gr.	
$Th_4(PO_4)_4P_2O_7$	Thorium phosphate/Diphosphate (TPD)	orthorhombic, Pbcm, Pcam	U, Np, Pu, Am, Cm
Tungstates $CaWO_4$	Scheelite	tetragonal, I4/c	Ca: Li, Na, K, Rb, Cs, Tl; Ca, Sr, Ba, Mn, Cu; Fe, Ce, La-Lu, Y; Th, U, Np, Pu; Nb, Ta; W: Mo, Re, I, V, Ge
Aluminates $X(Al,Fe)_{12}O_{19}$	Magnetoplumbite	hexagonal, $P6_3/mmc$	Na, Cs, Mg, Sr, Ba, Pb, Mn, Co, Cu, Al, Fe, Sc, Y, La, Ce, Sm, Gd, Yb, Lu, An, Si, Tl, Sn

Many of the compounds listed here have been studied and continue to be actively investigated by researchers led by the co-author of this work (Prof Orlova), including those with structures of garnet [185,189–194], P-pollucite [205–215], pollucite [214,215,293], monazite [141,352], sodium zirconium phosphate (NZP) [21,209,383,384,388,392–394,396,405,407–409,412–419], langbeinite [416–419], whitlockite [87,89,424–430] and scheelite [89,445,446]. Overall crystalline ceramics are characterized as much more durable compared with glasses of the same chemical composition e.g., the chemical durability of isomorph glasses is one to two orders of magnitude lower [458–460]. Nevertheless, the degree of the development of crystalline ceramics remains at the level of laboratory investigations rather than industrial use, except for SYNROC polyphase crystalline ceramic that is at the stage of the planned start of utilization by industry. Practically all structural forms developed (Table 1) are at the stage of obtaining compounds and their studies at the laboratory scale. The references [15,458–460] are also providing data on the acceptability of ionic size variability within the structure, and on chemical and radiation durability.

From the analysis of the presented data of various compounds with various compositions and structural forms it is clear that researchers in the field of materials for nuclear waste immobilization have many variants available for work. While materials are mineral-like the principle "from nature to nature" can be realized. Although many structures were included herewith, some could be missed, for example brannerite [15,99], which is currently considered for actinide immobilization [461]. Among most investigated structures one can note oxide ceramics. Some of crystalline ceramics such as monazite were synthesized using real (radioactive) actinides [15,235], whereas most of researchers use surrogate (non-radioactive) cations for investigations.

6. Conclusions

1. Ceramic waste-forms for nuclear waste immobilization are investigated in different countries with a focus on improving environmental safety during storage, transport and disposal.
2. Inorganic compounds of oxide and salt character, having structural analogs with natural minerals, are being studied as most perspective materials for the immobilization of radioactive waste.
3. Approaches based on crystallochemistry principles are used when choosing the most favorable structural forms. They are based on the materials science concept "composition-structure-method of synthesis-property" accounting for the real task to be achieved. The basic principle is the isomorphism of cations and anions in compounds when choosing a real structure. Possible isomorphic substitutions in both cationic and anionic structural sites were considered in the works analyzed.
4. Crystalline ceramic waste-forms are intended to increase the environmental safety barrier when isolating radioactive materials (containing both actinides and fission products) from the biosphere. Among the methods of obtaining ceramic waste-forms, special attention in recent years is paid to sintering methods which ensure the formation of ceramics that, first, are almost non-porous e.g., have a relative density of up to 99.0–99.9% of theoretical, and, second, can be obtained within a small processing time e.g., within a few minutes (i.e., 2–3 min). These requirements are met by high-speed electric pulse sintering processes e.g., so-called Spark Plasma Sintering (SPS), although hot pressing enables the synthesis of very dense ceramics as well.

Professor Albina Orlova is working in the field of new inorganic materials used in nuclear chemistry for the rad-waste immobilization of dangerous isotopes, for actinide transmutation, as well for construction materials. She uses the structure properties and physico-chemical principles for the elaboration of new ceramics with mineral-like crystal forms.

Professor Michael Ojovan is known for the connectivity-percolation theory of glass transition, the Sheffield model (two-exponential equation) of viscosity of glasses and melts, condensed Rydberg matter, metallic and glass-composite materials for nuclear waste immobilization, and self-sinking capsules to investigate Earth's deep interior.

Author Contributions: A.I.O. conceived the study, both A.I.O. and M.I.O. equally contributed to final paper preparation.

Funding: This research was funded by Russian Science Foundation grant number RSF-16-13-10464 (Scientific Supervisor A.I. Orlova).

Conflicts of Interest: The authors declare no conflict of interest.

References

1. Hatch, L.P. Ultimate Disposal of Radioactive Wastes. *Am. Sci.* **1953**, *41*, 410–421.
2. McCarthy, G.J. Quartz-Matrix Isolation of Radioactive Wastes. *J. Mater. Sci.* **1973**, *8*, 1358–1359. [CrossRef]
3. McCarthy, G.J.; Davidson, M.T. Ceramic Nuclear Waste Forms: I. Crystal Chemistry and Phase Formation. *Bull. Am. Ceram. Soc.* **1975**, *54*, 782–786.
4. Roy, R. Rationale Molecular Engineering of Ceramic Materials. *J. Am. Ceram. Soc.* **1977**, *60*, 350–363. [CrossRef]
5. Jantzen, C.M.; Lee, W.E.; Ojovan, M.I. Radioactive waste (RAW) conditioning, immobilization, and encapsulation processes and technologies: overview and advances. In *Radioactive Waste Management and Contaminated Site Clean-Up. Processes, Technologies and International Experience*; Lee, W.E., Ojovan, M.I., Jantzen, C.M., Eds.; Woodhead Published Limited: Oxford, UK; Cambridge, UK; Philadelphia, PA, USA; New Delhi, India, 2013; Chapter 6; pp. 171–272.
6. Schoebel, R.O. Stabilization of High Level Waste in Ceramic Form. *Bull. Am. Ceram. Soc.* **1975**, *54*, 459.
7. Ringwood, A.E.; Oversby, V.M.; Kesson, S.E. SYNROC: Leaching Peformance and Process Technology. In Proceedings of the Seminar on Chemistry and Process Engineering for High Level Liquid Waste Solidification, Jülichw, Germany, 1–5 June 1981; pp. 495–506.
8. Ringwood, A.E. *Safe Disposal of High Level Nuclear Reactor Wastes: A New Strategy*; Australian Nuclear University Press: Canberra, Australia, 1978; pp. 1–64.
9. Ringwood, A.E.; Kesson, S.E.; Ware, N.G.; Hibberson, W.O.; Major, A. The SYNROC Process: A Geochemical Approach to Nuclear Waste Immobilization. *Geochem. J.* **1979**, *13*, 141–165. [CrossRef]
10. Vance, E.R.; Moricca, S.A.; Stewart, M.W.A. Progress at ANSTO on a synroc plant for intermediate-level waste from reactor production of ^{99}Mo. *Adv. Sci. Technol.* **2014**, *94*, 111–114. [CrossRef]
11. McCarthy, G.J.; Pepin, J.G.; Pfoertsch, D.E.; Clarke, D.R. *Crystal Chemistry of the Synthetic Minerals in Current Supercalcine-Ceramics; U.S. DOE Report CONF-790420*; Battelle Pacific Northwest Labs.: Richland, WA, USA, 1979; pp. 315–320.
12. Lutze, W.; Ewing, R.C. *Radioactive Waste Forms for the Future*; North-Holland Publishers: Amsterdam, The Netherlands, 1988; p. 778.
13. Jantzen, C.M.; Flintoff, J.; Morgan, P.E.D.; Harker, A.B.; Clarke, D.R. Ceramic Nuclear Waste Forms. In Proceedings of the Seminar on Chemistry and Process Engineering for High-Level Liquid Waste Solidification, Jülichw, Germany, 1–5 June 1981; Volume 2, pp. 693–706.
14. Raison, P.E.; Haire, R.G.; Sato, T.; Ogawa, T. *Fundamental and Technological Aspects of Actinide Oxide Pyrochlores: Relevance for Immobilization Matrices*; MRS Online Proceedings Library Archive: Warrendale, PA, USA, 1999; p. 556.
15. Burakov, B.E.; Ojovan, M.I.; Lee, W.E. *Crystalline Materials for Actinide Immobilization*; Imperial College Press: London, UK, 2010.
16. Dacheux, N.; Clavier, N.; Podor, R. Monazite as a promising long-term radioactive waste matrix: Benefits of high-structural flexibility and chemical durability. *Am. Miner.* **2013**, *98*, 833–847. [CrossRef]
17. Clavier, N.; Podor, R.; Dacheux, N. Crystal chemistry of the monazite structure. *J. Eur. Ceram. Soc.* **2011**, *31*, 941–976. [CrossRef]
18. Orlova, A.I.; Lizin, A.A.; Tomilin, S.V.; Lukinykh, A.N. On the Possibility of Realization of Monazite and Langbeinite Structural Types in Complex Americium and Plutonium Phosphates. Synthesis and X-ray Diffraction Studies. *Radiochemistry* **2011**, *53*, 63–68. [CrossRef]
19. Alamo, J.; Roy, R. Revision of Crystalline Phases in the System ZrO_2-P_2O_5. *Commun. Am. Ceram. Soc.* **1984**, *67*, C80–C82. [CrossRef]

20. Hawkins, H.T.; Spearing, D.R.; Veirs, D.K.; Danis, J.A.; Smith, D.M.; Tait, C.D.; Runde, W.H. Synthesis and Characterization of Uranium(IV)-Bearing Members of the [NZP] Structural Family. *Chem. Mater.* **1999**, *11*, 2851–2857. [CrossRef]

21. Orlova, A.I. Isomorphism in Crystalline Phosphates of the $NaZr_2(PO_4)_3$ structural type and Radiochemical Problems. *Radiochemistry* **2002**, *44*, 423–445. [CrossRef]

22. Volkov, Y.F.; Tomlin, S.V.; Orlova, A.I.; Lizin, A.A.; Spirjakov, V.I.; Lukinikh, A.N. Actinide Phosphates $A^{I}M_2^{IV}(PO_4)_3$ (MIV = U, Np, Pu; AI = Na, K, Rb) with rombohedral structure. *Radiochemistry* **2003**, *46*, 319–328. [CrossRef]

23. Gregg, D.J.; Karatchevtseva, I.; Triani, G.; Lumpkin, G.R.; Vance, E.R. The thermophysical properties of calcium and barium zirconium phosphate. *J. Nucl. Mater.* **2013**, *441*, 203–210. [CrossRef]

24. Gregg, D.J.; Karatchevtseva, I.; Thorogood, G.J.; Davis, J.; Bell, B.; Jackson, M.; Dayal, P.; Ionescu, M.; Triani, G.; Short, K.; et al. Ion Beam Irradiation Effects in Strontium Zirconium Phosphate with NZP-structure type. *J. Nucl. Mater.* **2014**, *446*, 224–231. [CrossRef]

25. Orlova, A.I.; Koryttseva, A.K.; Loginova, E.E. A Family of Phosphates of Langbeinite Structure. Crystal-Chemical Aspect of Radioactive Waste Immobilization. *Radiochemistry* **2011**, *53*, 51–62. [CrossRef]

26. Orlova, A.I. Chemistry and structural chemistry of anhydrous tri- and tetravalent actinide orthophosphates. In *Structural Chemistry of Inorganic Actinide Compounds*; Chapter 8; Krivovichev, S.V., Burns, P.C., Tananaev, I.G., Eds.; Elsevier: Amsterdam, The Netherlands, 2007; pp. 315–340.

27. Huittinen, N.; Scheinost, A.C.; Ji, Y.; Kowalski, P.M.; Arinicheva, Y.; Wilden, A.; Neumeier, S.; Stumpf, T. A spectroscopic and computational study of Cm^{3+} incorporation in lanthanide phosphate rhabdophane $(LnPO_4·0.67H_2O)$ and monazite $(LnPO_4)$. *Inorg. Chem.* **2018**, *57*, 6252–6265. [CrossRef]

28. Clarke, D.R. Preferential Dissolution of an Intergranular Amorphous Phase in a Nuclear Waste Ceramic. *J. Am. Ceram. Soc.* **1981**, *64*, 89–90. [CrossRef]

29. Cooper, J.A.; Cousens, D.R.; Hanna, J.A.; Lewis, R.A.; Myhra, S.; Segall, R.L.; Smart, R.S.C.; Turner, P.S.; White, T.J. Intergranular Films and Pore Surfaces in Synroc C: Structure, Composition, and Dissolution characteristics. *J. Am. Ceram. Soc.* **1986**, *69*, 347–352. [CrossRef]

30. Buykx, W.J.; Hawkins, K.; Levins, D.M.; Mitamura, H.; Smart, R.S.C.; Stevens, G.T.; Watson, K.G.; Weedon, D.; White, T.J. Titanate Ceramics for the Immobilization of SodiumBearing High-Level Nuclear Waste. *J. Am. Ceram. Soc.* **1988**, *71*, 768–788. [CrossRef]

31. Dickson, F.J.; Mitamura, H.; White, T.J. Radiophase Development in Hot-Pressed Alkoxide-Derived Titanate Ceramics for Nuclear Waste Stabilization. *J. Am. Ceram. Soc.* **1989**, *72*, 1055–1059. [CrossRef]

32. Buykx, W.J.; Levins, D.M.; Smart, R.S.C.; Smith, K.L.; Stevens, G.T.; Watson, K.G.; Weedon, D.; White, T.J. Interdependence of Phase Chemistry, Microstructure, and Oxygen Fugacity in Titanate Nuclear Waste Ceramics. *J. Am. Ceram. Soc.* **1990**, *73*, 1201–1207. [CrossRef]

33. Buykx, W.J.; Levins, D.M.; St Smart, R.C.; Smith, K.L.; Stevens, G.T.; Watson, K.G.; White, T.J. Processing Impurities as Phase Assemblage Modifiers in Titanate Nuclear Waste Ceramics. *J. Am. Ceram. Soc.* **1990**, *73*, 217–225. [CrossRef]

34. Mitamura, H.; Matsumoto, S.; Hart, K.P.; Miyazaki, T.; Vance, E.R.; Tamura, Y.; Togashi, Y.; White, T.J. Aging Effects on Curium-Dopped Titantate Ceramics Containing Sodium-Bearing High-Level Nuclear Waste. *J. Am. Ceram. Soc.* **1992**, *75*, 392–400. [CrossRef]

35. Zhang, Z.; Carter, M.L. An X-Ray Photoelectron Spectroscopy Investigation of Highly Soluble Grain-Boundary Impurity Films in Hollandite. *J. Am. Ceram. Soc.* **2010**, *93*, 894–899. [CrossRef]

36. Harker, A.B.; Clarke, D.R.; Jantzen, C.M.; Morgan, P.E.D. *The Effect of Interfacial Material on Tailored Ceramic Nuclear Waste Form Dissolution, Surfaces and Interfaces in Ceramic and Ceramic-Metal Systems*; Pask, J., Evans, A., Eds.; Plenum Press: New York, NY, USA, 1981; Volume 14, pp. 207–216.

37. Roy, R. Hydroxylated Ceramic Waste Forms and the Absurdity of Leach Tests. In Proceedings of the International Seminar on Chemistry and Process Engineering for High-Level Liquid Waste Solidification, Julich, India, 1–5 June 1981; pp. 576–602.

38. Mason, J.B.; Oliver, T.W.; Carson, M.P.; Hill, G.M. *Studsvik Processing Facility Pyrolysis/Steam Reforming Technology for Volume and Weight Reduction and Stabilization of LLRW and Mixed Wastes*; American Institute of Chemical Engineers: New York, NY, USA, 1999.

39. Mason, J.B.; McKibbin, J.; Ryan, K.; Schmoker, D. *Steam Reforming Technology for Denitration and Immobilization of DOE Tank Wastes*; THOR Treatment Technologies, LLC.: Richland, WA, USA, 2003.

40. Jantzen, C.M.; Lorier, T.H.; Marra, J.C.; Pareizs, J.M. *Durability Testing of Fluidized Bed Steam Reforming (FBSR) Waste Forms WM06 Paper № 6373*; WWM Symposia: Phoenix, AZ, USA, 2006.
41. Jantzen, C.M.; Lorier, T.H.; Pareizs, J.M.; Marra, J.C. *Fluidized Bed Steam Reformed (FBSR) Mineral Waste Forms: Characterization and Durability Testing*; MRS Online Proceedings Library Archive: Warrendale, PA, USA, 2007; pp. 379–386.
42. Jantzen, C.M. *Mineralization of Radioactive Wastes By Fluidized Bed Steam Reforming (FBSR): Comparisons to Vitreous Waste Forms and Pertinent Durability Testing*; U.S. DOE Report WSRC-STI-2008-00268; Westinghouse Savannah River Company, Savannah River Site: Aiken, SC, USA, 2008.
43. Jantzen, C.M.; Crawford, C.L.; Burket, P.B.; Daniel, W.G.; Cozzi, A.D.; Bannochie, C.J. *Radioactive Demonstrations of Fluidized Bed Steam Reforming (FBSR) as a Supplementary Treatment for Hanford's Low Activity Waste (LAW) and Secondary Wastes (SW)*; Waste Management WM11, Paper № 11593; WM Symposia: Phoenix, AZ, USA, 2011.
44. Jantzen, C.M.; Crawford, C.L.; Burket, P.R.; Bannochie, C.J.; Daniel, W.G.; Nash, C.A.; Cozzi, A.D.; Herman, C.C. *Radioactive Demonstrations of Fluidized Bed Steam Reforming (FBSR) with Actual Hanford Low Activity Wastes: Verifying FBSR as a Supplemental Treatment*; WM12 Paper № 12317; WM Symposia: Phoenix, AZ, USA, 2012.
45. Jantzen, C.M.; Glasser, F.P. Crystallochemical Stabilization of Radwaste Elements in Portland Cement Clinker. In Proceedings of the International Symposium on Ceramics in Nuclear Waste Management; CONF-790420, Cincinnati, OH, USA, 30 April–2 May 1979; pp. 342–348.
46. Jantzen, C.M.; Glasser, F.P. Stabilization of Nuclear Waste Constituents in Portland Cement. *Am. Ceram. Soc. Bull.* **1979**, *58*, 459–466.
47. Bragg, L.; Claringbull, G.F.; Taylor, W.H. *Crystal Structures of Minerals*; G. Bell & Sons: London, UK, 1965; p. 409.
48. Shannon, R.D.; Prewitt, C.T. Effective Ionic Radii in Oxides and Fluorides. *Acta Cryst.* **1969**, *B25*, 925–946. [CrossRef]
49. Shannon, R.D.; Prewitt, C.T. Revised Values of Effective Ionic Radii. *Acta Cryst.* **1970**, *B26*, 1046–1048. [CrossRef]
50. Shannon, R.D. Revised Effective Ionic Radii and Systematic Studies of Interatomic Distances in Halides and Chalcogenides. *Acta Cryst.* **1976**, *A32*, 751–767. [CrossRef]
51. Felsche, J. Rare Earth Silicates with the Apatite Structure. *J. Solid State Chem.* **1972**, *5*, 266–275. [CrossRef]
52. Jantzen, C.M.; Glasser, F.P.; Lachowsli, E.E. Radioactive Waste-Portland Cement Systems: I. Radionuclide Distribution. *J. Am. Ceram. Soc.* **1984**, *67*, 668–673. [CrossRef]
53. Jantzen, C.M.; Glasser, F.P.; Lachowsli, E.E. Radioactive Waste-Portland Cement Systems: II. Leaching Characteristics. *J. Am. Ceram. Soc.* **1984**, *67*, 674–678. [CrossRef]
54. Ringwood, A.E.; Kesson, S.E. Immobilization of High-Level Wastes in Synroc Titanate Ceramic. In Proceedings of the International Symposium on Ceramics in Nuclear WaMc Management, CONF-790420, Cincinnati, OH, USA, 30 April–2 May 1979; pp. 174–178.
55. Cooper, J.A.; Cousens, D.R.; Lewis, R.A.; Myhra, S.; Segall, R.L.; Smart, R.S.C.; Turner, P.S.; White, T.J. Microstructural Characterization of Synroc C and E by Electron Microscopy. *J. Am. Ceram. Soc.* **1985**, *68*, 64–70. [CrossRef]
56. Aubin, V.; Caurant, D.; Gourier, D.; Baffier, N.; Advocat, F.; Bart, F.; Leturcq, G.; Costantini, J.M. Synthesis, Characterization and Study of the Radiation Effects on Hollandite Ceramics Developed for Cesium Immobilization. *Mater. Res. Soc. Symp. Proc.* **2004**, *807*, 315–320. [CrossRef]
57. Hart, K.P.; Vance, E.R.; Day, R.A.; Begg, B.D.; Angel, P.J. Immobilization of Separated Tc and Cs/Sr in SYNROC. *Mater. Res. Soc. Symp. Proc.* **1996**, *412*, 281–287. [CrossRef]
58. Buerger, M.J.; Klein, G.E.; Hamburger, G.E. Structure of Nepheline. *Geol. Soc. Am. Bull.* **1946**, *57*, 1182–1183.
59. Buerger, M.J.; Klein, G.E.; Hamburger, G.E. The structure of nepheline. *Am. Mineral.* **1947**, *32*, 197.
60. Simmons, W.B.; Peacor, D.R. Refinement of the crystal structure of a volcanic nepheline. *Am. Mineral.* **1972**, *57*, 1711–1719.
61. Rossi, G.; Oberti, R.; Smith, D.C. The crystal structure of a K-poor Ca-rich silicate with the nepheline framework, and crystal-chemical relationships in the compositional space $(K,Na,Ca,\_)_8(Al,Si)_{16}O_{32}$. *Eur. J. Mineral.* **1989**, *1*, 59–70. [CrossRef]

62. Tait, K.T.; Sokolova, E.V.; Hawthorne, F.C.; Khomyakov, A.P. The crystal chemistry of nepheline. *Can. Mineral.* **2003**, *41*, 61–70. [CrossRef]

63. Hassan, I.; Antao, S.M.; Hersi, A.A.M. Single-crystal XRD, TEM, and thermal studies of the satellite reflections in nepheline. *Can. Mineral.* **2003**, *41*, 759–783. [CrossRef]

64. Gatta, G.D.; Angel, R.J. Elastic behavior and pressure-induced structural evolution of nepheline: Implications for the nature of the modulated superstructure. *Am. Mineral.* **2007**, *92*, 1446–1455. [CrossRef]

65. Angel, R.J.; Gatta, G.C.; Ballaran, T.B.; Carpenter, M.A. The mechanism of coupling in the modulated structure of nepheline. *Can. Mineral.* **2008**, *46*, 1465–1476. [CrossRef]

66. Antao, S.M.; Nepheline, H.I. Structure of three samples from the Bancroft area, Ontario, obtained using synchrotron high-resolution powder x-ray diffraction. *Can. Mineral.* **2010**, *48*, 69–80. [CrossRef]

67. Chapman, K.W.; Chupas, P.J.; Nenoff, T.M. Radioactive Iodine Capture in Silver-Containing Mordenites through Nanoscale Silver Iodide Formation. *J. Am. Chem. Soc.* **2010**, *132*, 8897–8899. [CrossRef]

68. Matyas, J.; Fryxell, G.; Busche, B.; Wallace, K.; Fifield, L. Functionalised silica aerogels: Advanced materials to capture and immobilise radioactive iodine. In *Ceramic Engineering and Science Proceedings*; American Ceramic Society, Inc.: Columbus, OH, USA, 2011; pp. 23–32.

69. Riley, B.J.; Chun, J.; Ryan, J.V.; Matyáš, J.; Li, X.S.; Matson, D.W.; Sundaram, S.K.; Strachan, D.M.; Vienna, J.D. Chalcogen-based aerogels as a multifunctional platform for remediation of radioactive iodine. *RSC Adv.* **2011**, *1*, 1704–1715. [CrossRef]

70. Yang, J.H.; Cho, Y.-J.; Shin, J.M.; Yim, M.-S. Bismuth-embedded SBA-15 mesoporous silica for radioactive iodine capture and stable storage. *J. Nucl. Mater.* **2015**, *465*, 556–564. [CrossRef]

71. Subrahmanyam, K.S.; Sarma, D.; Malliakas, C.D.; Polychronopoulou, K.; Riley, B.J.; Pierce, D.A.; Chun, J.; Kanatzidis, M.G. Chalcogenide Aerogels as Sorbents for Radioactive Iodine. *Chem. Mater.* **2015**, *27*, 2619–2626. [CrossRef]

72. Matyas, J.; Canfield, N.; Silaiman, S.; Zumhoff, M. Silica-based waste form for immobilization of iodine from reprocessing plant off-gas streams. *J. Nucl. Mater.* **2016**, *476*, 255–261. [CrossRef]

73. Vienna, J.D.; Kroll, J.O.; Hrma, P.R.; Lang, J.B.; Crum, J.V. Submixture Model to Predict Nepheline Precipitation in Waste Glasses. *Int. J. Appl. Glass Sci.* **2017**, *8*, 143–157. [CrossRef]

74. Asmussen, R.M.; Matyáš, J.; Qafoku, N.P.; Kruger, A.A. Silver-functionalized silica aerogels and their application in the removal of iodine from aqueous environments. *J. Hazard. Mater.* **2018**. [CrossRef]

75. Asmussen, R.M.; Ryan, J.V.; Matyas, J.; Crum, J.V.; Reiser, J.T.; Avalos, N.; McElroy, E.M.; Lawter, A.R.; Canfield, N.C. Investigating the Durability of Iodine Waste Forms in Dilute Conditions. *Materials* **2019**, *12*, 686. [CrossRef]

76. Burghatz, M.; Matzke, H.; Leger, C.; Vambenepe, G.; Rome, M. Inert Matrices for the Transmuation of Actinides; Fabrication, Thermal Properties and Radiation Stability of Ceramic Materials. *J. Nucl. Mater.* **1998**, *271*, 544–548.

77. Sickafus, K.E.; Hanrahan, R.J.; McClellan, K.J.; Mitchell, J.N.; Wetteland, C.J.; Butt, D.P.; Chodak, P.; Ramsey, K.B.; Blair, T.H.; Chidester, K.; et al. Burn and Bury Option for Plutonium. *Bull. Am. Ceram. Soc.* **1999**, *78*, 69–74.

78. Gong, W.L.; Lutze, W.; Ewing, R.C. Zirconia Ceramics for Excess Weapons Plutonium Waste. *J. Nucl. Mater.* **2000**, *277*, 239–249. [CrossRef]

79. Zacate, M.O.; Minervini, L.; Bradfield, D.J.; Grimes, R.W.; Sickafus, K.E. Defect cluster formation in M_2O_3-doped cubic ZrO_2. *Solid State Ion.* **2000**, *128*, 243–254. [CrossRef]

80. Sickafus, K.E.; Minervini, L.; Grimes, R.W.; Valdez, J.A.; Ishimaru, M.; Li, F.; McClellan, K.J.; Hartmann, T. Radiation tolerance of complex oxides. *Science* **2000**, *259*, 748–751. [CrossRef]

81. Burakov, B.; Anderson, E.; Yagovkina, M.; Zamoryanskaya, M.; Nikolaeva, E. Behavior of [238]Pu-Doped Ceramics Based on Cubic Zirconia and Pyrochlore under Radiation Damage. *J. Nucl. Sci. A Technol.* **2002**, *3*, 733–736. [CrossRef]

82. Poinssot, C.; Ferry, C.; Kelm, M.; Grambow, B.; Martinez-Esparza, A.; Johnson, L.; Andriambololona, Z.; Bruno, J.; Cachoir, C.; Cavedon, J.M.; et al. *Final Report of the European Project: Spent Fuel Stability under Repository Conditions*; 2005; Available online: https://inis.iaea.org/collection/NCLCollectionStore/_Public/37/038/37038431.pdf?r=1&r=1 (accessed on 19 August 2019).

83. Rendtorff, N.M.; Grasso, S.; Hu, C.; Suarez, G.; Aglietti, E.F.; Sakka, Y. Zircon–zirconia ($ZrSiO_4$–ZrO_2) dense ceramic composites by spark plasma sintering. *J. Eur. Ceram. Soc.* **2012**, *32*, 787–793. [CrossRef]

84. Truphémus, T.; Belin, R.C.; Richaud, J.-C.; Reynaud, M.; Martinez, M.-A.; Félines, I.; Arredondo, A.; Miard, A.; Dubois, T.; Adenot, F.; et al. Structural studies of the phase separation in the UO_2–PuO_2–Pu_2O_3 ternary system. *J. Nucl. Mater. Vol.* **2013**, *432*, 378–387. [CrossRef]

85. Burakov, B.E.; Yagovkina, M.A. A study of accelerated radiation damage effects in PuO_2 and gadolinia-stabilized cubic zirconia, $Zr_{0.79}Gd_{0.14}Pu_{0.07}O_{1.93}$, doped with ^{238}Pu. *J. Nucl. Mater.* **2015**, *467*, 534–536. [CrossRef]

86. Diaz-Guillen, J.A.; Dura, O.J.; Diaz-Guillen, M.R.; Bauer, E.; Lopez de la Torre, M.A.; Fuentes, A.F. Thermophysical properties of $Gd_2Zr_2O_7$ powders prepared by mechanical milling: Effect of homovalent Gd^{3+} substitution. *J. Alloys Compd.* **2015**, *649*, 1145–1150. [CrossRef]

87. Orlova, A.I.; Chuvildeev, V.N. Chemistry, Crystal Chemistry and SPS technology for elaboration of perspective materials for nuclear wastes and minor actinides consolidation. *J. Nucl. Energy Sci. Rower Gener. Technol.* **2017**, *6*, 36.

88. Zhang, L.; Shelyug, A.; Navrotsky, A. Thermochemistry of UO_2-ThO_2 and UO_2-ZrO_2 Fluorite Solid Solutions. *J. Chem. Thermodyn.* **2017**, *114*, 48–54. [CrossRef]

89. Orlova, A.I.; Chuvildeev, V.N.; Nokhrin, A.V.; Boldin, M.S.; Potanina, E.A.; Mikhailov, D.A.; Golovkina, L.S.; Malanina, N.A.; Tokarev, M. Next Generation Ceramic Materials for Consolidation of radioactive alpha-wastes using the Innovative Technology Spark Plasma Sintering for their preparation. In Proceedings of the 3rd World Congress on Materials Science & Engineering, Barcelona, Spain, 24–26 August 2017.

90. Zubekhina, B.Y.; Burakov, B.E. Plutonium leaching from polycrystalline and monocrystalline PuO_2. *Radiochim. Acta* **2018**, *106*, 119–123. [CrossRef]

91. Shelyug, A.; Palomares, R.I.; Lang, M.; Navrotsky, A. Energetics of defect production in fluorite-structured CeO_2 induced by highly ionizing radiation. *Phys. Rev. Mater.* **2018**, *2*, 093607. [CrossRef]

92. Zhang, L.; Dzik, E.; Sigmon, G.; Szymanowski, J.; Navrotsky, A.; Burns, P. Experimental Thermochemistry of Neptunium Oxides: Np_2O_5 and NpO_2. *J. Nucl. Mater.* **2018**, *501*, 398–404. [CrossRef]

93. Yavo, N.; Sharma, G.; Kimmel, G.; Lubomirsky, I.; Navrotsky, A.; Yeheskel, O. Energetics of Bulk Lutetium Doped $Ce_{1-x}Lu_xO_{2-x/2}$ Compounds. *J. Am. Ceram. Soc.* **2018**, *101*, 3520–3526. [CrossRef]

94. Chakoumakos, B.C. Systematics of the Pyrochlore Structure Type. Ideal $A_2B_2X_6Y$. *J. Solid State Chem.* **1984**, *53*, 120–129. [CrossRef]

95. Chakoumakos, B.C.; Ewing, R.C. Crystal Chemical Constraints on the Formation of Actinide Pyrochlores. In Materials Research Society Symposium Proceedings. In *Scientific Basis for Nuclear Waste Management*; Jantzen, C.M., Stone, J.A., Ewing, R.C., Eds.; Materials Research Society: Pittsburgh, PA, USA, 1985; pp. 641–646.

96. Castro, A.; Rasines, I.; Turrillas, X.M. Synthesis, X-ray diffraction study, and ionic conductivity of new AB_2O_6 pyrochlores. *J. Solid State Chem.* **1989**, *80*, 227–234. [CrossRef]

97. Sobolev, I.A.; Stefanovsky, S.V.; Omelianenko, B.I.; Ioudintsev, S.V.; Vance, E.R.; Jostsons, A. Comparative Study of Synroc-C Ceramics Produced by Hotpressing and Inductive Melting. *Mater. Res. Soc. Symp. Proc.* **1997**, *465*, 371–378. [CrossRef]

98. Yudintsev, S.V.; Yudintseva, T.S. Nonstoichiometry of pyrochlore $Ca(U,Pu)Ti_2O_7$ and problem of brannerite $(U,Pu)Ti_2O_6$ in ceramic for actinide immobilization. In Proceedings of the 8th International Conference on Radioactive Waste Management and Environmental Remediation, Bruges, Belgium, 30 September–4 October 2001; ASME: New York, NY, USA, 2011; Volume 2, pp. 547–552.

99. Kar, T.; Choudhary, R.N.P. Structural, dielectric and electrical conducting properties of $CsB'B''O_6$ (B' = Nb, Ta; B'' = W, Mo) ceramics. *Mater. Sci. Eng.* **2002**, *B90*, 224–233. [CrossRef]

100. Ewing, R.C.; Weber, W.J.; Lian, J. Pyrochlore ($A_2B_2O_7$): A Nuclear Waste Form for the Immobilization of Plutonium and "Minor" Actinides, (Focus Review). *J. Appl. Phys.* **2004**, *95*, 5949–5971. [CrossRef]

101. Whittle, K.R.; Lumpkin, G.R.; Ashbrook, S.E. Neutron diffraction and MAS NMR of Cesium Tungstate defect pyrochlores. *J. Solid State Chem.* **2006**, *179*, 512–521. [CrossRef]

102. Thorogood, G.J.; Saines, P.J.; Kennedy, B.J.; Withers, R.L.; Elcombe, M.M. Diffuse scattering in the cesium pyrochlore $CsTi_{0.5}W_{1.5}O_6$. *Mater. Res. Bull.* **2008**, *43*, 787–795. [CrossRef]

103. Fukuda, K.; Akatsuka, K.; Ebina, Y.; Ma, R.; Takada, K.; Nakai, I.; Sasaki, T. Exfoliated Nanosheet Crystallite of Cesium Tungstate with 2D Pyrochlore. Structure: Synthesis, Characterization, and Photochromic Properties. *ACS Nano* **2008**, *2*, 1689–1695. [CrossRef]

104. Lukinykh, A.N.; Tomilin, S.V.; Lizin, A.A.; Yudintsev, S.V. Radiation and Chemical durability of Actinide Crystalline Matrices, In Book of Abstracts. In Proceedings of the III International Pyroprocessing Research Conference, Dimitrovgrad, Russia, 29 November–3 December 2010; Volume 42.

105. Laverov, N.P.; Yudintsev, S.V.; Livshits, T.S.; Stefanovsky, S.V.; Lukinykh, A.N.; Ewing, R.C. Synthetic Minerals with the Pyrochlore and Garnet Structures for Immobilization of Actinide-Containing Wastes. *Geochem. Int.* **2010**, *48*, 1–14. [CrossRef]

106. Laverov, N.P.; Urusov, V.S.; Krivovichev, S.V.; Pakhomova, A.S.; Stefanovsky, S.V.; Yudintsev, S.V. Modular Nature of the Polysomatic Pyrochlore-Murataite Series. *Geol. Ore Depos.* **2011**, *53*, 273–294. [CrossRef]

107. Hartmann, T.; Alaniz, A.; Poineau, F.; Weck, P.F.; Valdes, J.A.; Tang, M.; Jarvinen, G.D.; Czerwinski, K.R.; Sickafus, K.E. Structure studies on lanthanide technetium pyrochlores as prospective host phases to immobilize 99 technetium and fission lanthanides from effuents of reprocessed used nuclear fuels. *J. Nucl. Mater.* **2011**, *411*, 60–71. [CrossRef]

108. Krivovichev, S.V.; Urusov, V.S.; Yudintsev, S.V.; Stefanovsky, S.V.; Karimova, O.V.; Organova, N.N. Crystal Structure of Murataite Mu-5, a Member of the Murataite-Pyrochlore Polysomatic Series. In *Minerals as Advanced Materials II*; Springer: Berlin/Heidelberg, Germany, 2012; pp. 293–304.

109. Gregg, D.J.; Zhang, Y.; Middleburgh, S.C.; Conradson, S.D.; Lumpkin, G.R.; Triani, G.; Vance, E.R. The incorporation of plutonium in lanthanum zirconate pyrochlores. *J. Nucl. Mater.* **2013**, *443*, 444–451. [CrossRef]

110. Nash, K.L.; Lumetta, G.J.; Vienna, J.D. Irradiated nuclear Fuel management: resource versus waste. In *Radioactive Waste Management and Contaminated Siite Clean-up. Processes, Technologies and International Experience*; Lee, W.E., Ojovan, M.I., Jansen, M.C., Eds.; Woodhead Publishing Limited: Oxford, UK; Cambridge, UK; Philadelphia, PA, USA; New Delhi, India, 2013; Chapter 5; pp. 145–170.

111. Yudintsev, S.V.; Stefanovsky, S.V.; Nikonov, B.S. A Pyrochlore based matrix for isolation of the REE-actinide fraction of wastes from spent nuclear fuel reprocessing. *Dokl. Earth Sci.* **2014**, *454*, 54–58. [CrossRef]

112. Jafar, M.; Sengupta, P.; Achary, S.N.; Tuagi, A.K. Phase evolution and microstructural studies in CaZrTi$_2$O$_7$ (zirconolite)–Sm$_2$Ti$_2$O$_7$ (pyroclore) system. *J. Eur. Ceram. Soc.* **2014**, *34*, 4373–4381. [CrossRef]

113. Jafar, M.; Sengupta, P.; Achary, S.N.; Tuagi, A.K. Phase Evolution and Microstructural Studies in CaZrTi$_2$O$_7$ —Nd$_2$Ti$_2$O$_7$ System. *J. Am. Ceram. Soc.* **2014**, *97*, 609–616. [CrossRef]

114. Hollmann, D.; Merka, O.; Schwertmann, L.; Marschall, R.; Wark, M.; Brückner, A. Active Sites for Light Driven Proton Reduction in Y$_2$Ti$_2$O$_7$ and CsTaWO$_6$ Pyrochlore Catalysts Detected by In Situ EPR. *Top. Catal.* **2015**, *58*, 769–775. [CrossRef]

115. Kim, J.; Shih, P.-C.; Tsao, K.-C.; Pan, Y.-T.; Yin, X.; Sun, C.-J.; Yang, H. High-Performance Pyrochlore-Type Yttrium Ruthenate. Electrocatalyst for Oxygen Evolution Reaction in Acidic Media. *J. Am. Chem. Soc.* **2017**, *139*, 12076–12083. [CrossRef]

116. McMaster, S.A.; Ram, R.; Faris, N.; Pownceby, M.I. Radionuclide disposal using the pyrochlore supergroup of minerals as a host matrix-A review. *J Hazard Mater.* **2018**, *360*, 257–269. [CrossRef]

117. Kim, J.; Shih, P.-C.; Qin, Y.; Al-Bardan, Z.; Sun, C.-J.; Yang, H. A Porous Pyrochlore Y$_2$[Ru$_{1.6}$Y$_{0.4}$]O$_7$- δ Electrocatalyst for Enhanced Performance towards the Oxygen Evolution Reaction in Acidic Media. *Angew. Chem. Int. Ed.* **2018**, *57*, 13877–13881. [CrossRef]

118. Morgan, P.E.D.; Ryerson, F.J. A new "Cubic" Crystal Compound. *J. Mater. Sci. Lett.* **1982**, *1*, 351–352. [CrossRef]

119. Sobolev, I.A.; Stefanovsky, S.V.; Ioudintsev, S.V.; Nikonov, B.S.; Omelianenko, B.I.; Mokhov, A.V. Study of Melted Synroc Doped with Simulated High-level Waste. *Mater. Res. Soc. Symp. Proc.* **1997**, *465*, 363–370. [CrossRef]

120. Stefanovsky, S.V.; Yudintsev, S.V.; Nikonov, B.S.; Mokhov, A.V.; Perevalov, S.A.; Stefanovsky, O.I.; Ptashkin, A.G. Phase Compositions and Leach Resistance of Actinide-Bearing Murataite Ceramics. *Mater. Res. Soc. Symp. Proc.* **2006**, *893*, 0893-JJ05-23. [CrossRef]

121. Lian, I.; Yudintsev, S.V.; Stefanovsky, S.V.; Kirjanova, O.I.; Ewing, R.C. Ion Induced Amorphization of Murataite. *Mater. Res. Soc. Symp. Proc.* **2002**, *713*, 455–460. [CrossRef]

122. Urusov, V.S.; Organova, N.I.; Karimova, O.V.; Yudintsev, S.V.; Stefanovsky, S.V. Synthetic "Murataits" as Modular Members of Pyrochlore-Murataite Polysomatic Series. *Dokl. Earth Sci.* **2005**, *401*, 319–325.

123. Stefanovsky, S.V.; Yudintsev, S.V.; Nikonov, B.S.; Stefanovsky, O.I. Rare Earth-Bearing Murataite Ceramics. *Mater. Res. Soc. Symp. Proc.* **2007**, *985*, 175–180. [CrossRef]

124. Stefanovsky, S.V.; Ptashkin, A.G.; Knyazev, O.A.; Dimitriev, S.A.; Yudintsev, S.V.; Nikonov, B.S. Inductive Cold Crucible Melting of Actinide-bearning Murataite-based Ceramics. *J. Alloys Compd.* **2007**, *444*, 438–442. [CrossRef]
125. Stefanovsky, S.V.; Yudintsev, S.V.; Perevalov, S.A.; Startseva, I.V.; Varlakova, G.A. Leach Resistance of Murataite-based Ceramics Containing Actinides. *J. Alloys Compd.* **2007**, *444*, 618–620. [CrossRef]
126. Stefanovsky, S.V.; Ptashkin, A.G.; Knyazev, O.A.; Zen'kovskaja, M.S.; Stefanovsky, O.I.; Yudintsev, S.V.; Nikonov, B.S.; Lapina, M.I. Melted Murataite Ceramics Containing Simulated Actinide/Rere Earth Fraction of High Level Waste. In Proceedings of the WM2008 Conference, Phoenix, AZ, USA, 24–28 February 2008.
127. Krivovichev, S.V.; Yudintsev, S.V.; Stefanovsky, S.V.; Organova, N.I.; Karimova, O.V.; Urusov, V.S. Murataite–Pyrochlore Series: A Family of Complex Oxides with Nanoscale Pyrochlore Clusters. *Angew. Chem. Int. Ed.* **2010**, *49*, 9982–9984. [CrossRef]
128. Laverov, N.P.; Yudintsev, S.V.; Stefanovskii, S.V.; Omel'yanenko, B.I.; Nikonov, B.S. Murataite Matrices for Actinide Wastes. *Radiochemistry* **2011**, *53*, 229–243. [CrossRef]
129. Pakhomova, A.S.; Krivovichev, S.V.; Yudintsev, S.V.; Stefanovsky, S.V. Synthetic Murataite-3c, a complex form for long-term immobilization of nuclear waste: crystal structure and its comparison with natural analogue. *Z. Kristallogr* **2013**, *228*, 151–156. [CrossRef]
130. Poglyad, S.S.; Pryzhevskaya, E.A.; Lizin, A.A.; Tomilin, S.V.; Murasova, O.V. On possibility of the murataite fusion temperature lowering for radioactive waste immobilization. *J. Phys. Conf. Ser.* **2018**, *1133*, 012019. [CrossRef]
131. Lizin, A.A.; Tomilin, S.V.; Poglyad, S.S.; Pryzhevskaya, E.A.; Yudintsev, S.V.; Stefanovsky, S.V. Murataite: a matrix for immobilizing waste generated in radiochemical reprocessing of spent nuclear fuel. *J. Radioanal. Nucl. Chem.* **2018**, *318*, 2363–2372. [CrossRef]
132. Clinard, F., Jr.; Hobbs, L.W.; Land, C.C.; Peterson, D.E.; Rohr, D.L.; Roof, R.B. Alpha Decay Self-irradiation Damage in 238Pu-substituted Zirconolite. *J. Nucl. Mater.* **1982**, *105*, 248–256.
133. Clinard, F., Jr.; Peterson, D.E.; Rohr, D.L.; Hobbs, L.W. Self-irradiation Effects in 238Pu-substituted Zirconolite: I. Temperature Dependence of Damage. *J. Nucl. Mater.* **1984**, *126*, 245–254.
134. Boult, K.A.; Dalton, J.T.; Evans, J.P.; Hall, A.R.; Inns, A.J.; Marples, J.A.C.; Paige, E.L. *The Preparation of fully-active Synroc and its radiation stability–Final Report October 1988*; UKAEA: Harwell, UK, 1988.
135. Vance, E.R.; Ball, C.J.; Day, R.A.; Smith, K.L.; Blackford, M.G.; Begg, B.D.; Angel, P.J. Actinide and Rare Earth Incorporation in Zirconolite. *J. Alloys Compd.* **1994**, *213–214*, 406–409. [CrossRef]
136. Vance, E.R.; Angel, P.J.; Begg, B.D.; Day, R.A. Zirconolite-Rich Titanate Ceramics for High-Level Actinide Wastes. In *Scientific Basis for Nuclear Waste Management XVII*; van Konynenburg, R., Barkatt, A.A., Eds.; Materials Research Society: Pittsburgh, PA, USA, 1994; pp. 293–298.
137. Zhang, Y.; Stewart, M.W.A.; Li, H.; Carter, M.L.; Vance, E.R.; Moricca, S. Zirconolite-rich Titanate Ceramics for Immobilization of Actinides–Waste form/HIP Can Interactions and Chemical Durability. *J. Nucl. Mater.* **2009**, *395*, 69–74. [CrossRef]
138. Whittle, K.R.; Hyatt, N.C.; Smith, K.L.; Margiolaki, I.; Berry, F.J.; Knight, K.E.; Lumpkin, G.R. Combined neutron and X-ray diffraction determination of disorder in doped zirconolite-2M. *Am. Mineral.* **2012**, *97*, 291–298. [CrossRef]
139. Bohre, A.; Avasthi, K.; Shrivastava, O.P. Synthesis, Characterization, and Crystal Structure Refinement of Lanthanum and Yttrium Substituted Polycrystalline 2M Type Zirconolite Phases: $Ca_{1-x}Me_xZrTi_2O_7$ (Me = Y, La and x = 0.2). *J. Powder Technol.* **2014**, *2014*, 1–10. [CrossRef]
140. Lumpkin, G.R.; Gao, Y.; Giere, R.; Williams, C.T.; Mariano, A.N.; Geisler, T. The role of Th-U minerals in assessing the performance of nuclear waste forms. *Mineral. Mag.* **2014**, *78*, 1071–1095. [CrossRef]
141. Deschanels, K.; Seydonx-Guillaume, A.M.; Morgin, V.; Mesbah, A.; Tribet, M.; Moloney, M.P.; Serruys, Y.; Peuget, S. Swelling induced by alpha decay in monazite and zirconolite ceramics: A XRD and TEM comparative study. *J. Nucl. Mater.* **2014**, *448*, 184–194. [CrossRef]
142. Zhang, K.; Wen, G.; Zhang, H.; Teng, Y. Self-propagating high-temperature synthesis of CeO_2 incorporated zirconolite-rich waste forms and the aqueous durability. *J. Eur. Cream. Soc.* **2015**, *35*, 3085–3093. [CrossRef]
143. Zhang, K.; Wen, G.; Jia, Z.; Teng, Y.; Zhang, H. Self-Propagating High-Temperature Synthesis of Zirconolite Using CuO and MoO_3 as the Oxidants. *Int. J. Appl. Ceram. Technol.* **2015**, *12*, E111–E120. [CrossRef]
144. Wen, G.; Zhang, K.; Yin, D.; Zhang, H. Solid-state reaction synthesis and aqueous durability of Ce-doped zirconolite-rich ceramics. *J. Nucl. Mater.* **2015**, *466*, 113–119. [CrossRef]

145. Popa, K.; Cologna, M.; Martel, L.; Staicu, D.; Cambriani, A.; Ernstberger, M.; Raison, P.E.; Somers, J. CaTh(PO$_4$)$_2$ cheralite as a candidate ceramic nuclear waste form: Spark plasma sintering and physicochemical characterization. *J. Eur. Ceram.* **2016**, *36*, 4115–4121. [CrossRef]

146. Wen, J.; Sun, C.; Dholabhai, P.P.; Xia, Y.; Tang, M.; Chen, D.; Yang, D.Y.; Li, Y.H.; Uberuaga, B.P.; Wang, Y.Q. Temperature dependence of the radiation tolerance of nanocrystalline pyrochlores A$_2$Ti$_2$O$_7$ (A = Gd, Ho and Lu). *Acta Mater.* **2016**, *110*, 175–184. [CrossRef]

147. Clark, B.M.; Sundaram, S.K.; Misture, S.T. Polymorphic Transitions in Cerium-Substituted Zirconolite (CaZrTi$_2$O$_7$). *Sci. Rep.* **2017**, *7*, 5920. [CrossRef]

148. Sun, S.K.; Stennett, M.C.; Corkhill, C.L. Reactive spark plasma synthesis of CaZrTi$_2$O$_7$ zirconolite ceramics for plutonium disposition. *J. Nucl. Mater.* **2018**, *500*, 11–14. [CrossRef]

149. McCaughherty, S.; Crosvenor, A.P. Low-temperature synthesis of CaZrTi$_2$O$_7$ zirconolite-type materialss using ceramic, coprecipitation, and soil-gel methods. *J. Mater. Chem. C* **2019**, *7*, 177–187. [CrossRef]

150. Zhang, K.; Yin, D.; Xu, K.; Zhang, H. Self-Propagating Synthesis and Characterization Studies of Gd-Bearing Hf-Zirconolite Ceramic Waste Forms. *Materials* **2019**, *12*, 178. [CrossRef]

151. Boult, A.; Dalton, J.T.; Evans, J.P.; Hall, A.R.; Inn, A.J.; Marples, J.A.C.; Paige, E.L. The Preparation of Fully-active Synroc and its Radiation Stability. *Rep. Aere-R.* **1987**, *13*, 318.

152. Vance, E.R.; Carter, M.L.; Zhang, Z.; Finnie, K.S.; Thomson, S.J.; Begg, B.D. Uranium valences in perovskite, CaTiO$_3$. In *Environmental Issues and Waste Management Technologies in the Ceramic & Nuclear Industries IX*; The American Ceramic Society: Westeville, OH, USA, 2004; Volume 155, pp. 3–10.

153. Bozadjiev, L.; Georgiev, G.; Parashkevov, D. Synthesis of perovskites and perovskite based technical stones. In *National Conference "GEOSCIENSES 2006"*; Bulgarian Geophysical Society: Sofia, Bulgaria, 2006; pp. 127–129.

154. Bowles, J.F.W.; Howie, R.A.; Vaughan, D.J.; Zussman, J. *Rock-Forming Minerals, V5A: Non-Silicates: Oxides, Hydroxides and Sulphides*; The Geological Society: London, UK, 2011; p. 920.

155. Stewart, M.W.A.; Vance, E.R.; Moricca, S.A.; Brew, D.R.; Cheung, C.; Eddowes, T.; Bermudez, W. Immobilisation of Higher Activity Wastes from Nuclear Reactor Production of ^{99}Mo. *Sci. Technol. Nucl. Install.* **2013**, *2013*, 1–16. [CrossRef]

156. Ghosh, B.; Dutta, A.; Shannigrahi, S.; Sinha, T.P. Combined XPS and first principles study of double-perovskite Ca$_2$GdTaO$_6$. *J. Mater. Sci.* **2014**, *49*, 819–826. [CrossRef]

157. Bohre, A.; Avasthi, K.; Shrivastava, O.P. Structure Refinement of Polycrystalline Orthorombic Calcium Titanate Substituted by Lanthanum: Ca$_{1-x}$Ln$_x$TiO$_3$ (x = 0.1–0.4). *Crystallogr. Rep.* **2014**, *59*, 944–948. [CrossRef]

158. Livshits, T.S.; Zhang, J.; Yudintsev, S.V.; Stefanovsky, S.V. New titanate matrices for immobilization of REE-actinide high level waste. *J. Radioanal. Nucl. Chem.* **2014**, *304*, 47–52. [CrossRef]

159. Mahadik, P.S.; Sengupta, P.; Halder, R.; Abraham, G.; Dey, G.K. Perovskite-Ni composite: A potential route for management of radioactive metallic waste. *J. Hazard. Mater.* **2015**, *287*, 207–216. [CrossRef]

160. Carter, M.L.; Vance, E.R.; Mitchell, D.R.G.; Hanna, J.V.; Zhang, Z.; Loi, E. Fabrication, Characterization, and Leach Testing of Hollandite, (Ba,Cs)(Al,Ti)$_2$Ti$_6$O$_{16}$. *J. Mater. Res.* **2002**, *17*, 2578–2589. [CrossRef]

161. Carter, M.L.; Vance, E.R.; Li, H. Hollandite-rich Ceramic Melts for the Immobilization of Cs. *Mater. Res. Soc. Proc.* **2004**, *807*, 249–254. [CrossRef]

162. Whittle, K.R.; Ashbrook, S.; Redfem, S.; Lumpkin, G.R. Structural Studies of Hollandite-Based Radioactive Waste Forms Structural Studies of Hollandite-Based Radioactive Waste Forms. *Mater. Res. Soc. Symp. Proc.* **2004**, *87*, 1–7.

163. Nishiyama, N.; Rapp, R.P.; Irifune, T.; Sanehira, T.; Yamazaki, D.; Funakoshi, K. Stability and P-V-T equation of state KAlSi$_3$O$_8$-hollandite determined by in situ X-Ray observations and implications for dynamics of subducted continental crust material. *Phys. Chem. Miner.* **2005**, *32*, 627–637. [CrossRef]

164. Addelouas, A.; Utsunomiya, S.; Suzuki, T.; Grambow, B.; Advocat, T.; Bart, F.; Ewing, R.C. Effect of ionizing radiation on the hollandite structure-type: Ba$_{0.85}$Cs$_{0.26}$Al$_{1.35}$Fe$_{0.77}$Ti$_{5.90}$O$_{16}$. *Am. Mineral.* **2008**, *93*, 241–247. [CrossRef]

165. Shluk, L.; Niewa, R. Crystal Structure and Magnetic Properties of the Novel Hollandite Ba$_{1.3}$Co$_{1.3}$Ti$_{6.7}$O$_{16}$. *Z. Naturforsch.* **2011**, *66b*, 1097–1100. [CrossRef]

166. Shabalin, B.; Titov, Y.; Zlobenko, B.; Bugera, S. Ferric Titanous Hollandite Analogues–Matrices for Immobilization of Cs-Containing Radioactive Waste: Synthesis. *Mineral. J.* **2013**, *35*, 12–18.

167. Chen, T.-Y.; Maddrell, E.R.; Hyatt, N.C.; Hriljac, J.A. A potential wasteform for Cs immobilization: synthesis, structure determination, and aqueous durability of $Cs_2TiNb_6O_{18}$. *Inorg. Chem.* **2016**, *55*, 12686–12695. [CrossRef] [PubMed]

168. Zhao, M.; Xu, Y.; Shuller-Nickles, L.; Amoroso, J.; Frenkel, A.I.; Li, Y.; Gong, W.; Lilova, K.; Navrotsky, A.; Brinkman, K.S. Compositional control of radionuclide retention in hollandite-based ceramic waste forms for Cs immobilization. *J. Am. Ceram. Soc.* **2018**, *102*, 4314–4324. [CrossRef]

169. Grote, R.; Zhao, M.; Shuller-Nickles, L.; Amoroso, J.; Gong, W.; Lilova, K.; Navrotsky, A.; Tang, M.; Brinkman, K.S. Compositional control of tunnel features in hollandite-based ceramics: Structure and stability of $(Ba,Cs)_{1.33}(Zn,Ti)_8O_{16}$. *J. Mater. Sci.* **2018**, *54*, 1112–1125. [CrossRef]

170. Geller, S.; Miller, C.E. Silicate garnet yttrium-iron garnet solid solution. *Am. Mineral.* **1959**, *44*, 1115–1120.

171. Ito, J.; Frondel, C. Synthesis zirconium and titanium garnets. *Am. Mineral.* **1967**, *52*, 773–781.

172. Rickwood, P.C. On recasting analyses of garnet into end-member molecules. *Contrib. Mineral. Petrol.* **1968**, *18*, 175–198. [CrossRef]

173. Novak, G.A.; Gibbs, G.V. The crystal chemistry of the silicate garnets. *Am. Mineral.* **1971**, *56*, 791–825.

174. Dowty, E. Crystal chemistry of titanian and zirconian garnet: I. Review and spectral studies. *Am. Mineral.* **1971**, *56*, 1983–2009.

175. Kanke, Y.; Navrotsky, A. A calorimetric study of the lanthanide aluminum oxides and the lantha-nide gallium oxides: Stability of the perovskites and the garnets. *J. Solid State Chem.* **1998**, *141*, 424–436. [CrossRef]

176. Burakov, B.E.; Anderson, E.B.; Zamoryanskaya, M.V.; Petrova, M.A. Synthesis and Study of [239]Pu-Doped Gadolinium-Aluminum Garnet. In *Material Research Society Symposium Proceedings Scientific Basis for Nuclear Waste Management XXIII*; Materials Research Society: Warendalle, PA, USA, 2000; Volume 608, pp. 419–422.

177. Yudintsev, S.V. Incorporation of U, Th, Zr, and Gd into the Garnet-structured Host. In Proceedings of the 8th International Conference on Radioactive Waste Management and Environmental Remediation, The American Society of Mechanical Engineers, Bruges, Belgium, 30 September–4 October 2001.

178. Yudintsev, S.V.; Lapina, M.I.; Ptashkin, A.G. Accommodation of uranium into the garnet structure. *MRS Symp. Proc.* **2002**, *713*, 477–480. [CrossRef]

179. Utsunomiya, S.; Wang, L.M.; Yudintsev, S.; Ewing, R.C. Ion irradiation-induced amorphization and nano-crystal formation in garnets. *J. Nucl. Mater.* **2002**, *303*, 177–187. [CrossRef]

180. Yudintsev, S.V. A Structural-chemical Approach to Selecting Crystalline Matrices for Actinide Immobilization. *Geol. Ore Depos.* **2003**, *45*, 151–165.

181. Utsunomiya, S.; Yudintsev, S.V.; Ewing, R.C. Radiation effects in ferrate garnet. *J. Nucl. Mater.* **2005**, *336*, 251–260. [CrossRef]

182. Maslakov, K.I.; Teterin, U.A.; Vukchevich, L.; Udintseva, T.C.; Udintsev, S.V.; Ivanov, K.E.; Lapina, M.I. Issledovanie obraztsov keramiki $(Ca_{2.5}Th_{0.5})Zr_2Fe_3O_{12}$, $(Ca_{1.5}GdTh_{0.5})(ZrFe)Fe_3O_{12}$ i $(Ca_{2.5}Ce_{0.5})Zr_2Fe_3O_{12}$ so strukturoi granata. *Radiohimiya* **2007**, *49*, 31–37.

183. Suarez, M.; Fernandez, A.; Menendez, J.L.; Torrecillas, R. Transparent Yttrium Aluminium Garnet Obtained by Spark Plasma Sintering of Lyophilized Gels. *J. Nanomater.* **2009**, *2009*, 1–5. [CrossRef]

184. Tomilin, S.V.; Lizin, A.A.; Lukinykh, A.N.; Livshits, T.S. Radiation Resistance and Chemical of Yttrium Aluminium Garnet. *Radiochemistry* **2011**, *53*, 186–190. [CrossRef]

185. Golovkina, L.S.; Orlova, A.I.; Nokhrin, A.V.; Boldin, M.S.; Sakharov, N.V. Ceramics based on Yttrium Aluminium Garnet Containing Nd and Sm obtained by Spark Plasma Sintering. *Adv. Ceram. Sci. Eng.* **2013**, *2*, 261–265.

186. Hanc, E.; Zając, W.; Lu, L. On fabrication procedures of Li-ion conducting garnets. *J. Solid State Chem.* **2017**, *248*, 51–60. [CrossRef]

187. Yamada, H.; Ito, T.; Basappa, R.H.; Bekarevich, R.; Mitsuishi, K. Influence of strain on local structure and lithium ionic conduction in garnet-type solid electrolyte. *J. Power Sources* **2017**, *368*, 97–106. [CrossRef]

188. Selvi, M.M.; Chakraborty, D.; Venkateswaran, C. Magnetodielectric coupling in multiferroic holmium iron garnets. *J. Magn. Magn. Mater.* **2017**, *423*, 39–45. [CrossRef]

189. Golovkina, L.; Orlova, A.; Boldin, M.; Sakharov, N.; Chuvil'deev, V.; Nokhrin, A.; Konings, R.; Staicu, D. Development of composite ceramic materials with improved thermal conductivity and plasticity based on garnet-type oxides. *J. Nucl. Mater.* **2017**, *489*, 158–163. [CrossRef]

190. Golovkina, L.S.; Nokhrin, A.V.; Boldin, M.S.; Lantsev, E.A.; Orlova, A.I.; Chuvil'deev, V.N.; Murashov, A.A.; Sakharov, N.V. Preparation of Fine-Grained $Y_{2.5}Nd_{0.5}Al_5O_{12}$ + MgO composite ceramics for Inert Matrix Fuels by Spark Plasma Sintering. *Inorg. Mater.* **2018**, *54*, 1291–1298. [CrossRef]

191. Golovkina, L.S.; Orlova, A.I.; Nokhrin, A.V.; Boldin, M.S.; Lantsev, E.A.; Chuvil'deev, V.N.; Sakharov, N.V.; Shotin, S.V.; Zelenov, A.Y. Spark Plasma Sintering of fine-grained ceramic-metal composites YAG: Nd-(W,Mo) based on garnet-type oxide $Y_{2.5}Nd_{0.5}Al_5O_{12}$ for inert matrix fuel. *J. Nucl. Mater.* **2018**, *511*, 109–121. [CrossRef]

192. Golovkina, L.S.; Orlova, A.I.; Nokhrin, A.V.; Boldin, M.S.; Chuvil'deev, V.N.; Sakharov, N.V.; Belkin, O.A.; Shotin, S.V.; Zelenov, A.Y. Spark Plasma Sintering of fine-grain ceramic-metal composites based on garnet-structure oxide $Y_{2.5}Nd_{0.5}Al_5O_{12}$ for inert matrix fuel. *Mater. Chem. Phys.* **2018**, *214*, 516–526. [CrossRef]

193. Golovkina, L.S.; Orlova, A.I.; Chuvil'deev, V.N.; Boldin, M.S.; Lantcev, E.A.; Nokhrin, A.V.; Sakharov, N.V.; Zelenov, A.Y. Spark Plasma Sintering of high-density fine-grained $Y_{2.5}Nd_{0.5}Al_5O_{12}$ + SiC composite ceramics. *Mater. Res. Bull.* **2018**, *103*, 211–215. [CrossRef]

194. Golovkina, L.; Orlova, A.; Boldin, M.; Lantsev, E.; Sakharov, N.; Zelenov, A.; Chuvil'deev, V. Composite Ceramics Based on Garnet-type Oxide $Y_{2.5}Nd_{0.5}Al_5O_{12}$ and Silicon Carbide. *Mater. Res. Bull.* **2018**, *4*, 518–524.

195. Gong, W.L.; Ewing, R.C.; Wang, L.M.; Xie, H.S. Crichtonite Structure Type ($AM_{21}O_{38}$ and $A_2M_{19}O_{36}$) as a Host Phase in Crystalline Waste Form Ceramics. Scientific Basis for Nuclear Waste Management XVIII. T. Murakami and R.C. Ewing, Eds. *Proc. Mater. Res. Soc.* **1995**, *353*, 807–815. [CrossRef]

196. Grey, I.E.; Lloyd, D.J.; White, J.S., Jr. The Structure of Crichtonite and its Relationship to Senaite. *Am. Mineral.* **1976**, *61*, 1203–1212.

197. Wulser, P.-A.; Meisser, N.; Brugger, J.; Schenk, K.; Ansermet, S.; Bonin, M.; Bussy, F. Cleusonite, $(Pb,Sr)(U^{4+},U^{6+})(Fe^{2+},Zn)_2(Ti,Fe^{2+},Fe^{3+})_{18}(O,OH)_{38}$, a new mineral species of the crichtonite group from the western Swiss Alps. *Eur. J. Mineral.* **2005**, *17*, 933–942. [CrossRef]

198. Mills, S.J.; Bindi, L.; Cadoni, M.; Kampf, A.R.; Ciriotti, M.E.; Ferraris, G. Paseroite, $PbMn^{2+}(Mn^{2+},Fe^{2+})_2(V^{5+},Ti,Fe^{3+},\square)_{18}O_{38}$, a new member of the crichtonite group. *Eur. J. Mineral.* **2012**, *24*, 1061–1067. [CrossRef]

199. Rastsvetaeva, R.K.; Aksenov, S.M.; Chukanov, N.V.; Menezes, L.A.D. Crystal Structure of Almeidaite, a New Mineral of the Crichtonite Group. *Dokl. Chem.* **2014**, *455*, 53–57. [CrossRef]

200. Biagioni, C.; Orlandi, P.; Pasero, M.; Nestola, F.; Bindi, L. Mapiquiroite, $(Sr,Pb)(U,Y)Fe_2(Ti,Fe^{3+})_{18}O_{38}$, a new member of the crichtonite group from the Apuan Alps, Tuscany, Italy. *Eur. J. Mineral.* **2014**, *26*, 427–437. [CrossRef]

201. Menezes Filho, L.A.D.; Chukanov, N.V.; Rastsvetaeva, R.K.; Aksenov, S.M.; Pekov, I.V.; Chaves, M.L.S.C.; Richards, R.P.; Atencio, D.; Brandao, P.R.G.; Scholz, R.; et al. Almeidaite, $Pb(Mn,Y)Zn_2(Ti,Fe^{3+})_{18}O_{36}(O,OH)_2$, a new crichtonite-group mineral, from Novo Horizonte, Bahia. *Braz. Mineral. Mag.* **2015**, *79*, 269–283. [CrossRef]

202. Rezvukhin, D.I.; Malkovets, V.G.; Sharygin, I.S.; Kuzmin, D.V.; Gibsher, A.A.; Litasov, K.D.; Pokhilenko, A.P.; Sobolev, N.V. Inclusions of Crichtonite Group Minerals in Pyropes from the Internatsionalnaya Kimberlite Pipe. *Dokl. Akad. Nauk.* **2016**, *466*, 714–717. [CrossRef]

203. Ishiguro, T.; Tanaka, K.; Marumo, F.; Ismail, M.G.M.U.; Hirano, S. Somiya, Freudenbergite. *Acta Cryst.* **1978**, *B34*, 255–256. [CrossRef]

204. Vance, E.R.; Angel, P.J.; Cassidy, D.J.; Stewart, M.W.A.; Blackford, M.G.; McGlinn, P.A. Freudenbergite: A Possible Synroc Phase for Sodium-Bearing High-Level Waste. *J. Am. Ceram. Soc.* **1994**, *77*, 1576–1580. [CrossRef]

205. Ren, X.; Komarneni, S.; Roy, D.M. Novel $CsAl_2PO_6$ of pollucite structure: synthesis and characterization. *Mater. Res. Bull.* **1990**, *25*, 665–670. [CrossRef]

206. Komarneni, S.; Menon, V.C.; Li, Q.H.; Roy, R.; Ainger, F. Microwave-hydrothermal processing of $BiFeO_3$ and $CsAl_2PO_6$. *J. Am. Ceram. Soc.* **1996**, *75*, 1409–1412. [CrossRef]

207. Aloy, A.S.; Kol'tsova, T.I.; Trofimenko, A.V.; Tutov, A.G. New compound with pollucite structure forming in process of synthesis and crystallization of cesium-alumphosphate glasses. *Radiochemistry* **2000**, *42*, 273–274.

208. Hirst, J.P.; Claridge, J.B.; Rosseinsky, M.J.; Bishop, P. High temperature synthesis of a noncentrosymmetric site-ordered cobalt aluminophosphate related to the pollucite structure, The Royal Society of Chemistry. *Chem. Commun.* **2003**, *6*, 684–685. [CrossRef]

209. Orlova, A.I. Crystal Chemical View on Elaboration of Ecology Safe Materials for Immobilization of Alkaline elements of Radwaste. In *III International Pyroprocessing Research Conference*; NIIAR: Dimitrovgrad, Russia, 2010; pp. 27–28.

210. Loginova, E.E.; Orlova, A.I.; Mikhailov, D.A.; Troshin, A.N.; Borovikova, E.Y.; Samoilov, S.G.; Kazantsev, G.N.; Kazakova, A.Y.; Demarin, V.T. Phosphorus-Containing Compounds of Pollucite Structure and Radiochemical Problems. *Radiochemistry* **2011**, *53*, 593–603. [CrossRef]

211. Orlova, A.I.; Volgutov, V.Y.; Mikhailov, D.A.; Troshin, A.N.; Golovkina, L.S.; Skuratov, V.A.; Kirilkin, N.S.; Chuvil'diev, V.N.; Nokhrin, A.V.; Boldin, M.S.; et al. Mineral like compounds with NZP and Pollucite structures: synthesis of high density ceramic and radiation testing. In *European Congress on Advanced Materials and Processes*; EUROMAT 2011; FEMS: Montpellier, France, 2011.

212. Yanase, I.; Saito, Y.; Kobayashi, H. Synthesis and thermal expansion of (V, P, Nb)-replaced pollucite. *Ceram. Int.* **2012**, *38*, 811–815. [CrossRef]

213. Orlova, A.I.; Troshin, A.N.; Mikhailov, D.A.; Chuvil'deev, V.N.; Boldin, M.S.; Sakharov, N.V.; Nokhrin, A.V.; Skuratov, V.A.; Kirilkin, N.S. Phosphorus-Containing Cesium Compounds of Pollucite Structure. Preparation of High-Density Ceramic and Its Radiation Tests. *Radiochemistry* **2014**, *56*, 98–104. [CrossRef]

214. Klapshin, Y.P. Chemical and Phase Transformations during the Synthesis of $Cs[MgR_{0.5}P_{1.5}O_6]$ (R = B, Al, Fe) Complex Oxides from Metal Chlorides. *Russ. J. Inorg. Chem.* **2018**, *63*, 1156–1163. [CrossRef]

215. Klapshin, Y.P. Chemical and Phase Transformations during the Synthesis of $Cs[MgR_{0.5}P_{1.5}O_6]$ (R = B, Al, Fe) Complex Oxides from Metal Nitrates. *Russ. J. Inorg. Chem.* **2018**, *63*, 1381–1388. [CrossRef]

216. Townes, W.D.; Fang, J.H.; Perrotta, A.J. The Crystal Structure and Refinement of Ferromagnetic Barium Ferrite, $BaFe_{12}O_{19}$. *Z. Kristallogr.* **1967**, *125*, 437–449. [CrossRef]

217. Morgan, P.E.D.; Clarke, D.R.; Jantzen, C.M.; Harker, A.B. High Alumina Tailored Nuclear Waste Ceramics. *J. Am. Ceram. Soc.* **1981**, *64*, 249–258. [CrossRef]

218. Harker, A.B.; Jantzen, C.M.; Clarke, D.R.; Morgan, P.E.D. Tailored Ceramic Nuclear Waste Forms: Preparation and Characterization. In *Science Basis for Nuclear Waste Management, III*; Moore, J.G., Ed.; Plenum Press: New York, NY, USA, 1981; pp. 139–146.

219. Morgan, P.E.D.; Cirlin, E.H. The Magnetoplumbite Crystal Structure as a Radwaste Host. *J. Am. Ceram. Soc.* **1982**, *65*, C114–C115. [CrossRef]

220. Jantzen, C.M.; Clarke, D.R.; Morgan, P.E.D.; Harker, A.B. Leaching of Polyphase Nuclear Waste Ceramics: Microstructural and Phase Characterization. *J. Am. Ceram. Soc.* **1982**, *65*, 292–300. [CrossRef]

221. Bansal, N.P.; Zhu, D. Thermal properties of oxides with magnetoplumbite structure for advanced thermal barrier coatings. *Surf. Coat. Technol.* **2008**, *202*, 2698–2703. [CrossRef]

222. Men, D.; Patel, M.K.; Usov, I.O.; Pivin, J.C.; Porter, J.R.; Mecartney, M.L. Radiation Damage of $LaMgAl_{11}O_{19}$ and $CeMgAl_{11}O_{19}$ Magnetoplumbite. *J. Am. Ceram. Soc.* **2013**, *96*, 3325–3332.

223. Rakshit, S.K.; Parida, S.C.; Lilova, K.; Navrotsky, A. Thermodynamic studies of $CaLaFe_{11}O_{19}(s)$. *J. Solid State Chem.* **2013**, *2*, 68–74. [CrossRef]

224. Angle, J.P.; Nelson, A.T.; Men, D.; Mecartney, M.L. Thermal measurements and computational simulations of three-phase $(CeO_2–MgAl_2O_4–CeMgAl_{11}O_{19})$ and four-phase $(3Y-TZP–Al_2O_3–MgAl_2O_4–LaPO_4)$ composites as surrogate inert matrix nuclear fuel. *J. Nucl. Mater.* **2014**, *454*, 69–76. [CrossRef]

225. Ewing, R.C.; Lutze, W.; Weber, W.J. Zircon: A host-phase for the disposal of weapons plutonium. *J. Mater. Res.* **1995**, *10*, 243–246. [CrossRef]

226. Shi, Y.; Huang, X.; Yan, D. Fabrication of Hot-Pressed Zircon Ceramics: Mechanical Properties and Microstructure. *Ceram. Int.* **1997**, *23*, 457–462. [CrossRef]

227. Meldrum, A.; Boatner, L.A.; Ewing, R.C. A comparison of radiation effects in crystalline ABO_4-type phosphates and silicates. *Mineral. Mag.* **2000**, *64*, 185–194. [CrossRef]

228. Burakov, B.E.; Anderson, E.B.; Zamoryanskaya, M.V.; Yagovkina, M.A.; Strykanova, E.E.; Nikolaeva, E.V. Synthesis and Study of ^{239}Pu-Doped Ceramics Based on Zircon, $(Zr,Pu)SiO_4$, and Hafnon, $(Hf,Pu)SiO_4$. In *Material Research Society Symposium Proceedings Scientific Basis for Nuclear Waste Management XXIV*; Materials Research Society: Warrendale, PA, USA, 2001; Volume 663, pp. 307–313.

229. Hanchar, J.M.; Burakov, B.E.; Anderson, E.B.; Zamoryanskaya, M.V. Investigation of Single Crystal Zircon, (Zr,Pu)SiO$_4$, Doped with ^{238}Pu. In *Scientific Basis for Nuclear Waste Management XXVI, Materials Research Society Symposium Proceedings*; Finch, R.J., Bullen, D.B., Eds.; Materials Research Society: Warrendale, PA, USA, 2003; Volume 757, pp. 215–225.

230. Geisler-Wierwille, T.; Burakov, B.E.; Zirlin, V.; Nikolaeva, L.; Pöml, P. A Raman spectroscopic study of high-uranium zircon from the Chernobyl "Lava". *Eur. J. Mineral.* **2005**, *17*, 883–894. [CrossRef]

231. Burakov, B.E.; Domracheva, Y.V.; Zamoryanskaya, M.V.; Petrova, M.A.; Garbuzov, V.M.; Kitsay, A.A.; Zirlin, V.A. Development and synthesis of durable self-glowing crystals doped with plutonium. *J. Nucl. Mater.* **2009**, *385*, 134–136. [CrossRef]

232. Rendtorff, N.M.; Grasso, S.; Hu, C.; Suarez, G.; Aglietti, E.F.; Sakka, Y. Dense zircon (ZrSiO$_4$) ceramics by high energy ball milling and spark plasma sintering. *Ceram. Int.* **2012**, *38*, 1793–1799. [CrossRef]

233. IAEA. *Radiation Protection and NORM Residue Management in the Zircon and Zirconia Industries*; Safety Reports Series 51; IAEA: Vienna, Austria, 2007; 149p.

234. Pöml, P.; Burakov, B.; Geisler, T.; Walker, C.T.; Grange, M.L.; Nemchin, A.A.; Berndt, J.; Fonseca, R.O.C.; Bottomley, P.D.W.; Hasnaoui, R. Micro-analytical uranium isotope and chemical investigations of zircon crystals from the Chernobyl "lava" and their nuclear fuel inclusions. *J. Nucl. Mater.* **2013**, *439*, 51–56. [CrossRef]

235. Ojovan, M.I.; Burakov, B.E.; Lee, W.E. Radiation-induced microcrystal shape change as a mechanism of wasteform degradation. *J. Nucl. Mater.* **2018**, *501*, 162–171. [CrossRef]

236. Hayden, L.A.; Watson, E.B.; Wark, D.A. A thermobarometer for sphene (titanite). *Contrib. Mineral. Petrol.* **2008**, *155*, 529–540. [CrossRef]

237. Park, T.J.; Li, S.; Navrotsky, A. Thermochemistry of glass forming Y-substituted Sr-analogues of titanite (SrTiSiO$_5$). *J. Mater. Res.* **2009**, *24*, 3380–3386. [CrossRef]

238. Scanu, T.; Guglielmi, J.; Colomban, P. Ion exchange and hot corrosion of ceramic composites matrices: A vibrational and microstructural study. *Solid State Ion.* **1994**, *70–71*, 109–120. [CrossRef]

239. Weber, W.J. Radiation-induced Swelling and Amorphization in Ca$_2$Nd$_8$(SiO$_4$)$_6$O$_2$. *Radiat. Eff. Defects Solids* **1983**, *77*, 295–308. [CrossRef]

240. Fahey, J.A.; Weber, W.J.; Rotella, F.J. An X-ray and Neutron Powder Diffraction Study of the Ca$_{2+x}$Nd$_{8-x}$(SiO$_4$)$_6$O$_{2-0.5x}$ System. *J. Solid State Chem.* **1985**, *60*, 145–158. [CrossRef]

241. Weber, W.J. Alpha-decay-induced Amorphization in Complex Silicate Structures. *J. Am. Ceram. Soc.* **1993**, *76*, 1729–1738. [CrossRef]

242. Utsunomiya, S.; Yudintsev, S.; Wang, L.M.; Ewing, R.C. Ion-beam and Electron-beam Irradiation of Synthetic Britholite. *J. Nucl. Mater.* **2003**, *322*, 180–188. [CrossRef]

243. Leo'n-Reina, L.; Losilla, E.R.; Martı́nez-Lara, M.; Bruque, S.; Llobet, A.; Sheptyakov, D.V.; Aranda, M.A.G. Interstitial Oxygen in Oxygen-stoichiometric Apatites. *J. Mater. Chem.* **2005**, *15*, 2489–2498. [CrossRef]

244. Terra, O.; Dacheux, N.; Audubert, F.; Podor, R. Immobilization of tetravalent actinides in phosphate ceramics. *J. Nucl. Mater.* **2006**, *352*, 224–232. [CrossRef]

245. Malavasi, L.; Fisher, C.A.J.; Islam, M.S. Oxide-ion and proton conducting electrolyte materials for clean energy applications: Structural and mechanistic features. *Chem. Soc. Rev.* **2010**, *39*, 4370–4387. [CrossRef]

246. Knyazev, A.V.; Bulanov, E.N.; Korshunov, A.O.; Krasheninnikova, O.V. Synthesis and thermal expansion of some lanthanide-containing apatites. *Inorg. Mater.* **2013**, *49*, 1133–1137. [CrossRef]

247. Bulanov, E.N.; Knyazev, A.V. High-temperature in situ XRD investigations in apatites. Structural interpretation of thermal deformations. In *Apatite: Synthesis, Structural Characterization and Biomedical Applications*; Nova Science Publishers, Inc.: Hauppauge, NY, USA, 2014; Chapter 7; pp. 173–200.

248. Bulanov, E.N.; Wang, J.; Knyazev, A.V.; White, T.; Manyakina, M.; Baikie, T.; Lapshin, A.N.; Dong, Z. Structure and thermal expansion of calcium-thorium apatite, [Ca$_4$]F[Ca$_2$Th$_4$]T [(SiO$_4$)$_6$]O$_2$. *Inorg. Chem.* **2015**, *54*, 11356–11361. [CrossRef]

249. Knyazev, A.V.; Bulanov, E.N.; Smirnova, N.N.; Korokin, V.Z.; Shushunov, A.N.; Blokhina, A.G.; Xu, Z. Thermodynamic and thermophysics properties of synthetic britholite SrPr$_4$(SiO$_4$)$_3$O. *J. Chem. Thermodyn.* **2017**, *108*, 38–44. [CrossRef]

250. Cronstedt, A.F. Observation and Description of an Unknown Kind of Rock to be Named Zeolites. *Kong Vetenskaps Acad. Handl. Stockh.* **1756**, *12*, 120–123. (In Swedish)

251. Smith, J.V. Structural Classification of Zeolites. *Mineral. Soc. Am. Spec. Pap.* **1963**, *1*, 281.

252. Breck, D.A. *Zeolite Molecular Sieves: Structure, Chemistry and Use*; Wiley-Interscience: New York, NY, USA, 1974.

253. Smith, J.V. Origin and Structure of Zeolites. In *Zeolite Chemistry and Catalysis*; A.C.S. Series; Rabo, J.A., Ed.; American Chemical Society: Washington, DC, USA, 1976; Volume 171, pp. 1–79.

254. Barrer, R.M. *Hydrothermal Chemistry of Zeolites*; Academic Press: New York, NY, USA, 1982.

255. Harjula, R.; Lehto, J. Effect of sodium and potassium ions on cesium absorption from nuclear power plant waste solutions on synthetic zeolites. *Nucl. Chem. Waste Manag.* **1986**, *6*, 133–137. [CrossRef]

256. Higgins, F.M.; de Leeuw, N.H.; Parker, S.C. Modelling the Effect of Water on Cation Exchange in Zeolite A. *J. Mater. Chem.* **2002**, *12*, 124–131.

257. Lima, E.; Ibarra, A.; Bosch, P.; Bulbulian, S. Vitrification of CsA and CsX zeolites. *Stud. Surf. Sci. Catal.* **2004**, *154*, 1907–1911.

258. Tsukada, T.; Takahashi, K. Absorption Characteristics of Fission Product Elements on Zeolite. *Nucl. Technol.* **2008**, *162*, 229–243. [CrossRef]

259. Kaminski, M.D.; Mertz, C.J.; Ferrandon, M.; Dietz, N.L.; Sandi, G. Physical properties of an alumino-silicate waste form for cesium and strontium. *J. Nucl. Mater.* **2009**, *392*, 510–518. [CrossRef]

260. Cappelletti, P.; Rapisardo, G.; De Gennaro, B.; Colella, A.; Langella, A.; Fabio, S.; Lee, D.; De Gennaro, M. Immobilization of Cs and Sr in aluminosilicate matrices derived from natural zeolites. *J. Nucl. Mater.* **2011**, *414*, 451–457. [CrossRef]

261. Gatta, G.D.; Merlini, M.; Lotti, P.; Lausi, A.; Rieder, M. Microporous and Mesoporous Materials Phase stability and thermo-elastic behavior of $CsAlSiO_4$ (ABW): A potential nuclear waste disposal material. *Microporous Mesoporous Mater.* **2012**, *163*, 147–152. [CrossRef]

262. Yamagishi, I.; Nagaishi, R.; Kato, C.; Morita, K.; Terada, A.; Kamiji, Y. Characterization and storage of radioactive zeolite waste. *J. Nucl. Sci. Technol.* **2014**, *51*, 1044–1053. [CrossRef]

263. Gallis, F.S.; Ermanoski, I.; Greathouse, J.A.; Chapman, K.W.; Nenoff, T.M. Iodine Gas Adsorption in Nanoporous Materials: A Combined Experiment–Modeling Study. *Ind. Eng. Chem. Res.* **2017**, *56*, 2331–2338. [CrossRef]

264. Lee, H.Y.; Kim, H.S.; Jeong, H.-K.; Park, M.; Chung, D.-Y.; Lee, E.-H.; Lim, W.T. Selective Removal of Radioactive Cesium from Nuclear Waste by Zeolites: On the Origin of Cesium Selectivity Revealed by Systematic Crystallographic Studies. *J. Phys. Chem. C* **2017**, *121*, 10594–10608. [CrossRef]

265. Papynov, E.K. Spark plasma sintering of ceramic and glass-ceramic matrices for cesium radionuclides immobilization. In *Glas-Ceramics: Properties, Applications and Technolology*; Narag, K., Ed.; Nova Science Publisher, Inc.: New York, NY, USA, 2018; pp. 109–153.

266. Papynov, E.K.; Shichalin, O.O.; Mayorov, V.Y.; Kuryavyi, V.G.; Kaidalova, T.A.; Teplukhina, L.V.; Portnyagin, A.S.; Slobodyuk, A.B.; Belov, A.A.; Tananaev, I.G.; et al. SPS technique for ionizing radiation source fabrication based on dense cesium-containing core. *J. Hazard. Mater.* **2019**, *369*, 25–30. [CrossRef]

267. Strachan, D.M.; Schulz, W.W. Characterization of Pollucite as a Material for Long-Term Storage of Cesium-137. *Am. Ceram. Soc. Bull.* **1979**, *58*, 865–871.

268. Gallagher, S.A.; McCarthy, G.J. Preparation and X-ray Characterization of Pollucite ($CsAlSi_2O_6$). *Inorg. Nucl. Chem.* **1981**, *43*, 1773–1777. [CrossRef]

269. Komameni, S.; White, W.B. Stability of Pollucite in Hydrothermal Fluids. *Sci. Basis Nucl. Waste Manag.* **1981**, *3*, 387–396.

270. Yanagisawa, K.; Nishioka, M.; Yamasaki, N. Immobilization of Cesium into Pollucite Structure by Hydrothermal Hot-Pressing. *J. Nucl. Sci. Technol.* **1987**, *24*, 51–60. [CrossRef]

271. Mielearski, M. Preparation of [137]Cs Pollucite Source Core. *Isotopenpraxis* **1989**, *25*, 404–408. [CrossRef]

272. Mimura, H.; Shibata, M.; Akiba, K. Surface Alteration of Pollucite under Hydrothermal Conditions. *J. Nucl. Sci. Technol.* **1990**, *27*, 835–843. [CrossRef]

273. Anthony, R.G.; Phillip, C.V.; Dosch, R.G. Selective Adsorption and Ion Exchange of Metal Cations and Anions with Silico-Titanates and Layered Titanates. *Waste Manag.* **1993**, *13*, 503–512. [CrossRef]

274. Balmer, M.L.; Bunker, B.C. *Inorganic Ion Exchange Evaluation and Design-Silicotitanate Ion Exchange Waste Conversion*; PNL-10460; Pacific Northwest Laboratory: Richland, WA, USA, 1995.

275. Su, Y.; Balmer, M.L.; Bunker, B.C. Evaluation of Cesium Silicotitanates as an Alternate Waste Form. *Mater. Res. Soc. Symp. Proc.* **1996**, *465*, 457–464. [CrossRef]

276. McCready, D.E.; Balmer, M.L.; Keefer, K.D. Experimental and Calculated X-ray Diffraction Data for Cesium Titanium Silicate, $CsTiSi_2O_{6.5}$: A New Zeolite. *Powder Diffr.* **1997**, *12*, 40–46. [CrossRef]

277. Su, Y.; Balmer, M.L.; Wang, L.; Bunker, B.C.; Nyman, M.; Nenoff, T.; Navrotsky, A. Evaluation of Thermally Converted Silicotitanate Waste Forms. *Mater. Res. Soc. Symp. Proc.* **1999**, *556*, 77–84. [CrossRef]

278. Xu, H.; Navrotsky, A.; Nyman, M.D.; Nenoff, T.M. Thermochemistry of Microporous Silicotitanate Phases in the Na_2O-Cs_2O-SiO_2-TiO_2-H_2O System. *J. Mater. Res.* **2000**, *15*, 815–823. [CrossRef]

279. Xu, H.; Navrotsky, A.; Balmer, M.L.; Su, Y.; Bitten, E.R. Energetics of Substituted Pollucites along the $CsAlSi_2O_6$-$CsTiSi_2O_{6.5}$ Join: A High-Temperature Calorimetric Strudy. *J. Am. Ceram. Soc.* **2001**, *84*, 555–560. [CrossRef]

280. Bubnova, R.S.; Stepanov, N.K.; Levin, A.A.; Filatov, S.K.; Paufler, P.; Meyer, D.C. Crystal structure and thermal behavior of boropollucite $CsBSi_2O_6$. *Solid State Sci.* **2004**, *6*, 629–637. [CrossRef]

281. Richerson, D.W.; Hummel, F.A. Synthesis and Thermal Expansion of Polycrystalline Cesium Minerals. *J. Am. Ceram. Soc.* **2006**, *55*, 269–273. [CrossRef]

282. Rehspringer, J.-L.; Balencie, J.; Vilminot, S.; Burger, D.; Boos, A.; Estournes, C. Confining caesium in expanded natural Perlite. *J. Eur. Ceram. Soc.* **2007**, *27*, 619–622. [CrossRef]

283. Garino, T.G.; Nenoff, T.M.; Park, T.J.; Navrotsky, A. The Crystallization of Ba-Substituted $CsTiSi_2O_{6.5}$ Pollucite using $CsTiSi_2O_{6.5}$ Seed Crystals. *J. Am. Ceram. Soc.* **2009**, *92*, 2144–2146. [CrossRef]

284. Park, T.-J.; Garino, T.J.; Nenoff, T.M.; Rademacher, D.; Navrotsky, A. The Effect of Vacancy and Ba-Substitution on the Stability of the $CsTiSi_2O_{6.5}$ Pollucite. *J. Am. Ceram. Soc.* **2011**, *94*, 3053–3059. [CrossRef]

285. He, P.; Jia, D. Low-temperature sintered pollucite ceramic from geopolymer precursor using synthetic metakaolin. *J. Mater. Sci.* **2013**, *48*, 1812–1818. [CrossRef]

286. Garino, T.J.; Rademacher, D.X.; Rodriguez, M.; Nenoff, T.M. The Synthesis of Ba and Fe Substituted $CsAlSi_2O_6$ Pollucites. *J. Am. Ceram. Soc.* **2013**, *96*, 2966–2972.

287. Jing, Z.; Hao, W.; He, X.; Fan, J.; Zhang, Y.; Miao, J.; Jin, F. A novel hydrothermal method to convert incineration ash into pollucite for the immobilization of a simulant radioactive cesium. *J. Hazard. Mater.* **2016**, *306*, 220–229. [CrossRef]

288. Fan, J.; Jing, Z.; Zhang, Y.; Miao, J.; Chen, Y.; Jin, F. Mild hydrothermal synthesis of pollucite from soil for immobilization of Cs in situ and its characterization. *Chem. Eng. J.* **2016**, *304*, 344–350. [CrossRef]

289. Vance, E.R.; Gregg, D.J.; Griffiths, G.J.; Gaugliardo, P.R.; Grant, C. Incorporation of Ba in Al and Fe pollucite. *J. Nuc. Mater.* **2016**, *478*, 256–260. [CrossRef]

290. Henderson, C.M.B.; Charnock, J.M.; Bell, J.M.; van der Laan, G.C.M.B. X-ray absorption study of 3d transition-metals and Mg in glasses and analogue crystalline materials in $AFe^{3+}Si_2O_6$ and $A_2X^{2+}Si_5O_{12}$, where A = K, Rb, or Cs and X = Mg, Mn, Fe, Co, Ni, Cu, or Zn. *J. Non-Cryst. Solids* **2016**, *451*, 23–48. [CrossRef]

291. Omerasevic, M.; Matovic, L.; Ruzic, J.; Golubovic, Z.; Jovanovic, U.; Mentus, S.; Dondur, V. Safe trapping of cesium into pollucite structure by hot-pressing method. *J. Nuc. Mater.* **2016**, *474*, 35–44. [CrossRef]

292. Hamada, S.; Kishimura, H.; Matsumoto, H.; Takahashi, K.; Aruga, A. Effect of shock compression on luminescence properties of $CsAlSi_2O_6$:Eu^{2+} for white-light-emitting diodes. *Opt. Mater.* **2016**, *62*, 192–198. [CrossRef]

293. Orlova, A.I.; Chuvildeev, V.N.; Mikhailov, D.A.; Boldin, M.S.; Belkin, O.A.; Nokhrin, A.V.; Sakharov, N.V.; Skuratov, V.A.; Kirilkin, S. Preparation of ceramic materials with mineral-like structures by mean of SPS technology for the purpose of radwaste consolidation and radiation and hydrolytic investigations. *J. Nucl. Energy Sci. Power Gener. Technol.* **2017**, *6*, 33.

294. Berry, L.G.; Mason, B. *Mineralogy Concepts, Descriptions, Determinations*; W.H. Freeman & Co.: San Francisco, CA, USA, 1959.

295. Deer, W.A.; Howie, R.A.; Zussman, J. *Rock-Forming Minerals, Vol IV*; John Wiley & Sons, Inc.: New York, NY, USA, 1963.

296. Klingenberg, R.; Felsche, J. Interstitial Cristobalite-type Compounds (Na_2O)$\leq0.33Na[AlSiO_4]$). *J. Solid State Chem.* **1986**, *61*, 40–46. [CrossRef]

297. Kim, J.G.; Lee, J.H.; Kim, I.T.; Kim, E.H. Fabrication of a Glass-Bonded Zeolite Waste Form for Waste LiCl Salt. *J. Ind. Eng. Chem.* **2007**, *13*, 292–298.

298. Brookins, D.G. *Geochemical Aspects of Radioactive Waste Disposal*; Springer: Berlin, Germany, 1984.

299. Fleet, M.E. Structures of Sodium Alumino-Germanate Sodalites. *Acta Cryst.* **1989**, *C45*, 843–847.

300. McFarlane, H.F.; Goff, K.M.; Felicione, F.S.; Dwight, C.C.; Barber, D.B. Hot Demonstrations of Nuclear-Waste Processing Technologies. *JOM* **1997**, *49*, 14–21. [CrossRef]
301. Deer, W.A.; Howie, R.A.; Zussman, J. *Rock-Forming Minerals. Vol. 2A. Single-Chain Silicates*; Geological Society: London, UK, 1997.
302. Nakazawa, T.; Kato, H.; Okada, K.; Ueta, S.; Mihara, M. Iodine Immobilization by Sodalite Waste Form. *Mater. Res. Soc. Symp. Proc.* **2001**, *663*, 51–57. [CrossRef]
303. Olson, A.L.; Soelberg, N.R.; Marshal, D.W.; Anderson, G.L. *Fluidized Bed Steam Reforming of INEEL SBW Using THOR Mineralizing Technology*; INEEL/EXT-04-02564; Idaho National Laboratory: Idaho Falls, ID, USA, 2004.
304. Deer, W.A.; Howie, R.A.; Wise, W.S.; Zussman, J. *Rock-Forming Minerals, Vol. 4B, Framework Silicates: Silica Minerals, Feldspathoids and the Zeolites*; The Geological Society: London, UK, 2004.
305. Mattigod, S.V.; McGrail, B.P.; McCready, D.E.; Wang, L.; Parker, K.E.; Young, J.S. Synthesis and Structure of Perrhenate Sodalite. *J. Microporous Mesopourous Mater.* **2006**, *91*, 139–144. [CrossRef]
306. Angelis, G.D.; Capone, M.; Mannielo, A.; Mariani, M.; Maceratu, E.; Conti, C. Different methods for conditioning Chlorite Salt waste from Pyroprocesses. In *Book of Abstracts "III International Pyroprocessing Research Conference*; NIIAR: Dimitrovgrad, Russia, 2010; p. 48.
307. Simpson, M.F.; Allensworth, J.R.; Phongikaroon, S.; Williams, A.N.; Dunzik-Gourgar, M.L.; Ferguson, C. Immobilization of Salt from Zone Freezing Process in Zeolite-A. In *III International Pyroprocessing Research Conference*; NIIAR: Dimitrovgrad, Russia, 2010; pp. 30–31.
308. Akimkhan, A.M. Structural and Ion-Exchange Properties of Natural Zeolite. *Ion Exch. Technol.* **2012**, *10*, 261–283.
309. Tendeloo, L.V.; Blochouse, B.; Dom, D.; Vancluysen, J.; Snellings, R.; Martens, J.A.; Kirschhock, C.E.A.; Maes, A.; Breynaert, E. Cation Exchange Properties of Zeolites in Hyper Alkaline Aqueous Media. *Environ. Sci. Technol.* **2015**, *49*, 1729–1737. [CrossRef]
310. Olszewska, W.; Miśkiewicz, A.; Zakrzewska-Kołtuniewicz, G.; Lankof, L.; Pająk, L. Multibarrier system preventing migration of radionuclides from radioactive waste repository. *Nukleonika* **2015**, *60*, 557–563. [CrossRef]
311. Kim, H.S.; Park, J.S.; Lim, W.T. Site Competition of Ca^{2+} and Cs^+ Ions in the Framework of Zeolite Y (Si/Al = 1.56) and Their Crystallographic Studies. *J. Mineral. Soc. Korea* **2018**, *31*, 235–248. [CrossRef]
312. Ovhal1, S.; Butler, I.S.; Xu, S. The Potential of Zeolites to Block the Uptake of Radioactive Strontium-90 in Organisms. *Contemp. Chem.* **2018**, *1*, 1–13.
313. Dyer, A.; Hriljac, J.; Evans, N.; Stokes, I.; Rand, P.; Kellet, S.; Harjula, R.; Moller, T.; Maher, Z.; Heatlie-Branson, R.; et al. The use of columns of the zeolite clinoptilolite in the remediation of aqueous nuclear waste streams. *J. Radioanal. Nucl. Chem.* **2018**, *318*, 2473–2491. [CrossRef]
314. Grundy, H.D.; Hassan, I. The crystal structure of a carbonate-rich cancrinite. *Can. Mineral.* **1982**, *20*, 239–251.
315. Zhao, H.; Deno, Y.; Harsh, J.B.; Flury, M.; Boyle, J.S. Alteration of Kaolinite to Cancrinite and Sodalite by simulated Hanford tank waste and its inpact on cesium retention. *Clay Mater.* **2004**, *52*, 1–13.
316. Mon, J.; Deng, Y.; Flury, M.; Harsh, J.B. Cesium incorporation and diffusion in cancrinite, sodalite, zeolite, and allophone. *Microporous Microporous Mater.* **2005**, *86*, 277–286. [CrossRef]
317. Hassan, I.; Antao, S.M.; Parise, J.B. Cancrinite: Crystal structure, phase transitions, and dehydration behavior with temperature. *Am. Mineral.* **2006**, *91*, 1117–1124. [CrossRef]
318. Deng, Y.; Flury, M.; Harsh, J.B.; Felmy, A.R.; Qafoku, O. Cancrinite and sodalite formation in the presence of cesium, potassium, magnesium, calcium and strontium in Hanford tank waste stimulant. *Appl. Geochem.* **2006**, *21*, 2049–2063. [CrossRef]
319. Dickson, J.O.; Harsh, J.B.; Flury, M.; Lukens, W.W.; Pierce, E.M. Immobilization and Exchange of Perrhenate in Sodalite and Cancrinite. *Microporous Mesoporous Mater.* **2015**, *214*, 115–120. [CrossRef]
320. Zheng, Z.; Anthony, R.G.; Miller, J.E. Modeling Multicomponent Ion Exchange Equilibrium Utilizing Hydrous Crystalline Silicotitanates by a Multiple Interactive Ion Exchange Site Model. *Ind. Eng. Chem. Res.* **1997**, *36*, 2427–2434. [CrossRef]
321. Miller, J.E.; Brown, N.E. *Development and Properties of Crystalline Silicotitanate (CST) Ion Exchangers for Radioactive Waste Applications*; SAND97-0771; Sandia National Laboratories: Albuquerque, NM, USA, 1997.
322. Andrews, M.K.; Harbour, J. *Glass Formulation Requirements for Hanford Coupled Operations Using Crystalline Silicotitanates (CST)*; WSRC-RP-97-0265; Westinghouse Savannah River Company, Savannah River Site: Aike, SC, USA, 1997.

323. Yu, B.; Chen, J.; Song, C. Crystalline Silicotitanate: A New Type of Ion Exchange for Cs Removal from Liquid Waste. *J. Mater. Sci. Technol.* **2002**, *18*, 206–210.

324. Tripathi, A.; Medvedev, D.G.; Nyman, M.; Clearfield, A. Selectivity for Cs and Sr in Nb-Substituted Titanosilicate with Sitinakite Topology. *J. Solid State Chem.* **2003**, *175*, 72–83. [CrossRef]

325. Keppler, H. Ion Exchange Reactions Between Dehydroxylated Micas and Salt Melts and the Crystal Chemistry of the Interlayer Cation in Micas. *Am. Mineral.* **1990**, *75*, 529–538.

326. Fleet, M.E. *Rock-Forming Minerals: Sheet Silicates: Micas, V. 3A*; The Geological Society: Bath, Englad, 2003.

327. Jantzen, C.M.; Williams, M.R.; Bibler, N.E.; Crawford, C.L.; Jurgensen, A.R. *Fluidized Bed Steam Reformed (FBSR) Mineral Waste Forms: Application to Cs-137/Sr-90 Wastes for the Global Nuclear Energy Partnership (GNEP)*; U.S. DOE Report WSRC-MS; Savannah River National Laboratory: Jackson, MS, USA, 2008.

328. Jantzen, C.M.; Williams, M.R. *Fluidized Bed Steam Reforming (FBSR) Mineralization for High Organic and Nitrate Waste Streams for the Global Nuclear Energy Partnership (GNEP)*; Waste Management 08, Paper #8314 (2008); WM Symposia: Phoenix, AZ, USA, 2008; p. 8314.

329. Neeway, J.J.; Qafoku, N.P.; Peterson, R.A.; Brown, C.F. *Characterization and Leaching Tests of the Fluidized Bed Steam Reforming (FBSR) Waste Form for LAW Immobilizatio*; Waste Management 2013, Paper 13400; WM Symposia: Phoenix, AZ, USA, 2013.

330. Kumar, A.; Singh, Y.P.; Pradhan, G.; Dhawan, N. Utilization of Mica for Potassium Recovery. In Proceedings. *Mater. Today* **2018**, *5*, 17030–17034.

331. Boatner, L.A.; Sales, B.C. Monazite. In *Radiation Waste Forms for the Future*; Lutze, W., Ewing, R.C., Eds.; North-Holland Press: Amsterdam, The Netherlands, 1988; pp. 495–564.

332. Ewing, R.C.; Weber, W.J.; Lutze, W. Crystalline Ceramics: Waste Forms for the Disposal of Weapons Plutonium. In *Disposal of Ex-weapons Plutonium as Waste*; NATIO ASI Series; Merz, E.R., Walter, C.E., Eds.; Kluwer Academic Publishers: Dordrecht, The Netherlands, 1996; pp. 65–83.

333. Merz, E.R.; Walter, C.E. *Disposal of Ex-weapons Plutonium as Waste, NATIO ASI Series*; Kluwer Academic Publishers: Dordrecht, The Netherlands, 1996; pp. 65–83.

334. Van Emden, B.; Thornber, M.R.; Graham, J.; Lincoln, F.J. The incorporation of actinides in monazite and xenotime from placer deposits in Western Australia. *Can. Mineral.* **1997**, *35*, 95–104.

335. Chang, L.L.Y.; Howie, R.A.; Zussman, J. Rock-Forming Minerals, V.5B Non-Silicates. In *Sulphates, Carbonates, Phosphates, Halides*, 2nd ed.; The Geological Society: London, UK, 1998; Volume 383, ICBN 978-1897799901.

336. Genet, M.; Dacheux, N.; Thomas, A.C.; Chassigneux, B.; Pichot, E.; Brandel, V. Thorium phosphate-diphosphate as a ceramic for the immobilization of tetravalent uranium, neptunium and plutonium. *Waste Manag.* **1999**, *99*, 38.

337. Clavier, N.; Dacheux, N.; Podor, R.; Le Coustumer, P. Study of Actinides Incorporation in Thorium Phosphate-Diphosphate/Monazite Based Ceramics. *Mater. Res. Soc. Symp. Proc.* **2004**, *802*, DD3.6.1–DD3.6.6. [CrossRef]

338. Burakov, B.E.; Yagovkina, M.A.; Garbuzov, V.M.; Kitsay, A.A.; Zirlin, V.A. SelfIrradiation of Monazite Ceramics: Contrasting Behavior of PuPO4 and (La,Pu)PO$_4$ Doped with Pu-238. In *Scientific Basis for Nuclear Waste Management XXVIII*; Hanchar, J.M., Stroes-Gascoyne, S., Browning, L., Eds.; Materials Research Society Symposium Proceedings: San Francisco, CA, USA, 2004; Volume 824, pp. 219–224.

339. Montel, J.M.; Glorieux, B.; Seydoux-Guilaume, A.M.; Wirth, R. Synthesis and Sintering of a Monazite-brabantite Solid Solution Ceramic for Nuclear Waste Storage. *J. Phys. Chem. Solids* **2006**, *67*, 2489–2500. [CrossRef]

340. Zhang, Y.J.; Vance, E.R. Plutonium in Monazite and Brabanite: Diffuse Reflectance Spectroscopy Study. *J. Nucl. Mater.* **2008**, *375*, 311–314. [CrossRef]

341. Glorieux, B.; Montel, J.M.; Matecki, M. Synthesis and Sintering of a Monazite-brabantite Solid Solution Ceramics Using Metaphosphate. *J. Eur. Ceram. Soc.* **2009**, *29*, 1679–1686. [CrossRef]

342. Brandt, F.; Neumeier, S.; Schuppik, T.; Arinicheva, Y.; Bukaemskiy, A.; Modolo, G.; Bosbach, D. Conditioning of minor actinides in lanthanum monazite ceramics: A surrogate with Europium. *Prog. Nucl. Energy* **2014**, *72*, 140–143. [CrossRef]

343. Arinicheva, Y.; Bukaemskiy, A.; Neumeier, S.; Modolo, G.; Bosbach, D. Studies on thermal and mechanical properties of monazite-type ceramics for the conditioning of minor actinides. *Prog. Nucl. Energy* **2014**, *72*, 144–148. [CrossRef]

344. Mezentseva, L.P.; Kruchinina, I.Y.; Osipov, A.V.; Kuchaeva, S.K.; Ugolkov, V.L.; Popova, V.F.; Pugachev, K.E. Nanopowders of Orthophosphate LaPO$_4$–YPO$_4$–H$_2$O System and Ceramics Based on Them. *Glass Phys. Chem.* **2014**, *40*, 356–361. [CrossRef]

345. Ma, J.; Teng, Y.; Huang, Y.; Wu, L.; Zhang, K.; Zhao, X. Effects of sintering process, pH and temperature on chemical durability of Ce$_{0.5}$Pr$_{0.5}$PO$_4$ ceramics. *J. Nucl. Mater.* **2015**, *465*, 550–555. [CrossRef]

346. Zhao, X.; Teng, Y.; Yang, H.; Huang, Y.; Ma, J. Comparison of microstructure and chemical durability of Ce$_{0.9}$Gd$_{0.1}$PO$_4$ ceramics prepared by hot-press and pressureless sintering. *Ceram. Int.* **2015**, *41*, 11062–11068. [CrossRef]

347. Teng, Y.; Zeng, P.; Huang, Y.; Wu, L.; Wang, X. Hot-pressing of monazite Ce$_{0.5}$Pr$_{0.5}$PO$_4$ ceramic and its chemical durability. *J. Nucl. Mater.* **2015**, *465*, 482–487. [CrossRef]

348. Zhao, X.; Teng, Y.; Wu, L.; Huang, Y.; Ma, J.; Wang, G. Chemical durability and leaching mechanism of Ce$_{0.5}$Eu$_{0.5}$PO$_4$ ceramics: Effects of temperature and pH values. *J. Nucl. Mater.* **2015**, *466*, 187–193. [CrossRef]

349. Teng, Y.; Wang, X.; Huang, Y.; Wu, L.; Zeng, P. Hot-pressure sintering, microstructure and chemical durability of Ce$_{0.5}$Eu$_{0.5}$PO$_4$ monazite ceramics. *Ceram. Int.* **2015**, *41*, 10057–10062. [CrossRef]

350. Meng, C.; Ding, X.; Zhao, J.; Ren, C.; Fu, H.; Yang, H. Phase evolution and microstructural studies of Gd$_{1-x}$Yb$_x$PO$_4$ (0≤x≤1) ceramics for radioactive waste storage. *J. Eur. Ceram. Soc.* **2016**, *36*, 773–779. [CrossRef]

351. Potanina, E.; Golovkina, L.; Orlova, A.; Nokhrin, A.; Boldin, M.; Sakharov, N. Lanthanide (Nd, Gd) compounds with garnet and monazite structures. Powders synthesis by "wet" chemistry to sintering ceramics by Spark Plasma Sintering. *J. Nucl. Mater.* **2016**, *473*, 93–98. [CrossRef]

352. Ji, Y.; Kowalski, P.M.; Neumeier, S.; Deissmann, G.; Kulriya, P.K.; Gale, J.D. Atomistic modeling and experimental studies of radiation damage in monazite-type LaPO$_4$ ceramics. *Nucl. Instrum. Methods Phys. Res.* **2017**, *393*, 54–58. [CrossRef]

353. Babelot, C.; Bukaemskiy, A.; Neumeier, S.; Modolo, G.; Bosbach, D. Crystallization processes, compressibility, sinterability and mechanical properties of La-monazite-type ceramics. *J. Eur. Ceram. Soc.* **2017**, *37*, 1681–1688. [CrossRef]

354. Guo, L.; Yan, Z.; Li, Z.; Yu, J.; Wang, Q.; Li, M.; Ye, F. GdPO$_4$ as a novel candidate for thermal barrier coating applications at elevated temperatures. *Surf. Coat. Technol.* **2018**, *349*, 400–406. [CrossRef]

355. Arinicheva, Y.; Gausse, C.; Neumeier, S.; Brandt, F.; Rozov, K.; Szenknect, S.; Dacheux, N.; Bosbach, D.; Deissmann, G. Influence of temperature on the dissolution kinetics of synthetic LaPO$_4$-monazite in acidic media between 50 and 130 °C. *J. Nucl. Mater.* **2018**, *509*, 488–495. [CrossRef]

356. Arinicheva, Y.; Clavier, N.; Neumeier, S.; Podor, R.; Bukaemskiy, A.; Klinkenberg, M.; Roth, G.; Dacheux, N.; Bosbach, D. Effect of powder morphology on sintering kinetics, microstructure and mechanical properties of monazite ceramics. *J. Eur. Ceram. Soc.* **2018**, *38*, 227–234. [CrossRef]

357. Zhao, X.; Li, Y.; Teng, Y.; Wu, L.; Bi, P.; Wang, L.; Wang, S. The structure, sintering process, and chemical durability of Ce$_{0.5}$Gd$_{0.5}$PO$_4$ ceramics. *Ceram. Int.* **2018**, *44*, 19718–19724. [CrossRef]

358. Zhao, X.; Li, Y.; Teng, Y.; Wu, L.; Bi, P.; Yang, X.; Wan, L. The effect of Ce content on structure and stability of Gd$_{1-x}$Ce$_x$PO$_4$: Theory and experiment. *J. Eur. Ceram. Soc.* **2019**, *39*, 1555–1563. [CrossRef]

359. Milligan, W.O.; Mullica, D.F.; Beall, G.W.; Boatner, L.A. Structural investigation of YPO$_4$, ScPO$_4$, and LuPO$_4$. *Inorg. Chim. Acta* **1982**, *60*, 39–43. [CrossRef]

360. Boatner, L. Synthesis, Structure, and Properties of Monazite, Pretulite, and Xenotime. *Rev. Mineral. Geochem.* **2002**, *48*, 87–121. [CrossRef]

361. Hetherington, C.J.; Harlov, D.E.; Budzyn, B. Experimental metasomatism of monazite and xenotime: Mineral stability, REE mobility and fluid composition. *Miner. Petrol.* **2010**, *99*, 165–184. [CrossRef]

362. Ji, Y.; Beridze, G.; Bosbach, D.; Kowalski, P.M. Heat capacities of xenotime-type ceramics: An accurate ab initio prediction. *J. Nucl. Mater.* **2017**, *494*, 172–181. [CrossRef]

363. Kondrat'eva, O.N.; Nikiforova, G.E.; Tyurin, A.V.; Ryumin, M.A.; Gurevich, V.M.; Kritskaya, A.P.; Gavrichev, K.S. Calorimetric study of ytterbium orthovanadate YbVO$_4$ polycrystalline ceramics. *Ceram. Int.* **2018**, *44*, 18103–18107. [CrossRef]

364. Weber, W.J.; Turcotte, R.P.; Bunnell, L.R.; Roberts, F.P.; Westsik, J.H., Jr. Radiation Effects in Vitreous and Devitrified Simulated Waste Glass (Contains Apatite). In Proceedings of the International Symposium on Ceramics in Nuclear Waste Management, Cincinnati, OH, USA, 28 April–3 May 1979; pp. 294–299.

365. Weber, W.J. Radiation Damage in Rare-earth Silicate with the Apatite Structure. *J. Am. Ceram. Soc.* **1982**, *65*, 544–548. [CrossRef]

366. Bros, R.; Carpens, J.; Sere, V.; Beltritti, A. Occurrence of Plutonium and Fissiogenic REE in Hydrothermal Apatites from the Nuclear Reactor 16 at Oklo (Gabon). *Radiochim. Acta* **1996**, *74*, 277–282. [CrossRef]

367. Weber, W.J.; Ewing, R.C.; Meldrum, A. The Kinetics of Alpha-decay-induced Amorphization in Zircon and Apatite Containing Weapons-grade Plutonium or Other Actinides. *J. Nucl. Mater.* **1997**, *250*, 147–155. [CrossRef]

368. Audubert, F.J.; Lacout, J.L.; Tetard, F. Elaboration of an Iodine-Bearing Apatite Iodine Diffusion into a $Pb_3(VO_4)_2$ Matrix. *Solid State Ion.* **1997**, *95*, 113–119. [CrossRef]

369. Boyer, L.; Carpena, J.; Lacout, J.L. Synthesis of Phosphate-Silicate Apatites at Atmospheric Pressure. *Solid State Ion.* **1997**, *95*, 121–129. [CrossRef]

370. Carpena, J.; Donazzon, B.; Ceraulo, E.; Prene, S. Composite Apatitic Cement as a Material to Retain Cesium and Iodine. *Comptes Rendus de L Academie Des Sciences Serie II Fascicule C—Chimie* **2001**, *4*, 301–308.

371. Park, H.S.; Kim, I.T.; Kim, H.Y.; Lee, K.S.; Ryu, S.K.; Kim, J.H. Application of Apatite Waste Form for the Treatment of Water-soluble Wastes Containing Radioactive Elements. Part 1: Investigation on the Possibility. *J. Ind. Eng. Chem.* **2002**, *8*, 318–327.

372. Elliott, J.C.; Wilson, R.M.; Dowker, S.E.P. Apatite Structures. *Adv. X-ray Anal.* **2002**, *45*, 172–181.

373. Kim, J.Y.; Dong, Z.L.; White, T.J. Model Apatite Systems for the Stabilization of Toxic Metals: II, Ccation and Metalloid Substitutions in Chlorapatites. *J. Am. Ceram. Soc.* **2005**, *88*, 1253–1260. [CrossRef]

374. Carpena, J.; Lacout, J.L. Calcium Phosphate Nuclear Materials: Apatitic Ceramics for Separated Wastes. *Actual. Chim.* **2005**, 66–71.

375. Jothinathan, E.; Vammeesel, K.; Vleugels, J. Apatite type lanthanum silicate and composite anode half cells. *Solid State Ion.* **2011**, *192*, 419–423. [CrossRef]

376. Knyazev, A.V.; Chernorukov, N.G.; Bulanov, E.N. Apatite-structured compounds: Synthesis and high-temperature investigation. *Mater. Chem. Phys.* **2012**, *132*, 773–781. [CrossRef]

377. Wang, J. Incorporation of iodine into apatite structure: a crystal chemistry approach using Artificial Neural Network. *Front. Earth Sci.* **2015**, *3*, 1–11. [CrossRef]

378. Kirkland, C.L.; Yakymchuk, C.; Szilas, K.; Evans, N.; Hollis, J.; McDonald, B.; Gardiner, N.J. Apatite: A U-Pb thermochronometer or geochronomete. *Lithos* **2018**, *318–319*, 143–157. [CrossRef]

379. Hong, H.Y.-P. Crystal structures and crystal chemistry in the system $Na_{1+x}Zr_2SixP_3$-xO_{12}. *Mater. Res. Bull.* **1976**, *11*, 173–182. [CrossRef]

380. Goodenough, J.B.; Hong, H.Y.-P.; Kafalas, J.A. Fast Na+-ion transport in skeleton structures. *Mater. Res. Bull.* **1976**, *11*, 203–220. [CrossRef]

381. Boilot, J.P.; Salanié, J.P.; Desplanches, G.; Le Potier, D. Phase transformation in Na1+xSixZr2P3−xO12 compounds. *Mater. Res. Bull.* **1979**, *14*, 1469–1477. [CrossRef]

382. de la Rochère, M.; d'Yvoire, F.; Collin, G.; Comès, R.; Boilot, J.P. NASICON type materials—$Na_3M_2(PO_4)_3$ (M = Sc, Cr, Fe): Na^+-Na^+ correlations and phase transitions. *Solid State Ion.* **1983**, *9–10 Pt 2*, 825–828. [CrossRef]

383. Manthiram, A.; Goodenough, J.B. Lithium Insertion into Fe2(MO4)3 Frameworks: Comparison of M = W with M = Mo. *J. Solid State Chem.* **1987**, *71*, 349–360. [CrossRef]

384. Roy, R.; Vance, E.R.; Alamo, J. [NZP], a new radiophase for ceramic nuclear waste forms. *Mater. Res. Bull.* **1982**, *17*, 585–589. [CrossRef]

385. Alamo, R.; Roy, R. Crystal chemistry of the $NaZr_2(PO_4)_3$, NZP or CTP, structure family. *J. Mater. Sci.* **1986**, *21*, 444–450. [CrossRef]

386. Scheetz, B.E.; Roy, R. Novel Waste Forms. In *Radioactive Waste Forms for the Future*; Lutze, W., Ewing, R.C., Eds.; North-Holland: Amsterdam, The Netherlands, 1988; pp. 596–599.

387. Orlova, A.I.; Zyryanov, V.N.; Kotel'nikov, A.R.; Demarin, V.T.; Rakitina, E.V. Ceramic phosphate matrices for high level waste. Behaviour in hydrothermal conditions. *Radiokhimiya* **1993**, *35*, 120–126.

388. Orlova, A.I.; Volkov, Y.F.; Melkaya, R.F.; Masterova, L.Y.; Kulikov, I.A.; Alferov, V.A. Synthesis and Radiation Stability of NZP Phosphates Containg F-elements. *Radiochemistry* **1994**, *36*, 322–325.

389. Scheetz, B.E.; Agrawal, D.K.; Breval, E.; Roy, R. Sodium Zirconium-phosphate (NZP) as a Host Structure for Nuclear Waste Immobilization—A Review. *Waste Manag.* **1994**, *14*, 489–505. [CrossRef]

390. Hawkins, H.T.; Scheetz, B.E.; Guthrie, G.D., Jr. Preparation of Monophasic [NZP] Radiophases: Potential Host Matrices for the Immobilization of Reprocessed Commercial High-Level Wastes. In *Scientific Basis for Nuclear Waste Management. XX*; Gray, W.J., Triay, I.R., Eds.; Material Research Society: Pittsburgh, PA, USA, 1997; pp. 387–394.

391. Zyryanov, V.N.; Vance, E.R. Comparison of Sodium Zirconium Phosphate-Structured HLW forms and Synroc for High-Level Nuclear Waste Immobilization. In *Scientific Basis for Nuclear Waste Management. XX*; Gray, W.J., Triay, I.R., Eds.; Material Research Society: Pittsburgh, PA, USA, 1997; pp. 409–416.

392. Miyajima, Y.; Miyoshi, T.; Tamaki, J.; Matsuoka, M.; Yamamoto, Y.; Masquelier, C.; Tabuchi, M.; Saito, Y.; Kageyama, H. Solubility range and ionic conductivity of large trivalent ion doped $Na_{1+x}M_xZr_{2-x}P_3O_{12}$ (M: In, Yb, Er, Y, Dy, Tb, Gd) solid electrolytes. *Solid State Ion.* **1999**, *124*, 201–211. [CrossRef]

393. Orlova, A.I.; Charlamova, A.A.; Volkov, Y.F. Investigation of Plutonium, Americium and Curium Phosphates as a Basis for Inclusion into Kosnarite-type Ceramic Waste Worms. In *Review of Excess Weapons Plutonium Disposition, LLNL Contract Work in Russia, Proceedings of the 3rd Annual Meeting for Coordination and Review of LLNL Contract Work, St. Petersburg, Russia, 14–18 January 2002*; Lawrence Livermore National Laboratory: Livermore, CA, USA, 2002; pp. 407–418.

394. Orlova, A.I.; Orlova, V.A.; Buchirin, A.V.; Beskrovnyi, A.I.; Kurazhkovskaya, V.S. Cesium and Its Analogs, Rubidium and Potassium, in Rhombohedral [$NaZr_2(PO_4)_3$ Type] and Cubic (Langbeinite Type) Phosphates: 1. Crystal-Chemical Studies. *Radiochemistry* **2005**, *47*, 225–234. [CrossRef]

395. Bykov, D.M.; Orlova, A.I.; Tomilin, S.V.; Lizin, A.A.; Lukinykh, A.N. Americium and Plutonium on Trigonal Phosphates (NZP Type) $AM_{1/3}Zr_2(PO_4)_3$ and $Pu_{1/4}Zr_2(PO_4)_3$. *Radiochemistry* **2006**, *48*, 234–239. [CrossRef]

396. Bykov, D.M.; Konings, R.J.M.; Orlova, A.I. High Temperature Investigations of the rare earth NZP phosphates $R_{1/3}Zr_2(PO_4)_3$ (R = La, Nd, Eu, Lu) by drop calorimetry. *J. Alloy. Compd.* **2007**, *439*, 376–379. [CrossRef]

397. Nalk, A.H.; Deb, S.B.; Chalke, A.B.; Saxena, M.K.; Ramakumar, K.L.; Venugopal, V.; Dharwadkar, S.R. Microwave-assisted low temperature synthesis of sodium zirconium phosphate (NZP) and the leachability of some selected fission products incorporated in its structure—A case study of leachability of cesium. *J. Chem. Sci.* **2010**, *122*, 71–82.

398. Orlova, A.I.; Lizin, A.A.; Tomilin, S.V.; Lukinykh, A.N.; Kanunov, A.E.; Chuvil'deev, V.N.; Boldin, M.S.; Sakharov, N.V.; Nokhrin, A.V. Actinide Phosphates with $NaZr_2(PO_4)_3$ Structure. High-Speed Production of Dense Ceramics. In *Book of Abstracts: The 49-th Conference on Hot Laboratories and Remote Handling "HOTLAB 2012"*; CEA: Marcoule, France, 2012.

399. Orlova, A.I.; Volgutov, V.Y.; Mikhailov, D.A.; Bykov, D.M.; Skuratov, V.A.; Chuvil'deev, V.N.; Nokhrin, A.V.; Boldin, M.S.; Sakharov, N.V. Phosphate $Ca_{1/4}Sr_{1/4}Zr_2(PO_4)_3$ of the $NaZr_2(PO_4)_3$ type: Synthesis of a dense ceramic material and its radiation testing. *J. Nucl. Mater.* **2014**, *441*, 232–239. [CrossRef]

400. Bohre, A.; Shrivastava, O.P.; Awasthi, K. Crystal Chemistry of Immobilization of Tetravalent Ce and Se in Ceramic Matrix of Sodium Zirconium Phosphate. *Phys. Chem. Res.* **2014**, *2*, 21–29.

401. Pet'kov, V.; Asabina, E.; Loshkarev, V.; Sukhanov, M. Systematic investigation of the strontium zirconium phosphate ceramic form for nuclear waste immobilization. *J. Nucl. Mater.* **2016**, *471*, 122–128. [CrossRef]

402. Pet'kov, V.I.; Dmitrienko, A.S.; Sukhanov, M.V.; Koval'skii, A.M.; Borovikova, E.Y. Synthesis, phase formation, and thermal expansion of sulfate phosphates with the $NaZr_2(PO_4)_3$ structure. *Russ. J. Inorg. Chem.* **2016**, *61*, 623–629. [CrossRef]

403. Glukhova, I.O.; Asabina, E.A.; Pet'kov, V.I.; Borovikova, E.Y.; Koval'skii, A.M. Phase Formation, Structure, and Thermal Expansion of Phosphates $M_{0.5(1+x)}Fe_xTi_{2-x}(PO_4)_3$ (M = Mn, Zn). *Russ. J. Inorg. Chem.* **2016**, *61*, 681–687. [CrossRef]

404. Liu, T.; Wang, B.; Gu, X.; Wang, L.; Ling, M.; Liu, G.; Wang, D.; Zhang, S. All-climate sodium ion batteries based on the NASICON electrode materials. *Nano Energy* **2016**, *30*, 756–761. [CrossRef]

405. Kim, Y.; Kim, H.; Park, S.; Seo, I.; Kim, Y. Na ion-conducting ceramic as solid electrolyte for rechargeable seawater batteries. *Electrochim. Acta* **2016**, *191*, 1–7. [CrossRef]

406. Wang, J.; Zhang, Z.J. Luminescence properties and energy transfer studies of color tunable Tb^{3+}-doped $RE_{1/3}Zr_2(PO_4)_3$ (RE = Y, La, Gd and Lu). *J. Alloys Compd.* **2016**, *685*, 841–847. [CrossRef]

407. Ribero, D.; Seymour, K.C.; Kriven, W.M.; White, M.A. Synthesis of $NaTi_2(PO_4)_3$ by the inorganic–organic steric entrapment method and its thermal expansion behavior. *J. Am. Ceram. Soc.* **2016**, *99*, 3586–3593. [CrossRef]

408. Ananthanarayanan, A.; Ambashta, R.D.; Sudarsan, V.; Ajithkumar, T.; Sen, D.; Mazumder, S.; Wattal, P.K. Structure and short time degradation studies of sodium zirconium phosphate ceramics loaded with simulated fast breeder (FBR) waste. *J. Nucl. Mater.* **2017**, *487*, 5–12. [CrossRef]

409. Kanunov, A.; Glorieux, B.; Orlova, A.; Borovikova, E.; Zavedeeva, G. Synthesis, structure and luminescence properties of phosphates $A_{1-3x}Eu_xZr_2(PO_4)_3$ (A-alkali metal). *Bull. Mater. Sci.* **2017**, *40*, 7–16. [CrossRef]

410. Orlova, A. Next Generation Ceramic Materials for Consolidation of radioactive alpha-wastes using the Innovative Technology Spark Plasma Sintering for their preparation. In *Book of Abstracts: 3rd World Congress on Materials Science, Engineering, Oil, Gas and Petrochemistry*; StatNano: Dunbai, UAE, 2018.

411. Kanunov, A.E.; Orlova, A.I. Phosphors Based on Phosphates of $NaZr_2(PO_4)_3$ and Langbeinite Structural Families. *Rev. J. Chem.* **2018**, *8*, 1–33. [CrossRef]

412. Hallopeau, L.; Bregiroux, D.; Rousse, G.; Portehault, D.; Stevens, P.; Toussaint, G.; Laberty-Robert, C. Microwave-assisted reactive sintering and lithium ion conductivity of $Li_{1.3}Al_{0.3}Ti_{1.7}(PO_4)_3$ solid electrolyte. *J. Power Sources* **2018**, *378*, 48–52. [CrossRef]

413. Wang, H.; Okubo, K.; Inada, M.; Hasegawa, G.; Enomoto, N.; Hayashi, K. Low temperature-densified NASICON-based ceramics promoted by Na_2O-Nb_2O_5-P_2O_5 glass additive and spark plasma sintering. *Solid State Ion.* **2018**, *322*, 54–60. [CrossRef]

414. Savinkh, D.O.; Khainakov, S.A.; Orlova, A.I.; Garcia-Granda, S. New Phosphate-Sulfates with NZP Structure. *Russ. J. Inorg. Chem.* **2018**, *63*, 714–724. [CrossRef]

415. Savinkh, D.O.; Khainakov, S.A.; Orlova, A.I.; Garcia-Granda, S. Preparation and Thermal Expansion of Calcium Iron Zirconium Phosphates with the $NaZr_2(PO_4)_3$ Structure. *Inorg. Mater.* **2018**, *54*, 591–595. [CrossRef]

416. Orlova, A.; Khainakov, S.; Alexandrov, A.; Garcia-Granda, S.; Savinykh, D. Crystallographic studies of $NaZr_2(PO_4)_3$ phosphates at high temperatures. In *Book of Abstracts: 31-st European Crystallographic Meeting "ECM31"*; European Crystallographic Association: Oviedo, Spain, 2018.

417. Savinkh, D.O.; Khainakov, S.A.; Boldin, M.S.; Orlova, A.I.; Aleksandrov, A.A.; Lantsev, E.A.; Sakharov, N.V.; Murashov, A.A.; Garcia-Granda, S.; Nokhrin, A.V.; et al. Preparation of NZP-Type $Ca_{0.75+0.5x}Zr_{1.5}Fe_{0.5}(PO_4)_{3-x}(SiO_4)_x$ Powders and Ceramic, Thermal Expansion Behavior. *Inorg. Mater.* **2018**, *54*, 1267–1273. [CrossRef]

418. Orlova, A.I.; Loginova, E.E.; Logacheva, A.A.; Demarin, V.T.; Shmidt, O.V.; Nikolaev, A.Y. A Crystal-Chemical Approach in the Development of Phosphate Materials as Environmentally Safe Chemical Forms of Utilization of Spent Cs-Containing Ferrocyanide Sorbents. *Radiochemistry* **2010**, *52*, 462–468. [CrossRef]

419. Lizin, A.A.; Tomilin, S.V.; Gnevashov, O.E.; Lukinykh, A.N.; Orlova, A.I. Orthophosphates of Langbeinite Structure for Immobilization of Alkali Metal Cations of Salt Wastes from Pyrochemical Processes. *Radiochemistry* **2012**, *54*, 542–548. [CrossRef]

420. Martynov, K.V.; Nekrasov, A.N.; Kotel'nikov, A.R.; Tananaev, I.G. Synthesis and study of the chemical stability and strength of zirconium phosphates with the structure of langbeinite with imitators of high-level radioactive waste (HLRW). *Glass Phys. Chem.* **2017**, *43*, 75–82. [CrossRef]

421. Mold, P.; Bull, R.K.; Durrani, S.A. Constancy of ^{244}Pu distribution in chondritic whitlockite. *Nuclear Tracks* **1981**, *5*, 27–31. [CrossRef]

422. Nakamura, S.; Otsuka, R.; Aoki, H.; Akao, M.; Miura, N.; Yamamoto, T. Thermal expansion of hydroxyapatite-β-tricalcium phosphate ceramics. *Thermochim. Acta* **1990**, *165*, 57–72. [CrossRef]

423. Belik, A.A.; Morozov, V.A.; Grechkin, S.V.; Khasanov, S.S.; Lazoryak, B.I. Crystal Structures of Double Vanadates, $Ca_9R(VO_4)_7$. III. R = Nd, Sm, Gd, or Ce. *Crystallogr. Rep.* **2000**, *45*, 798–803.

424. Orlova, A.I.; Orlova, M.P.; Solov'eva, E.N.; Loginova, E.E.; Demarin, V.T.; Kazantsev, G.N.; Samojlov, S.G.; Stefanovsky, S.V. Lanthanides in Phosphates with the structure of whitlockite mineral (analog of β-$Ca_3(PO_4)_2$). *Radiochemistry* **2006**, *48*, 561–567. [CrossRef]

425. Orlova, M.; Glorieux, B.; Orlova, A.; Montel, J.M.; Kazantsev, G.; Samoilov, S. Phosphates with structure of mineral whitlockite (beta-$Ca_3(PO_4)_3$). In *Book of Abstracts: Engineering Conf. International ECI "Alternative Materials for Radioactive Waste Stabilization and Nuclear Materials Containment"*; ECI: Brooklyn, NY, USA, 2007.

426. Orlova, A.I.; Orlova, M.P.; Loginova, E.E.; Lizin, A.A.; Tomilin, S.V.; Lukinykh, A.N.; Khainakov, S.A.; Garcia-Granda, S.; Demarin, V.T.; Oleneva, T.A.; et al. Thorium, Plutonium, Lanthanides and Some 1- and 2-Valent Elements in the New Orthophosphates with the Structure of Mineral Whitlockite. Chemistry, Structure, Stability. In *Abstracts Booklet "Plutoniun Future—"The Science", Proceedings of the Topical Conference on Plutonium and Actinides, Dijon, France, 7–11 July 2008*; Elsevier: Amsterdam, Netherlands, 2008; pp. 313–314.

427. Orlova, A.I.; Khainakov, S.A.; Loginova, E.E.; Oleneva, T.A.; Garcia Granda, S.; Kurazhkovskaya, V.S. Calcium Thorium Phosphate (whitlockite type Mineral). Synthesis and Structure Refinement. *Crystallogr. Rep.* **2009**, *54*, 591–597. [CrossRef]

428. Benhamou, R.A.; Bessiere, A.; Wallez, G.; Viana, B.; Elaatmani, M.; Daoud, M.; Zegzouti, A. New insight in the structure-luminescence relationship of Ca9Eu(PO4)7. *J. Solid State Chem.* **2009**, *182*, 2319–2325. [CrossRef]

429. Orlova, A.I.; Orlova, M.P.; Loginova, E.E.; Khainakov, S.; Garcia-Granda, S.; Lizin, A.A.; Tomilin, S.V.; Lukinikh, A.N.; Kurazshkovskay, V.S. Nature "experience" and experimental data on immobilization of actinides into calcium phosphate with whitelochite mineral structure. In *Book of Abstracts: 6th Russian Conference on Radiochemistry "Radiochemistry-2009"*; Lomonosov Moscow State University: Moscow, Russia, 2009; p. 294.

430. Orlova, A.I.; Malanina, N.V.; Chuvil'deev, V.N.; Boldin, M.S.; Sakharov, N.V.; Nokhrin, A.V. Praseodymium and neodymium phosphates Ca9Ln(PO4)7 of whitlockite structure. Preparation of a ceramic with a high relative density. *Radiochemistry* **2014**, *56*, 380–384. [CrossRef]

431. Adcock, C.T.; Tschauner, O.; Hausrath, E.M.; Udry, A.; Luo, S.N.; Cai, Y.; Ren, M.; Lanzirotti, A.; Newville, M.; Kunz, M.; et al. Shock-transformation of whitlockite to merrillite and the implications for meteoritic phosphate. *Nat. Commun.* **2017**, *8*, 14667. [CrossRef]

432. Carrasco, I.; Piccinelli, F.; Bettinelli, M. Optical Spectroscopy of Ca9Tb1-xEux(PO4)7 (x = 0, 0.1, 1): Weak Donor Energy Migration in the Whitlockite Structure. *J. Phys. Chem. C* **2017**, *121*, 16943–16950. [CrossRef]

433. Bénard, P.; Brandel, V.; Dacheux, N.; Jaulmes, S.; Launay, S.; Lindecker, C.; Genet, M.; Louër, D.; Quarton, M. Th4(PO4)4P2O7, a New Thorium Phosphate: Synthesis, Characterization, and Structure Determination. *Chem. Mater.* **1996**, *8*, 181–188. [CrossRef]

434. Dacheux, N.; Podor, R.; Chassigneux, B.; Brandel, V.; Genet, M. Actinides Iimmobilization in New Matrices Bbased on Solid Solutions: Th4-xMxIV(PO4)(4)P2O7, (M-IV = U-238, Pu-239). *J. Alloy. Compd.* **1998**, *271*, 236–239. [CrossRef]

435. Dacheux, N.; Podor, R.; Brandel, V.; Genet, M. Investigations of systems ThO2-MO2-P2O5 (M = U, Ce, Zr, Pu). Solid Solutions of Thorium-Uranium(IV) and ThoriumPlutonium(IV) Phosphate-diphosphates. *J. Nucl. Mater.* **1998**, *252*, 179–186. [CrossRef]

436. Pichot, E.; Dacheux, N.; Brandel, V.; Genet, M. Investigation of Cs-137(+), Sr-85(2+) and Am-241(3+) Ion Exchange on Thorium Phosphate Hydrogenphosphate and their Immobilization in the Thorium Phosphate Diphosphate. *New J. Chem.* **2000**, *24*, 1017–1023. [CrossRef]

437. Clavier, N.; Dacheux, N.; Martinez, P.; Du Fou de Kerdaniel, E.; Aranda, L.; Podor, R. Sintering of β-Thorium−Uranium(IV) Phosphate−Diphosphate Solid Solutions from Low-Temperature Precursors. *Chem. Mater.* **2004**, *16*, 3357–3366. [CrossRef]

438. Clavier, N.; Dacheux, N.; Podor, R. Synthesis, characterization, sintering, and leaching of β-TUPD/monazite radwaste matrices. *Inorg. Chem.* **2006**, *45*, 22. [CrossRef]

439. Clavier, N.; Du Fou de Kerdaniel, E.; Dacheux, N.; Le Coustumer, P.; Drot, R.; Ravaux, J.; Simoni, E. Behavior of thorium–uranium (IV) phosphate–diphosphate sintered samples during leaching tests. Part II. Saturation processes. *J. Nucl. Mater.* **2006**, *349*, 304–316. [CrossRef]

440. Morozov, V.A.; Mironov, A.V.; Lazoryak, B.I.; Khaikina, E.G.; Basovich, O.M.; Rossell, M.D.; Van Tendeloo, G. Ag1/8Pr5/8MoO4: An incommensurately modulated scheelite-type structure. *J. Solid State Chem.* **2006**, *179*, 1183–1191. [CrossRef]

441. Maček Kržmanc, M.; Logar, M.; Budič, B.; Suvorov, D. Dielectric and Microstructural Study of the SrWO4, BaWO4, and CaWO4 Scheelite Ceramics. *J. Am. Ceram. Soc.* **2011**, *94*, 2464–2472. [CrossRef]

442. Cheng, J.; He, J. Electrical properties of scheelite structure ceramic electrolytes for solid oxide fuel cells. *Mater. Lett.* **2017**, *209*, 525–527. [CrossRef]

443. Zhang, B.; Zhao, Q.; Zhao, C.; Chang, A. Comparison of structure and electrical properties of vacuum-sintered and conventional-sintered Ca1-xYxCeNbWO8 NTC ceramics. *J. Alloys Compd.* **2017**, *698*, 1–6. [CrossRef]

444. Xiao, M.; Sun, H.; Zhou, Z.; Zhang, P. Bond ionicity, lattice energy, bond energy, and microwave dielectric properties of $Ca_{1-x}Sr_xWO_4$ ceramics. *Ceram. Int.* **2018**, *44*, 20686–20691. [CrossRef]

445. Potanina, E.A.; Orlova, A.I.; Nokhrin, A.V.; Boldin, M.S.; Sakharov, N.V.; Belkin, O.A.; Chuvil'deev, V.N.; Tokarev, M.G.; Shotin, S.V.; Zelenov, A.Y. Characterization of $Na_x(Ca/Sr)_{1-2x}Nd_xWO_4$ complex tungstates fine-grained ceramics obtained by Spark Plasma Sintering. *Ceram. Int.* **2018**, *44*, 4033–4044. [CrossRef]

446. Potanina, E.A.; Orlova, A.I.; Mikhailov, D.A.; Nokhrin, A.V.; Chuvil'deev, V.N.; Boldin, M.S.; Sakharov, N.V.; Lantcev, E.A.; Tokarev, M.G.; Murashov, A.A. Spark Plasma Sintering of fine-grained $SrWO_4$ and $NaNd(WO_4)_2$ tungstates ceramics with the scheelite structure for nuclear waste immobilization. *J. Alloys Compd.* **2019**, *774*, 182–190. [CrossRef]

447. Pang, L.-X.; Zhou, D.; Yue, Z.-X. Temperature independent low firing $[Ca_{0.25}(Nd_{1-x}Bi_x)_{0.5}]MoO_4$ ($0.2 \leq x \leq 0.8$) microwave dielectric ceramics. *J. Alloys Compd.* **2019**, *781*, 385–388. [CrossRef]

448. Hanusa, J. Raman scattering and ifra-red spectra of tungstates $KLn(WO_4)_2$-family (Ln: La–Lu). *J. Mol. Struct.* **1984**, *114*, 471–474. [CrossRef]

449. Pages, M.; Freundlich, W. Phases of scheelite structure in the neptunium molybdate and sodium or litium molybdate systems. *J. Inorg. Nucl. Chem.* **1972**, *34*, 2797–2801. [CrossRef]

450. Lee, M.R.; Mahe, P. Molybdates et tungstates d'uranium IV et de sodium. *C. R. Acad. Sci. Paris.* **1974**, *279*, 1137–1170.

451. Tabuteau, A.; Pages, M. Identification and crystal chemistry of double molybdates of alkalimetals (K, Rb, Cs) and transuranium elements (Np, Pu, Am). *J. Inorg. Nucl. Chem.* **1980**, *42*, 401–403. [CrossRef]

452. Tabuteau, A.; Pages, M.; Freundlich, W. Sur les phases de structure sheelite dans les systems molybdate deplutonium-molybdate de litium du sodium. *Mater. Res. Bull.* **1972**, *7*, 691–697. [CrossRef]

453. Müller-Buschbaum, H.; Gallinat, S. Synthese und Röntgenstrukturanalyse von $KCuGd_2Mo_4O_{16}$ und $CuTb_2Mo_4O_{16}$. *Z. Naturforsch.* **1995**, *50*, 1794–1798. [CrossRef]

454. Basovich, M.; Khaikina, E.G.; Vasil'ev, E.V.; Frolov, A.M. Phase formation in the Li_2MoO_4—Rb_2MoO_4—$Ln_2(MoO_4)_3$ systems and the poperties of $LiRbLn_2(MoO_4)_4$. *Zh. Neorg. Khim.* **1995**, *40*, 2047-251.

455. Basovich, O.M.; Khaikina, E.G.; Solodovnikov, S.F.; Tsyrenova, G.D. Phase formation in the systems Li_2MoO_4—K_2MoO_4—$Ln_2(MoO_4)_3$ (Ln = La, Nd, Dy, Er) and properties of triple molybdates $LiKLn_2(MoO_4)_4$. *J. Solid State Chem.* **2005**, *178*, 1580–1588. [CrossRef]

456. Szillat, H.; Müller-Buschbaum, H. Synthese und Kristallstructur von $KCuHoMo_4O_{16}$. *Z. Nat.* **1994**, *49*, 350–354.

457. Wang, Y.; Wu, W.; Fu, X.; Liu, M.; Cao, J.; Shao, C.; Chen, S. Metastable scheelite $CdWO_4$:Eu^{3+} nanophosphors: Solvothermal synthesis, phase transitions and their polymorph-dependent luminescence properties. *Dyes Pigment.* **2017**, *147*, 283–290. [CrossRef]

458. Donald, I.W. *Waste Immobilisation in Glass and Ceramic Based Hosts*; Wiley: Chichester, UK, 2010; 507p.

459. Kinoshita, H. Development of ceramic matrices for high level radioactive waste. In *Handbook of Advanced Radioactive Waste Conditioning Technologies*; Ojovan, M., Ed.; Woodhead: Cambridge, UK, 2011; Chapter 10; p. 293.

460. Ojovan, M.I.; Lee, W.E.; Kalmykov, S.N. *An Introduction to Nuclear Waste Immobilisation*, 3rd ed.; Elsevier: Amsterdam, The Netherlands, 2019; p. 497.

461. Zhang, Y.; Wei, T.; Zhang, Z.; Kong, L.; Dayal, P.; Gregg, D.J. Uranium brannerite with Tb(III)/Dy(III) ions: Phase formation, structures, and crystallizations in glass. *J. Am. Ceram. Soc.* **2019**, 1–11. [CrossRef]

MDPI

St. Alban-Anlage 66

4052 Basel

Switzerland

Tel. +41 61 683 77 34

Fax +41 61 302 89 18

www.mdpi.com

Materials Editorial Office

E-mail: materials@mdpi.com

www.mdpi.com/journal/materials

www.ingramcontent.com/pod-product-compliance
Lightning Source LLC
Chambersburg PA
CBHW051842210326
41597CB00033B/5751